Praise for

The Way of a Ship

"A tribute to the seamen of the Age of Sail."
—JONATHAN YARDLEY, *Washington Post Book World*

"A fantastic ride through one of the greatest moments in the history of adventure . . . Lundy is a master of tension and storytelling."
—*Seattle Times*

"The wealth and authority of this book make it a worthy companion to the very best histories on seafaring. . . . This masterly documentary becomes a page-turner to rank with the very best of Patrick O'Brian."
—PETER NICHOLS in the *Sunday Times* (London)

"A saga of life under sail that touches to the quick . . . Lundy is particularly good at evoking the most dangerous situations, recounting the interplay between heavy weather and the captain's decisions with grim realism, yet lyrically portraying the ship as a living thing that must work, if not in harmony then at least in concert with the riotous elements that surround it."
—*Kirkus Reviews* (starred review)

" . . . readers will surface with a strong sense of seagoing history, a knowledge of the specialized skills involved in keeping square-riggers afloat and a respect not only for the fierce power of the elements but also for Lundy's considerable talent as a writer."
—*Publishers Weekly*

"An exceptionally rich and satisfying weave . . . making the reader feel the sting of the wind and the vertiginous plunge of the wave . . . heir to the tradition of Dana, Melville, Conrad . . . a descendant of such classics."
—JONATHAN RABAN, author of *Bad Land* and *Passage to Juneau*

"Fascinating. I don't think I've ever read anything that so authoritatively brings to life what it was like to sail a square-rigged vessel."

—NATHANIEL PHILBRICK, author of *In the Heart of the Sea*

"Derek Lundy's new, and marvelous, *The Way of a Ship* attempts an unusual trick. After declaring itself a work of non-fiction, it co-opts the form of the novel, with narrative arc and wholly imagined characters . . . It is indeed a work of nonfiction, but in the way that the best novels always are. The setting—the sea—ultimately becomes the principal character in this gorgeous book and, as a protagonist, Lundy's ocean is as real and nuanced and true as Emma Bovary."

—*Toronto Globe and Mail*

"A unique and masterful book that offers both intellectual and visceral thrills as it explores the history of the last great windships and the men who sailed them. Lundy gives us an informed and well-researched discussion of late nineteenth-century merchant seafaring; and shows us the mind-numbing heroism and bravery of the common forecastle sailors—the despised of the Earth—performing feats and enduring hardship and privation that we can scarce imagine, in what is a riveting account of a white-knuckle voyage around Cape Horn."

—JAMES L. NELSON, author of *The Pirate Round* and *Blackbirder*

"Even those familiar with the lore of the sea will be impressed with the immediacy [Lundy] brings to his . . . narrative of one of the hardest passages of the days of sail. . . . You can . . . feel the numbing fatigue of half-frozen men beating themselves half to death fisting tattered canvas in the middle of the night, 40, or 80, or 100 feet above a deck that disappears under huge ocean swells."

—*San Diego Union-Tribune*

"Unique. . . . a chronicle of seafaring life in the late nineteenth century. . . . I became engrossed [by] Lundy's attention to detail and his ability to balance the romance of deepwater-sail . . . with often nail-biting action."

—*The Times-Colonist* (Victoria, BC)

The Way of a Ship

Also by DEREK LUNDY

Godforsaken Sea:
A True Story of a Race Through
the World's Most Dangerous Waters

Scott Turow: Meeting the Enemy

Barristers and Solicitors in Practice (General Editor)

The Way of a Ship

A Square-Rigger Voyage in the Last Days of Sail

DEREK LUNDY

ecco

An Imprint of HarperCollins*Publishers*

First published in Canada in 2002 by Knopf Canada.

Permissions may be found on page 351.
The drawings on pages viii, ix, and x are by C.S. Richardson.
The photograph on page xii is courtesy of the author.

HarperCollins books may be purchased for educational,
business, or sales promotional use. For information please write:
Special Markets Department, HarperCollins Publishers Inc.,
10 East 53rd Street, New York, NY 10022.

First Ecco paperback edition published 2004

Designed by Mia Risberg

The Library of Congress has catalogued the hardcover editions as follows:
Lundy, Derek, 1946–
The way of a ship : a square-rigger voyage in the last days of sail /
Derek Lundy.—1st ed.
p.   cm.
ISBN 0-06-621012-7
1. Seafaring life.   2. Lundy, Benjamin—Journeys.
3. Square-riggers.   I. Title.
G540 .L96 2003
910.4'5—dc21   2002035246

ISBN 0-06-093537-5 (pbk.)

04 05 06 07 08 OS/RRD 10 9 8 7 6 5 4 3 2 1

TO MY MOTHER

AND TO

ROBERT AND DIANE HELE

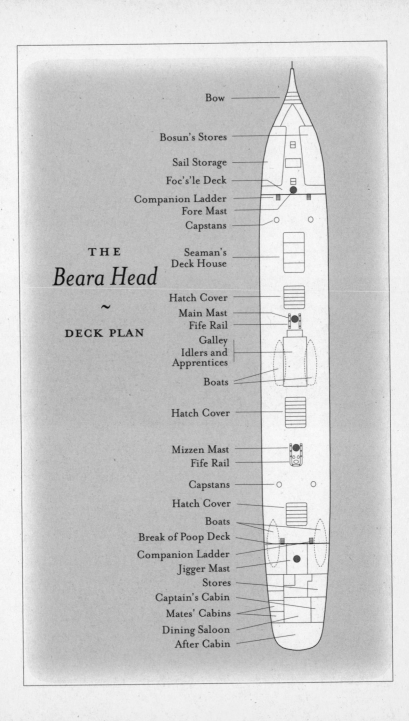

Bow

Bosun's Stores

Sail Storage

Foc's'le Deck

Companion Ladder

Fore Mast

Capstans

THE

Beara Head

~

DECK PLAN

Seaman's
Deck House

Hatch Cover

Main Mast

Fife Rail

Galley

Idlers and
Apprentices

Boats

Hatch Cover

Mizzen Mast

Fife Rail

Capstans

Hatch Cover

Boats

Break of Poop Deck

Companion Ladder

Jigger Mast

Stores

Captain's Cabin

Mates' Cabins

Dining Saloon

After Cabin

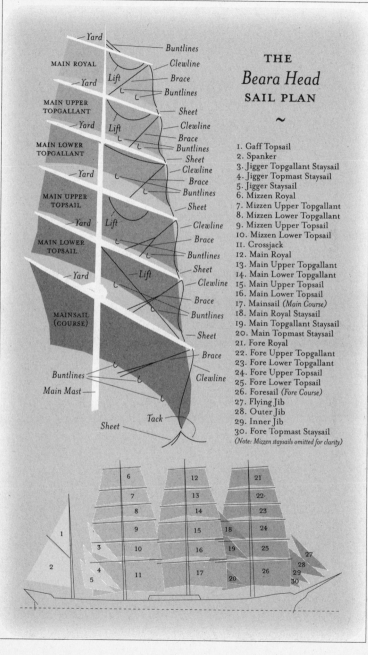

THE
Beara Head
SAIL PLAN

~

1. Gaff Topsail
2. Spanker
3. Jigger Topgallant Staysail
4. Jigger Topmast Staysail
5. Jigger Staysail
6. Mizzen Royal
7. Mizzen Upper Topgallant
8. Mizzen Lower Topgallant
9. Mizzen Upper Topsail
10. Mizzen Lower Topsail
11. Crossjack
12. Main Royal
13. Main Upper Topgallant
14. Main Lower Topgallant
15. Main Upper Topsail
16. Main Lower Topsail
17. Mainsail *(Main Course)*
18. Main Royal Staysail
19. Main Topgallant Staysail
20. Main Topmast Staysail
21. Fore Royal
22. Fore Upper Topgallant
23. Fore Lower Topgallant
24. Fore Upper Topsail
25. Fore Lower Topsail
26. Foresail *(Fore Course)*
27. Flying Jib
28. Outer Jib
29. Inner Jib
30. Fore Topmast Staysail

(Note: Mizzen staysails omitted for clarity)

Diagram labels: MAIN ROYAL, Yard, Buntlines, Clewline, Brace, Lift, Buntlines, MAIN UPPER TOPGALLANT, Yard, Sheet, Clewline, Brace, Lift, Buntlines, MAIN LOWER TOPGALLANT, Yard, Sheet, Clewline, Brace, Buntlines, MAIN UPPER TOPSAIL, Yard, Sheet, Clewline, Brace, Lift, Buntlines, MAIN LOWER TOPSAIL, Yard, Sheet, Clewline, Lift, Brace, Buntlines, MAINSAIL (COURSE), Sheet, Brace, Clewline, Buntlines, Main Mast, Tack, Sheet

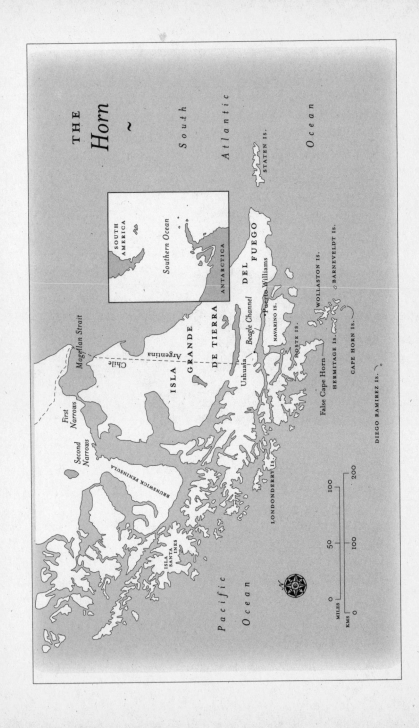

THE *Horn* ~

South Atlantic Ocean

SOUTH AMERICA

Southern Ocean

ANTARCTICA

STATEN IS.

DEL FUEGO

Puerto Williams

Beagle Channel

NAVARINO IS.

HOSTE IS.

WOLLASTON IS.

BARNEVELDT IS.

HERMITAGE IS.

CAPE HORN IS.

False Cape Horn

DIEGO RAMIREZ IS.

Magellan Strait

Chile Argentina

ISLA GRANDE DE TIERRA

Ushuaia

First Narrows

Second Narrows

BRUNSWICK PENINSULA

ISLA SANTA INES

LONDONDERRY IS.

Pacific Ocean

MILES   0   50   100   200

KMS   0   100   200

CONTENTS

Them was the days, sonnies,
Them was the men,
Them was the ships
As we'll never see again.
C. FOX SMITH, *"What the Old Man Said"*

PROLOGUE

Baltazar is anchored in a small bay on the east side of Isla Herschel, on the Paso al Mar del Sur, almost exactly seven miles north-northeast of Cape Horn. We're recording a sustained wind of more than sixty knots at deck level and gusts of close to eighty—although they're as much as 30 per cent stronger at the masthead. The anchor is dug into the sand bottom, good holding, with two hundred feet of chain out and two nylon snubbing lines to absorb the shock of the waves.

Before the front crossed over, the wind was out of the northeast. We were almost wide open in that direction, out into the Bahia Arquistade and beyond it, to the great Southern Ocean itself. For twelve hours, we pitched into seas eight to ten feet high before the wind backed first to the northwest and then west, rising to near-hurricane strength as it clocked round. By then, the low, grassy hills astern and on our sides gave us the protection we had counted on.

Apart from the anchor, we have three lines out to shore, two secured to wind-carved dwarf trees one hundred feet off our stern and one running off our port side, shackled to a cable we have wrapped around a large rock. The lines are fouled with hundreds of pounds of cleaving kelp, broken away from its beds by the storm. The weight adds strain to the lines but also dampens down their surges as the wind slams into the hull and rigging of our fifty-foot steel boat.

Two of us dragged the lines ashore in the rubber dinghy, paddling in frantic haste like commandos storming a beach. *Baltazar* was difficult to control in the rising and gusty wind, and we had to get the stabilizing lines secured in a hurry. Our skipper, Bertrand, worked the engine to keep the boat off the close, encircling rocks. Twice, I scaled the low cliff behind the beach and tied off two lines. Such exertion was unusual for me in my sedentary life, and afterwards, I slumped on the rocks gasping, heart thumping much too fast. I wondered if it was my destiny to die on this stony shore.

Weatherfax maps limned the growth of the unfolding storm. Twenty-four hours earlier, it had been an unremarkable, loose-structured low-pressure system. We would keep an eye on it as we rounded the Horn, but we wouldn't worry too much about it, maybe even using it to make a fast run across the Bahia Nassau and back to the Beagle Channel and shelter in this corner of inhospitable Tierra del Fuego. Then the barometer began to drop fast, going down at a sixty-degree angle until its line disappeared off the graph. The next weatherfax disclosed that the system had tightened up, its isobars bunching together until they almost merged, air pressure down to a frightening 950 millibars at the centre. The Chilean navy broadcast a *securité*, warning all vessels to get to shelter immediately. We began to see the cirrus and cirrocumulus clouds that signalled the depression— the "mares' tail and mackerel sky" that makes any sailor apprehensive. After clearing the eastern tip of Isla Hornos, we ran hard for a haven.

Bertrand is a fifteen-year veteran of these waters, and of many voyages across Drake Passage to Antarctica. He's never seen a storm that looks like this one, he tells us. And when the worst of the wind hits us, it is the strongest he's ever experienced. That's when the barometer begins to rise again, its tracing line reappearing on the graph and shooting up almost vertically. I didn't know a barometer could do that.

Fast rise after low foretells a stronger blow. With anxious fascination, we watch the wind lay our boat over on its side as if it was sailing close-hauled into a strong headwind.

The Horn has lived up to its reputation again. In twelve hours, its malign influences have transformed an innocuous summer low coming in out of the Southern Ocean into the most dangerous of storms: what the old square-rigger sailors used to call a Cape Horn snorter.

On deck for ten minutes to check shore lines for chafe and take photographs, dressed in my modern warm and impermeable foul-weather gear, I can, nevertheless, feel the windchill, fifteen or twenty, or more, below zero. The weight of wind is like a soft yet powerful, un-yielding wall moulding itself to my body. It's impossible to keep my eyes open looking to windward; raindrops are tiny, blinding missiles. I must concentrate on not getting flipped off the deck and into the sea. Later, from our snug, dry cabin, I look out at the horizontal rain and hail, the fog of sea water as the wind lashes the sea's surface into the air.

I often think of the nineteenth-century square-rigger men during the two days we wait out the storm in our little bay of refuge. I say to Bertrand: "How could they have done it?"

It's the question I've been asking myself since the storm began. It's the question I have come to Cape Horn to try to answer.

Day after day, week after week, summer or winter, wind-ship sailors endured just the sort of battering wind and deluge we were comfortably observing. They went aloft a hundred feet or more on icy ratlines and footropes, up masts that could whip to and fro through ninety degrees of arc in a few seconds, to grapple with homicidal sails, certain death just one small mistake, a slip, away. In leaky oilskins, al-ways soaked, no heat or light in their squalid fo'c's'les, malnourished, scurvy—the sailor's ancient bane—still a possibility even at the end of the nineteenth century.

One writer, a square-rigger sailor himself, coined the phrase "the Cape Horn breed" to describe the men who worked the beautiful, widow-making deep-sea sailing ships in their dying days. It felt apt to me. Those seamen's work was fraught with so much danger, their plane of discomfort such true suffering, that the men who matter-of-factly did it seemed remote and alien, like shadowy warriors in old and vanished wars.

I had a personal interest in these sailors. Some of my ancestors had been Cape Horn seamen. One of them was my great-great-uncle Ben-jamin Lundy, at sea in the 1880s. I had some of his letters and I knew what he looked like; I had met his descendants and become friends with them. I wanted to write about his voyage around the Horn. In that way, I thought I would come to better understand the men who sailed the last square-riggers, and what the experience had been like for them. Maybe I could answer the questions that had bubbled up with such urgency in our Cape Horn refuge.

South from our storm anchorage, past the low sheltering headland, lay the Horn, and beyond it, the Southern Ocean. That's where the wind ships would have been a century ago: fifty or a hundred miles out, or several hundred, close to the Antarctic drift ice, beating endlessly into contrary and hostile wind and seas, mothering their cargoes—the only reason they were there at all—struggling to make their westing before they could finally turn north, clear of the continent's lethal lee shore, towards benign seas, warmth and harbour.

I'm looking at an old photograph, or to be more precise, a photograph of an old photograph, since the one I'm holding shows the wide, decorated wooden frame forming an oval around the original. The recent one was taken by my father on Salt Spring Island, British Columbia, in 1985; the original dates from around 1895, and it was taken on Salt Spring Island as well. It shows my great-great-uncle, a lean man with a moustache and sloping shoulders wearing a dark shirt buttoned to the neck and khaki-coloured pants, standing beside a deer he's just shot, an antlered buck strung up beside him by a single hind leg from an overhead beam. Benjamin is looking directly at the camera without a discernible expression, his big hands holding a rifle whose barrel is cradled in the crook of his left arm; a black dog, which is also staring at the camera, sits beside him.

Benjamin settled on this green, and motionless, island to grow apples and vegetables, raise chickens and whatever else would feed his wife, Annie, a part-aboriginal woman, their two young daughters and a son. The most remarkable thing to me about the photograph is that Benjamin in 1895 so much resembles my father, his great-nephew, at about the same age sixty or seventy years later. It's as if someone had kitted out Alexander Lundy, low-level Cold War cipher clerk and former naval petty officer, with a rifle, stetson and bushy *bandido* moustache, stood him up next to a dead deer and a live dog, and taken his picture.

Among other things, the photograph demonstrates the startling persistence of living form through time. It made me feel connected to Benjamin through the intervening generations, truly part of a family line (as an only child of a scattered immigrant family, I don't often feel that), and also part of a tradition of "following the sea." Benjamin was a seaman more than a century ago, but my dead father's face in his, clear and heart-wrenching in an old photograph, seems to bring Benjamin and the ships he sailed on much closer.

It was quite a journey, when you came to think of it. Immigrants to North America, including members of my own family, did it all the time—it was the quintessential immigrant experience—and that made it seem commonplace. But what an alchemy! The voyage away from the confines of European class, accent, religion, imperial diktat and the claustrophobic "close-togetherness" of everything to the space and light of the New World, its even-handed presentation of the possibility of success and failure. It was like the first true deep breath of a person's life. Although he hadn't followed the immigrant's usual route—at first, perhaps, hadn't even intended to immigrate at all—Benjamin had eventually made that leap too. I wanted to find out more about the man in the photograph, and at least part of the story of his trek from a two-up, two-down worker's house in the Irish Quarter of Carrickfergus, under the shadow of a Norman-English castle in occupied Ireland, to become a landowner on an Edenic island in the Northwest rainforest.

It seemed to me entirely apt that Benjamin's self-displacement from one species of existence to another had been accomplished by means of a sea voyage under sail. He changed his life, made it new, by crossing oceans to a new world. At the same time, his journey of six months on a wind ship, like all such passages, was a sea change in itself.

In literature, the sea voyage has often been an overarching metaphor: "That unsounded ocean you gasp in, is life," Melville wrote. And Conrad: ". . . there are those voyages that seem ordered for the illustration of life, that might stand for a symbol of existence." Melville and Conrad were seamen before they became writers, and that early experience provided ample material for the writerly affinities they later created—or whose reality, perhaps, they merely confirmed—between the sea and human life. A third nineteenth-century seaman, who wrote a deceptively plain narrative of his voyage, completes a triumvirate of sailing informants. Richard Henry Dana's *Two Years before the Mast*, his account of his experiences as an ordinary seaman on a voyage around the Horn from Boston to California and back, was the unadorned voice from the fo'c's'le. These three pre-eminent sea writers—"sea brothers," as Melville described himself and Dana after reading the latter's book—were instrumental in telling shoreside society about seafaring culture. Without their writing, and the tradition of voyage narratives they inspired, life under sail would be an obscure, barely known world.

There's a mundane purpose to sea voyages in either sailing ships or steamers. The ship and its company have a destination, and the rationale for their venture is to arrive there with people and cargo intact. But steamer passages are also homogeneous and predictable: the vessel merely slogs along a kind of watery highway, the care and feeding of its engines the only real concern. Melville, Dana and Conrad affirmed that a sea voyage under sail meant much more.

Each one was unique. From the moment the sailing ship upanchored or unmoored, or dropped its tow, and began to move under the force of wind on sails alone, everything was thrown into the balance. No one could foretell the incidence or shape of the great things to come: storms, fire, stranding, collision, ice, Cape Horn's disposition, the severity and duration of the inevitable struggle ahead. Nor was it possible to predict from moment to moment what claims, burdens, ultimatums the wind and waves would bring down on the ship and its crew. Every decision to take in or set more sail, each turn of the wheel in heavy seas, the speed and skill with which seamen hauled or furled, spliced or lashed, all the ways of devotion hour by hour, or even minute by minute, by which the ship was continually made able to sail on, or indeed to survive, in the endless chaos of the sea—all these were subject to chance and laden with the possibility of failure or ruin.

Each sailing ship's voyage, before it began, was an unknowable adventure, a great contingency that could be completed, as one seaman-writer said, only "by the sea-cunning of men, not by the strength of machinery." The archetypical sea voyage is *The Odyssey*, and its lineaments of knowledge-seeking and adventure have been reproduced in every deep-sea passage under sail. Homer often describes Odysseus himself as cunning. He had to be to get through his trials on land, but he also needed a sailor's cunning to outwit and endure the hazards of the sea. On passages under sail, it's always the men that count: what they do, what they have in them and what the voyage teaches them along the way.

Benjamin Lundy was at sea during the last twenty years of the nineteenth century. That was Conrad's time. The young Polish exile shipped aboard his first square-rigger—the French wooden barque *Mont Blanc*, from Marseille to Martinique—as a passenger in 1874 and as an apprentice seaman the following year. He signed off the steamer *Adowa* in 1894, although it never sailed anywhere—the owners' company failed. Conrad's last real sea-berth was aboard the British wind ship *Torrens*, from which he was officially discharged as first mate in 1893, after two voyages to Australia.

The story of Benjamin's Cape Horn passage begins in Liverpool in May 1885. A month earlier, the twenty-seven-year-old Conrad signed on to the iron ship *Tilkhurst* as second mate for five pounds a month. The vessel carried coal from Cardiff to Singapore. Someone from the Bible Society gave Conrad a miniature copy of the New Testament before the ship sailed. The small, flimsy paper was just right for rolling cigarettes, and Conrad used it for that purpose throughout the voyage. But first, before he smoked them, he read the pages, studied them, in fact. In this way, he absorbed into his unfolding English the prose of King James. The day before the ship left Singapore on its subsequent passage to Calcutta, one of the seamen got a severe head-blow in a fight. He became delirious and, nine days later, committed suicide by jumping overboard. A year earlier, Conrad was second mate aboard the wind ship *Narcissus*, and one of its crewmen, a black man from Georgia, died of illness during the passage from Bombay to Dunkirk. Conrad later drew upon both deaths for *The Nigger of the "Narcissus."*

Sailors face similar difficulties and dangers at sea in any time: a storm is a storm, a lee shore equally deadly, Cape Horn the worst of places, the gales and ice that Dana encountered in the Southern Ocean in the 1830s the same today. Yet each epoch is different in some ways, and each sea writer looks to his own horizons and meanings. Melville, the metaphysician of whales and men, has little interest in describing the practical or psychological aspects of handling a ship, or surviving storms at sea—neither of which is done convincingly, if at all, in *Moby Dick*. The eighteenth-century experience of beating through unknown waters in a small, hemp-rigged wooden caravel, racing for some landfall before scurvy got its hooks in deep, was unfamiliar to a sailor of Conrad's time. He sailed aboard iron ships with polyglot crews carrying passengers or bulk cargoes when steam was in its inevitable ascendency. Conrad could see that the future of sail was no future. His sea writing is saturated with poignancy and nostalgia. When Conrad writes about the sea and ships, they are the stuff of Benjamin's voyage too, or of any man's on any ship in those last days of sail. The sea experiences of Melville and Dana are intriguing context, but Conrad's own life at sea, and his recollections of it, is the most useful set of minor variations on the theme of Benjamin's story.

For years after he left the Royal Navy, my father referred to closets as lockers, walls as bulkheads. He would clean the heads and sweep the deck. He kept his gear squared away. Whenever there was any sewing to be done, he did it. He made a long, narrow bag for me to carry my recorder in when I took group lessons in music class. The stitches were small, precise, evenly spaced—the work of a man who knew that each job, however small, had to be done absolutely right. Any shoddiness or carelessness could begin the slow, or quick, often irreversible, slide into disaster. So the bag was strong, overbuilt for its purpose; it had redundancy written all over it. I could have carried a recorder made out of lead in the bag, and it would still have held together. It was an offshore, deep-sea bag. Even aboard a diesel-powered, twentieth-century warship, my father had learned the lesson that, at sea, there was no margin for error.

I can't help thinking of my father and Benjamin Lundy in the same mental breath. It's partly because of their close physical resemblance, the ghost of one a lingering apparition in the other. But it's also because they are sea brothers of a sort themselves—part not of an immortal literary lineage like Dana and Melville, but of a modest family's tradition, two separated generations out of many that followed the sea. The line stretches up to me from a darkening past: a great-grandfather aboard HMS *Thunderer* 120 years ago, whose certification as a skilled shipwright in any vessel of the Queen-Empress hangs on the wall beside me; before him, intimations of other ancestors at sea aboard warships, or merchantmen round the Horn—the details petering out in the usual faded trail of anonymous lives—or building ships in Belfast in the Protestant shipyard there (my grandfather hammered rivets into the *Titanic*); another who sailed aboard one of the Shamrocks of Sir Thomas Lipton, the Irish tea magnate who challenged four times for the America's Cup; Great-great-uncle Benjamin himself and his namesake nephew, my great-uncle, shot dead in 1921 on his brother's doorstep in the uniform of the English king's navy during the Irish civil war; my father hunting submarines and picking up near-drowned merchant seamen in the Atlantic and Indian oceans in HMS *Nigella*, one of the eccentrically designated Flower-class corvettes; my young uncle, a long-service naval petty officer telegraphist; two cousins serving aboard British warships for the last twenty years or more; my own time under sail in small boats, and one passage in a twentieth-century square-rigger as well.

What a bloodline there is in that long usage of the sea! I was aware of this family record, of course, but in a vague way and incompletely. In fact, the depth and duration of my family's absorption with seagoing did not truly become part of my consciousness until after I first saw the photograph of Benjamin on Salt Spring and found out that he had been a seaman. Later, I saw his sea chest as well, now used as a general storage box in a hallway of his grandson's house on the island. It might have been with him aboard his first ship, lashed down next to his bunk, the small cache of all his belongings, table for salt horse and weevily biscuit, and for card games with shipmates.

The photographic image and its startling likeness is an eloquent, if ambiguous, memento. So is the chest, with its smooth amber surface, the expert symmetry of its joints, the solid, sober longevity of the century-old wood. They seem to squeeze together the generations between Benjamin and me in a kind of time-lapse; with such incontrovertible, conspicuous evidence of his existence, the man himself cannot be long gone. At the same time, the testimony is, in truth, paltry, conveying some impressions, making a few suggestions, but emphasizing how little time it takes for our ancestors to all but disappear from sight. Depending on the angle of my regard, I can feel close to Benjamin and his life at sea or remote from it, a past lost and done for.

In fact, Benjamin's trail was almost cold. There isn't near enough of it in the vague and fragmentary memories of living people, either in Ireland or on Salt Spring Island; it's the same with most families. There are records—yards of shelves, rooms of boxes—but they had been "organized" in a way that made them unhelpful to me. I spent time at the Public Records Office in Kew, on the border of London (where my father began his Cold War career, and an underground stop away from where I lived for six years as a child and became, briefly, a little English boy). I searched in the National Maritime Museum in Greenwich and corresponded with Newfoundland's Memorial University, which took most of the merchant-seaman records when the British government wanted to destroy them. (They took up too much space, and who cared?) If I knew the name of a ship that Benjamin had sailed on, or an exact year, then it was possible I might eventually find him among the thousands of stored crew lists—although maybe not; the records are far from complete.

Even if I found his name, it would tell me only that he had been aboard that ship between the noted ports. Most of the official logbooks kept aboard sailing ships have been destroyed, unless they recorded a birth or a death, and anyway they were notoriously stingy with details. Except for the odd voyage record by some seaman or officer who kept a journal that became public, or who published a book—and there is only a handful of those from the nineteenth century—the stories of the thousands upon thousands of square-rigger passages are lost. All I had was a man's name.

A book about the exact circumstances of a specific voyage made by young Benjamin Lundy, or about any of the anonymous seamen of the past, can never be written. It's a book that's impossible to write about almost all human beings a few generations, even one or two, after their lives, their stories, end. Robert Foulke writes about *The Odyssey* and its descendants in *The Sea Voyage Narrative*, and he concludes: "Clearly, historical and literary voyage narratives are often nearly identical in structure and substance: Usually no clear demarcation exists between fact and fiction, experience and imagination." So a different kind of book, then: take the fragments of what I know about Benjamin, and the great deal that's known about square-riggers and life aboard them in his time, and create a voyage. It will be typical in its incidents, suffering and accomplishment. Its officers and crew will be representative seamen of that time and of those ships. The ship itself: an imaginary one, the *Beara Head*—the name of an Irish promontory—but an actual sister to the big iron Cape Horners in all other respects. I will imagine the tale of Benjamin's voyage; not the voyage itself—that's unrecoverable—but as it might have been and emphatically, could have been. *The Way of a Ship* is the biography of a particular kind of journey, and in the telling, it is also the story of a man.

Benjamin's passage as a sailor before the mast aboard the *Beara Head* is, in part, the mere account of a young man learning the ropes; standing his watches; following orders; enduring cold, exhaustion and danger; helping to save the ship and himself; becoming a seaman. He is also a young man who, in the process of doing all that, learns the eternal lessons of the sea, which is to say that he finds out the sort of man he is, and that he is capable of doing things that, before or even after he did them, seemed almost unimaginably difficult and perilous. And although he is unaware of it, Benjamin is on a voyage freighted with the meanings and burdens of a whole world giving way to another.

CHAPTER ONE

Down to the Seas

*. . . as a simple sailor, right before the mast, plumb
down into the forecastle, aloft there to the royal mast-head.*
HERMAN MELVILLE, Moby Dick

O ff the pitch of the Horn. Wind at full-gale strength, waves as
high as the maintop, sometimes hail and then snow coming
down thick, clouds so low they enfold the mastheads, spume
and sky indistinguishable. The laden barque, down on its marks like a
half-tide rock, labours to windward under three lower topsails and fore
and mizzen staysails, seas sweeping the main deck like grapeshot. The
hull twists, pitches and rolls; its iron plates grind and groan. The wind
whistles, screams, as it encounters the vessel's four masts and their
dense network of standing and running rigging. The ship is close-
hauled, heading as close as possible to the direction of the wind; it
must contend for every inch to the west, although in this gale, these
seas, all it can manage is a stubborn retreat, a slow, grudging slide to
leeward, losing as little ground as it can until things improve.

Like the other men of the off-watch, ordinary seaman Benjamin
Lundy lies sleeping like a dead man in his sopping berth below. He
turned in "all standing"—that is, still wearing his soaked oilskins and
sea boots—two hours ago, after his watch on deck. His body heat under
blankets has begun to dry out his clothes a little; a light mist rises off
him, like steam. Seas boarding the main deck spurt through the door,
and a foot of water sloshes from one side of the deck house to the other

as the ship rolls, sometimes splashing up onto Benjamin in his lower bunk. But he sleeps through this and the deafening clamour of the storm.

Only one thing can drag him out of his near coma, and it happens now. The mate's whistle sounds above the wind's racket; someone hammers on the fo'c's'le door.

A hoarse roar: "All hands on deck to shorten sail!"

No matter what its state of exhaustion, Benjamin's brain long ago incorporated these words as an irresistible stimulus. Almost before he's awake, he has rolled out of his bunk onto the flooded floor. Colliding with other shapes in the cold, pitch darkness, he reels out onto the deck and the familiar wallop of the wind and driven spray.

"Clew up main and mizzen tops'ls!"

The awakened men join the watch already on deck, and with an unerring orientation in the murky night, they find the lines they need to haul on, a few out of hundreds in the precise, universal order of a wind ship's running rigging. With the seaman's coordinated jerk, they sweat up the sails, no shanties now for this hard-pressed crew. Other lines—the sheets—must be slackened off as the sails are gathered into looping folds, flogging in the wind with thunderclap explosions. Several times, the men must interrupt their hauling, belay the lines in a deft flash and jump for the lifelines, or the rigging, anything that's made fast to the ship, as a Southern Ocean greybeard breaks aboard, flooding the deck with tons of overpowering water. In these conditions, it's as dangerous on the main deck as it is aloft.

"Now lay aloft and furl 'em!"

Ben joins his mates as they climb the weather ratlines on the mainmast; the other watch will deal with the mizzen topsail. The wind sometimes presses against his body with such force that he must strain to bring his leg back against it to take the next step up. His hands, already numb with cold, get colder as they grip the rigging wire. Now he's aware of the deep cuts and cracks in the skin, ravaged knuckles, the salt water burning away at them. He's lost four of his fingernails, ripped away in earlier bloody skirmishes with recalcitrant sails. His oilskins, hard as metal, rub against the raw-chafed skin at his wrists and neck, the salt-water boils throbbing. The men climb up seventy feet and fan out along the windward side of the yard, slithering on the icy footrope, dodging the flailing number-one storm canvas and gear of the topsail. A mere tip or nudge from the heavy sail and a man could be flicked off the yard like an insect, to die on the deck or in the sea.

The crew of the *Beara Head* have been working the ship off the Horn for thirty-nine days, trying to get past this most troublesome of capes. They call it Cape Stiff, an evocative nickname of unknown origin. They began to win the war only a few days ago. In this austral winter of 1885, the westerly gales have come on them one after the other, no breathers in between. The six-week campaign has changed the seamen. From a sometimes fractious and ill-sorted gang, the unending storms have sculpted a desperate but united and determined unit of men who suffer and work without complaint. The equation is simple: they will live only if the ship survives, and so they fight to save the ship. The ship is everything, and in the fierce tumult of wind and waves, paradoxically, they have learned one of the sailor's most profound, and tender, lessons: what Conrad called the "serious relation in which a man stands to his ship." They know in their guts, unconsciously and without articulation, that they owe the ship the fullest measure of their thought, skill, self-love. They may be, as Conrad described all crews, "a small knot of men upon the great loneliness of the sea," but they will be saved by their instinctive devotion to the ship. Only then can it do its best for them, which may be good enough, maybe not.

On the main lower topsail yard, Benjamin and his soaked, frozen, half-starved watchmates begin their battle with a third of a ton of wet, flogging canvas. The object is to gather it up and lash it to the yard before it blows itself to rags or kills someone. This would take five minutes in a calm harbour; in a gale off the Horn, it's a long, intense battle. Sometimes, it's impossible to get even a grip on the material as the wind bellies it out, stiff as wood. When the ship does a heavy roll, the seamen must pause and hang on, with their bellies, eyelids, anything remotely flexible, as the yard end dips seaward at a forty-five-degree angle. They could slide off and down with ease, their fall a graceful, effortless parabola into the sea. A week ago, a man did just that. The ship rolled, and when it came back, he was gone. No one heard a sound from him. It was truly, as Melville described it, "the speechlessly quick chaotic bundling of a man into eternity." The unavoidable memory is in all their minds as they wrestle with the topsail; it adds a superfluous pungency to the sweat they're working up.

The fight to control the sail becomes nightmarish toil without end.

They get it half muzzled, but a gust rips it away from their fingers, many bleeding now. Several times, they almost have it and the precarious balance is disrupted by the wind or an awkward roll of the ship. The sail breaks free, and they must begin again. They fist the stiff canvas, trying to pound it into graspable shape; fresh blood oozes from their knuckles to join old stains. Each man has tried to secure his oilskins with soul and body lashings—light lines around wrists, ankles and waists—to try to keep some water out, and to stop the wind from blowing their clothes around their heads or even right off their bodies. But the wind is about sixty knots or so, driving sleet and snow hard at them and jumbling the leaking oilskins around them anyway, hampering their movement, hog-tying them. When the men lean far down over the yard to grasp the canvas, their feet, jammed down onto the footrope, swing up high into the air behind them, higher than their bodies. It looks as if they're diving down the forward side of the sail towards the deck. It's a sensation Benjamin has long since become accustomed to, although the first dozen times it happened, he nearly fainted with fear, indeed barely controlled his sphincter.

"Keep yer arse tight," his shipmates advised him.

No one's quiet now about the job he's doing. Grunts, oaths, imprecations, curses, exclamations of encouragement or frustration, vigorous expressions of anger at shipmates whose clumsiness or weakness has allowed the goddamned, son of a bitch of a sail to break free again—all fill the air about the small co-operative of men absorbed in their work.

As with all the battles they've fought aloft, in the end they win this one too. In fact, it's inevitable that they will furl the sail eventually, or that it will blow itself out before they do. (Even then, they'll gather in the remnants; nothing is wasted on a sailing ship.) It's unthinkable for them to come down from aloft without one or the other of these things happening. This time, it takes nearly one and a half hours. When the main lower topsail on the barque *Beara Head*, 110 days out of Liverpool, bound for Valparaiso with a cargo of coal, is finally secure, the port watch climbs down to the sea-swept deck and musters at the break of the poop.

The mate looks them over without emotion; the storms seem to have leached the viciousness out of him. The mizzen topsail was sorted out fifteen minutes before. The vessel is happier now under its three remaining sails.

"That'll do the watch," he shouts over the wail of wind and growl of waves.

Benjamin and the other seamen dodge forward along the main deck to the fo'c's'le. They have half an hour before the four-hour change of watch, when they must turn out again to relieve the men officially on duty. In the still-pitch darkness, Benjamin climbs back into his bunk. He falls asleep in less than two minutes. The ship still drifts to the east, the direction they don't want to go in, but refreshed by its crew's most recent devotion, it gives ground more slowly, more grudgingly, than before.

The first time Benjamin Lundy saw the *Beara Head*, his heart did not leap for joy, nor were his eyes dazzled by the beauty of a deep-sea wind ship.

The vessel was trussed up against one of the stone walls of the Bramley-Moore dock, surrounded by the rattle and screech of rail cars on the overhead line, its yards cock-a-bill to clear the obstructions of the hydraulic cranes overhanging the ship. Coal dust coated everything and rose in billows from the hatches as new wagonloads of coal were tipped in; the ship's decks and the nearby dock walls swarmed with grimy, yelling Irish coal-whippers. All the varieties of rubbish of a nineteenth-century industrial port littered the surrounding water. The smoke of thousands of coal fires and coal-fed boilers had blackened the warehouses and dock buildings. The line of rail cars stretched away towards the Sandhills station, where the Lancashire and Yorkshire Railway offloaded its cargoes of soft, sooty coal, which, like salt and cotton, was one of the city's raw commodities. It was a short way down the nearby Regent and Waterloo roads to Princes Dock and the ferry landing stage. Behind it squatted the jumbled houses, alleys, taverns, boardinghouses and brothels of Sailortown. Up and down the river, the line of docks stretched into the smoky mist and watery sun of a Liverpool spring day. Opposite lay the smaller, mirror-image portlands of Birkenhead, the two cities squeezing the Mersey in a cankerous red-brick ring before it widened out into its estuary. In the midst of all this, the wind ship looked like a graceful but grubby and reluctant guest at a noisy, cramped party, waiting a decent interval before making a run for it.

Standing on deck beside his sea chest, dressed in his only suit and a duncher cap, Benjamin had the same first impression as all novice sailors do of one of the modern four-masted iron barques: he was astonished, overwhelmed, cowed by the scale of the ship. He had some experience in small coastal vessels, ketches and smacks forty- and fifty-feet long. He had often seen the big ocean-going square-riggers in dock in Belfast and Kingstown. But none of that prepared him for the gut-heaving wrench produced by the *Beara Head*'s magnitude. It was like the sight of a seven-foot-tall man in a room of ordinary people; Benjamin had a queasy sense that the scale of things in the world had suddenly shifted, that normal dimensions had become unstable and unruly.

From his position just off the gangplank near the poop, the raised aftermost deck, he could look up and down the three-hundred-foot sweep of the ship and, forward of that, the forty-foot spike bowsprit soaring out and up from the bow. Between the raised decks of the poop at the stern and the fo'c's'le at the barque's forward end, the open main deck was broken up by hatches to the cargo hold below, by a deck-house aft of the foremast, containing the crew's quarters, and by the half-deck, another iron house between the main and mizzen masts, home to the galley, the idlers and the boy apprentices. All around him was the web of the standing and running rigging, the shrouds—iron wire supporting the masts—and the lines, hundreds upon hundreds of them, that raised, furled, trimmed, checked, eased and tweaked the vessel's yards and sails. It was like being inside a loose-woven cocoon; he looked out at the docks and cranes and shouting, filthy men through a screen of rope and wire, the world of the shore already seeming to recede from him inside his enclosure. And aloft, good God! That was where the eye was drawn, especially the eye of a man who knew he would very soon have to climb up there for the first time. The trucks—the mastheads—seemed distant, impressionistic in the haze of moist spring air, smoke and coal dust.

Benjamin had the same queasy feeling about the yards, their sails neatly furled in a harbour stow. The lower yard on the mainmast looked to be a relatively secure working area; in the middle, it was as big around as a barrel. In fact, it weighed close to four tons and was eighty-four feet long. When squared—that is, trimmed straight across the ship's beam—it stuck out twenty feet or so over the sea on each side of the hull. Later, Benjamin would find out that a whole watch was needed to move the monster, the men hauling on the braces (the lines leading from the yard's ends) when the mainsail (or main course) was full of wind. It seemed like hard labour until they had to do the even harder work of getting the sail in and furled. The other yards, which bore the lower and upper topsails, the lower and upper topgallants and the royals, stretched up in a ladder of cross-spars, becoming, like the masts, progressively diminished and abstract in their misty height.

In fact, Benjamin knew that climbing the masts was not the worst of his worries. Eventually, very soon, he would have to lay out along those yards, belly pressed against them, hands grasping the jackstay and feet on the footropes. He feared the footropes the most. Square-rigger men he had encountered had told him about them and their bouncy, swinging sway and drop. With two or three, or a dozen, men on the same side of the yard, each man's movements were transmitted to the footrope. As the seamen moved out on the yard or swung down the forward slope of the sail to gather in and furl it, the rope twitched and jumped and snapped in a spastic dance that chilled the blood of every man new to it. It got worse the farther up he went. There had to be some sort of quantifiable progression involved, an existential equation, something like the fear of the man increases in proportion to the height of the yard times the strength of the wind and the size of the seas. To Benjamin, peering aloft from the dirty deck, the possibilities of disgracing himself in the rigging of the *Beara Head* seemed limitless, and the chance of finding his death there not a small one.

Before submitting himself to these trials, Benjamin had to make a more immediate adjustment. He was used to the relaxed democracy of the coastal vessels he had sailed on for a year or so, easy fore-and-aft rigs with a handful of crew and captains who stood watches and mucked in with the men, hauling up muddy anchor chains or hoisting the heavy gaff-rigged sails, and who chatted with seamen amiably and unselfconsciously during the cold, black hours of the gravy-eye watches. Moving from those vessels to a deep-sea limejuicer was a displacement like Billy Budd's coming off the merchantman *Rights of Man* into the murderous, hard-labour regime of a man-of-war. Benjamin found out right away how different things would be on his first, and long-anticipated, square-rigger passage.

He heard the mate of the *Beara Head* before he saw him. Benjamin thought that the man had to be shouting through a megaphone, so powerful and commanding was the raspy bass voice. It sounded like the voice of God, in a very bad mood, chastizing unworthy men. That illusion passed when he recognized the words, or most of them: "you fuckin' paddies . . . pick up the goddamn pace down there. . . . I'll gut you, you sons of bitches . . . worthless organs home to your mammies in a sack, you bastards. So help me, I will knock down the next ape . . . you damned dogan slackers. . . ."

And more words to the same effect in a tirade of inventive obscenity and insult that a more detached man than Benjamin might have admired, even laughed at. But he foresaw that the frenzy of invective would soon and inevitably fall on him, and that didn't seem funny at all. The words had an American accent too, and he had heard all he wanted to hear about Yankee "bucko" mates, their violent discipline. They beat men up, sometimes killed them, and none that he'd heard of had been brought to account for any of it. In one of the favourite shanties seamen sang as they hauled up sails, the "cruel and hard treatment of every degree" had been handed out by a Yankee officer.

The mate appeared at the top of the poop steps still shouting, and Benjamin was surprised to see that there was no megaphone; it was the man's own natural voice sounding out in freakish amplification, so dominant a force that Benjamin didn't register the mate's appearance at all. It was illogical, but he just stood and watched the sound billow in noisy gusts from the mouth of the *Beara Head*'s first officer.

Benjamin shouldn't have been surprised. Vocal power was almost a necessity for officers on sailing ships, a job requirement. A man needed at least the sort of voice one often hears rising above the general din of large crowds—at sports events or during riots—with a distinctive timbre, perhaps, but above all, more shrill or raucous than the rest. A windship mate had to be able to make himself heard over the shriek and roar of gale-force wind and seas. He had to make known his suggestions and wishes to small groups of struggling men a hundred or 150 feet up in the air—direct them, threaten them, calm and comfort them with the familiarity of his ferocious voice, out-hollering the wind that could, at any moment, undo them all.

The mate fell silent, looked down at his new seaman, who looked back. The mate saw a wiry young man, medium height, dark brown hair, pale blue eyes, slightly protruding ears, a high forehead, a shabby tweed suit, old shiny leather shoes, a well-made sea chest. Benjamin saw a short, thick, bull-necked man with bloodshot blue eyes, stiff sandy hair, a hard burnished face that looked as if it had been scraped clean by wind-blasted salt, filthy dungarees and vest, an expression of vindictive rage and body posture that signified an imminent assault.

None came, however. Instead, the mate shouted: "New hand?"

The voice again. Dazed by its ferocity, and confused by the lazy American accent, Benjamin nodded.

"Anderson!" the mate yelled, the word clear above the racket of steam cranes and cascading coal.

A muscular man, much dirtier than the mate, with a crooked broken nose, climbed out of the cloud of coal dust hanging over the nearest cargo hatch and double-timed to the poop steps.

"Take this man to the fo'c's'le and get him out of that damned suit. Then you"—he glared at Benjamin; it was apparently the man's permanent expression—"lay aloft to the mainmast truck. Enjoy the fuckin' view, and then lay down here and do some work."

The seamen each took an end of the sea chest and lugged it forward to the deck house. Benjamin was used to sailors' accommodations and so wasn't surprised by the look of the *Beara Head*'s fo'c's'le. It was a rectangular iron house, riveted to the deck, with an iron partition running through it fore and aft. Although there was a doorway to allow movement between the two halves of the house, the partition would separate the members of the traditional port and starboard watches, when they were chosen. The bulkheads were lined with double-decker bunks. Each side of the house contained a single oil lamp hanging from an overhead hook. There was no table. A scattering of ring-eyes were bolted to the deck, some of them already used to lash down the crewmen's sea chests, which would, in this fo'c's'le, act as the men's only horizontal surfaces for eating, card playing, writing (if they knew how). There were four portholes on either outside bulkhead. And that was it. In this space, twenty-two men would live for whatever time it took to get back to England, or to convince them it was time to jump ship. It had the severe, brutal look of an institutional dormitory, a place for punishment or incarceration. Although it was populated by nominally free citizens of the British Empire, the United States of America and some enlightened European countries, the fo'c's'le would become a place of hard confinement, considerable suffering and scant shelter in the storms to come.

In his dungarees and workboots, skirting the pandemonium of the loading, Benjamin was taken by Anderson, another Ulsterman, and a Protestant too, as he had quickly found out, to the foot of the mainmast rigging. He watched Anderson disappear into the choking dust of the hold. Benjamin looked up. He had to strain neck muscles and bend back at the waist to see the truck, as if he was trying to catch a glimpse of a star directly overhead. Once again, he had the impression that the mast actually faded into low cloud. Probably an illusion. But it was a mistake to look up, as unnerving as looking down was reputed to be once you got up there. It was either the mast's unworldly height or the squeezing of blood vessels in his neck as he compressed it to look up, but he felt dizzy already. The bottom section, the one-piece iron lower mast and the topmast (these used to be separate, and although they were now joined into one pole mast, they retained their names), was reassuringly solid; above it, the wooden topgallant and royal masts, attached to the topmast at the cross-trees, in a single wooden section beginning about 105 feet off the deck, had a distinct ethereal quality to them that would be, he speculated, expressed in a discouraging willingness to bend and sway in any breeze. The topmost part in particular, the royal, appeared positively flimsy. It had the look of a mast in the planning stage, a theoretical spar that one day would become a solid reality but was now anything but. Each section was encased in its supporting structure of wire shrouds. Looking behind him, he could see the aftermost, fourth mast, the jigger, much shorter and in only two sections since it carried no square sails, only the fore-and-aft gaff spanker and a topsail above it. Together with the other fore-and-aft sails on board—four jibs and nine staysails set up between the masts— the sails on the jigger would provide considerable sailing power but were most useful for balancing and steering. The great mass of square sails on their yards were the barque's real engine.

Seeing the mate at the edge of the poop turning towards him, Benjamin knew his time had come. He wiped his (already) sweaty hands on his (already) coal-dust-grimed dungarees and pulled himself up onto the top of the bulwarks, the ship's side. No more looking up, he began to climb, placing his feet with careful precision on the wooden batten ladder of the ratlines, inspecting as he did so their attachment points, where they were seized to the shrouds. He knew enough to keep his hands on the big wire of the standing rigging so that if one of the ratlines gave way, there was less chance he'd go plunging down to the deck and die, or maim himself for good and wind up living off resentful relatives or begging on the hard streets of Carrickfergus, no safety nets, social or otherwise.

As he got up under the overhanging platform of the maintop, fifty feet up, the space between the shrouds narrowed so that he couldn't get his square-toed boots fully between them; in effect, he was forced to climb on his tip-toes. But apart from that, so far so good. It was higher, but no worse, than going aloft to the gaff to set a topsail on a ketch or smack. Benjamin wasn't comfortable with heights, but he'd adjusted to that after a while. He'd learned enough about square-riggers to know that he now had a choice. To get onto the platform of the maintop, he could squeeze up through the lubbers hole, a space between the platform and the mast that allowed the lower-mast shrouds to pass through, or he could climb, fly-like, out and around the edge of the platform. To accomplish that, he would have to take most of his weight with his arms and spread-eagle himself on the ratlines with his body at a forty-five-degree angle to the deck five and a half storeys down, like a rock climber getting around an overhang, and pull himself up. The idea terrified him, but not as much as the shame of using the lubbers hole. As its name clearly signified, only landlubbers—landsmen who happened to be at sea—used it. Sailors went out and around. The mate's voice reached him as if the mate himself was only a few feet away: "What are ya waiting for, ya slackin' son of a bitch? Lay aloft, I said. . . ."

And more words to the same effect as Benjamin, sweating and trembling, managed to drag himself up and onto the top. The semicircular wooden platform was laid on iron supports and had the dual purpose of providing a better angle for the topmast rigging, which was attached to the rim of the top, and acting as a working surface for sailors aloft. It was the pushing-off spot for men heading out along the massive main yard, which stretched out more than forty feet from each side of the mast.

The mate's supply of oaths and disparagement was unlimited and drove Benjamin up the topmast ratlines to the cross-trees another fifty feet up. There was no platform here, just an iron framework of trees, braces and spreaders. He had to squeeze himself up through the various components and, hanging onto the topgallant shrouds, balance on the narrow metal beams. Now he could look straight down past the main and topsail yards more than a hundred feet to the distorted, elongated shape of the barque's hull; from that height, it looked far too slender and fragile to carry anything, let alone three thousand tons of coal. The deck was shrouded in a mist of coal dust, which added to his sense of the vessel's lack of substance. He thought it was like a ghost ship sinking from under him, ebbing away.

He began a slow, deliberate climb up the topgallant ratlines and past the lower and then upper topgallant yards, the height now an almost-overpowering physical presence, a weight of fear. The ratlines themselves were narrow and unstable here. His hands and knees shook; he could see them wobble each time he changed position. Without surprise, he realized that he could still hear the mate's voice; it was stronger and clearer than the clanking cranes tipping coal into the ship. He reached the royal yard and climbed up onto it, both arms embracing the mast, which was more substantial than it had looked from the deck, but not nearly enough comfort. As he looked down now, the superstructure of the *Beara Head* had an abstract quality, like a complex geometric shape, but nothing he could recognize any longer as a familiar object.

Before the construction of tall buildings or the mundane carrying aloft of people in airplanes, seeing the surface of the earth from 165 feet up was a strange sensation that could be experienced only on the edges of cliffs or mountain chasms. Benjamin had seen little or nothing of either, and his position now, his perspective of the ground, was almost entirely new to him. He had expected that extreme height would be like distance: objects would look the same from the masthead as they did if both he and they were 165 feet away from each other on the ground. Not so; things weren't the same at all. He was discovering that height, the square-rigger sailor's dimension, was a unique category of distance with its own rules of perception, its special calculus of fear. He remembered a veteran deep-water sailor telling him that the height of the yards and masts on a square-rigger meant nothing. Past thirty feet or so, the fall would kill you as surely as from 150 feet. The only difference was the amount of time you had to think about things on the way down. Maybe it was even safer higher up; there were more things in your way as you fell, more chance of having your flight broken by a sail or a yard, the possibility of a desperate grab for a line or a piece of wire.

Not good for the hands, but you'd live on with the stripped-off flesh repairing itself in time. Benjamin didn't subscribe to this comforting theory, however. The higher he went, the harder things were, the less familiar the world looked, the more his fear grew, the more he shook. His left knee drubbed against the mast with a quick drumbeat.

From the royal yard, he could see through the haze the faint outlines of the Mersey opening out into Liverpool Bay and, behind Birkenhead, the green fields of the Wirral Peninsula. He looked down on the sprawl of the red-brick city and on Paradise Street in Sailortown. It was like a toy town. In fact, like all Sailortowns, it was rife with what Melville described as "all the variety of land-shark, land-rats and other vermin, which make the hapless mariner their prey." Beneath him, Benjamin saw the miniature trains and coal cars surrounded by moving tiny people. All around the docks and up and down the river, he saw ships, barques, brigs, barquentines, ketches, yawls, smacks, schooners, tugs and steamers, a part of that vast British imperial merchant fleet—many more ships than the rest of the world combined—doing its lawful business in the port of Liverpool, former slave emporium and point of embarkation for the Irish diaspora, now the second city of the empire.

He remained in this position for some time, gazing outwards, not downwards, trying to forget where he was, and that a return trip was involved. Indeed, he tried to follow the mate's suggestion and enjoy the fuckin' view, but evidently the mate had not been serious; his loud threats and orders, which had formed background noise for Benjamin for the past several minutes, became so frantic that even a man frozen with fear on a royal yard couldn't ignore them. Apparently his climb wasn't yet finished; apparently when the mate said the mainmast truck, he meant the mainmast truck, even if there were no steps with which to reach it. No reason to reach it either, other than the need to terrorize. Without really thinking about it any longer, Benjamin wrapped his arms and legs around the royal mast and, clamped onto the wood like a mobile limpet, inched his way up eight feet until he could plant his hand on the truck, the topmost point on the ship, nothing above him but air. He counted to five and then slid back down to the yard. Without hesitation, maintaining his momentum as if he was in a slow, controlled fall, he kept going down the royal and topgallant ratlines, through the crosstrees, down the topmast ratlines and once again to the maintop.

From there, the scale of things below him looked less alien again, more accustomed; his fear diminished. Once you'd stood on the royal yard, the maintop was easy. He was aware of the beginning of an idea that he might be able to get used to the strange world of height after all. He felt proud he had overcome his fear. That was one of the reasons he was here at all, he thought—to do things that would make a man out of him. Perhaps there was method in the mate's vitriol. Maybe he wasn't just a vicious bastard but was actually following a plan for the new hand, beginning his induction into the brotherhood.

Nevertheless, it took some time for Benjamin to figure out where his feet should go and what he should hang on to with great care as he swung down and around the maintop platform and onto the lower mast rigging. Then down through the dust and noise to the bulwarks, the pin rail and the deck once again, and to the mate's forceful opinion, delivered, with spittal, six inches from Benjamin's face, that he was, without any question, the least useful and legitimate human being to tread the deck of a ship, or indeed the earth itself, since the dawn of time.

If Benjamin had just climbed close to heaven, now he descended to hell.

"Lay below the main hatch," the mate shouted. "Lend a hand with trimming."

Benjamin joined a group of half-clothed men—whether sailors or longshoremen coal-whippers he couldn't tell—so completely layered with sweat and coal dust they looked to him like *faux* negroes in a music-hall skit. The loading was about half complete, and they were standing on top of the cargo under the deck and out of the way of the incoming coal, which was being tipped out of big iron boxes and through the hatch by one of the dockside cranes. After each load, the men continued their work with shovels and rakes to throw and push it away from the hatch, spreading it out, levelling it off. Each dumped load of the soft coal threw off a dense fog of dust. The men wore kerchiefs and bandanas over their mouths and noses, but these gave little protection. Within a few minutes, Benjamin's mouth was coated with soot, his nose clogged. He had to keep blowing hard out his nostrils to keep them clear. It was hot, claustrophobic, difficult to breathe deeply enough to meet his strained body's oxygen demands. His saliva was black when he spat, although after a while, he had no spit left. They worked for hours; once in a while, water was passed down to them, and Benjamin was able to swallow the wads of dust that had built up in his mouth and throat. He imagined his whole body inside slowly turning as black as his skin.

Sometimes when a load was tipped in, it kicked up only a small amount of dust. Benjamin felt a momentary relief. But there were experienced hands in the work crew, seamen who had made passages with coal before, and they didn't like the look of this. No dust meant that some of the box-loads were wet, or at least not dry. Wet coal had one very unpleasant characteristic: over time, it was much more likely than dry coal to burn by spontaneous combustion. Time was plentiful in sailing-ship voyages, and fire at sea one of the greatest dangers faced by any ship. For the veteran sailors of the coal-carriage trade, the intermittent clear air under the main hatch was a baleful sign.

When Melville arrived in Liverpool in 1839 as a seaman before the mast, he was fascinated by the city's enclosed docks, which were so much more organized and substantial than the haphazard wooden affairs of the New York waterfront. Each Liverpool dock was like a walled town, "or rather it is a small archipelago, an epitome of the world, where all the nations of Christendom, and even those of Heathendom, are represented. . . . Here, under the beneficient sway of the Genius of Commerce, all climes and countries embrace; and yard-arm touches yard-arm in brotherly love." By 1885, however, it had become difficult to recognize Melville's soothing conceit in the concussive racket of an industrial port with its specialized loading operations and purpose-built machinery.

The big dockside cranes in the Bramley-Moore dock had been installed nearly thirty years earlier as part of a coal-loading facility. They were typically ingenious products of the Industrial Revolution, each one with its own set of chains and hydraulic rams for lifting, lowering and tipping. In theory, a crane could load about three hundred tons an hour, but in practice, it never even approached this amount. The railway couldn't deliver coal wagons fast enough, for one thing, and the hatches of most vessels were too small to keep pace with the cranes or were not properly aligned with them. A ship was lucky to receive much more than six hundred tons of coal a day from the two cranes that could actually be used.

There wasn't much incentive to improve the system in the handful of Liverpool coal-loading docks. Across the muddy, fast-flowing river, Birkenhead was the dominant coal port because it was the railhead for the Great Western Railway, which brought the best-quality hard steam coal from mines in the Welsh valleys. Steamers could travel farther burning hard, rather than soft, coal in their boilers, a boon when every ton carried as fuel for the vessel was a ton of payload lost for profit.

Further loading delays were caused by the Bramley-Moore dock's inadequate provision for disposing of ballast discharged by sailing ships before loading; this was hauled away by pre-industrial horses and carts. A top-heavy square-rigger like the *Beara Head* couldn't be moved, even from one side of a dock to another, without several hundred tons of ballast—sand or rock, anything dense and heavy—in the hold to prevent capsize. The elaborate design and construction of the wind ship's tophamper, its tons of masts, yards, sails and rigging, required a good deal of countervailing weight below the waterline to lower the centre of gravity enough that the whole vessel didn't turn turtle under the assault of a gentle breeze or an unassuming wave.

Nothing more ably demonstrated the single-mindedness of the square-riggers' existence as a means of moving things across the sea. The ship lived in symbiotic contentment with what it carried. Without a cargo, or its ballast substitute, the ship couldn't live; it was a cripple that would fold up and roll over with pathetic ease. The wind ship might have been an elegant machine, but it was a mere happy accident that the organization of the materials involved in converting the energy of the wind into forward motion had produced a thing of beauty. The heart of the matter was cargo, freight, time, profit. If these elements of successful commerce could be better served by other means—by steamships, for example—then sailing ships would simply lose their reason for existence. Their comeliness would not save them when the market made its hard, impersonal calculation.

This was happening already. Square-riggers that carried coal, for example, were caught up in an ironic traffic. The *Beara Head* was bound for Valparaiso, Chile. Its cargo was fuel for the steamers that had usurped sailing vessels in the South American coastal trade. It was the same everywhere: steam more and more displaced sail; more steamers required more coal; the most economical means of transporting the coal to the steamers around the world was by sailing ship; as coal was more economically and efficiently delivered by sail, the use of steam became more feasible and widespread; that encouraged the technical improvement of steam engines that made much more efficient use of their coal fuel, which in turn enabled steamers to carry more cargo farther and spurred their even more widespread use; and so on. Even at the time, the end was clear to many people. The sailing ships served their up-and-coming replacements, helping to ensure their own obsolescence, *working the dying fall.*

At the end of his first day as a square-rigger seaman, Benjamin found out which of his fellow trimmers were shipmates; sluiced the coal grime off with buckets of very cold water from a dockside pipe; ate a supper of pea soup, salt meat, coffee and, to his great surprise, fresh soft bread. He saw the captain briefly on the poop, a lean, lithe, red-bearded man in his mid-thirties, accompanied by a little bristling, bouncing terrier. In port, the ship's routine and well-being were the mate's job; captains left them to it and were often invisible until departure. Benjamin found out that the captain's name was McMillan. He was also an Ulsterman; the ship's clannish Belfast owners preferred one of their own. There were rumours that he used to be a dancer—some kind of Irish jigging—until he got tired of it, or his legs gave out, and he belatedly acquiesced in his family's tradition and went to sea. The mate, whom the men detested already, wasn't American. It was even worse than that: he was from Nova Scotia, a Bluenose. They were reputed to be harder men even than Yankees when it came to driving crews and the laying on of fists and rope-end. The *Beara Head* promised to be what old hands called a "hot ship." The term used to mean a vessel that was well armed, but by Benjamin's time, it meant a strict-discipline merchantman subject to much "belaying-pin soup." There was a scale from warm to fairly hot to very hot to red-hot. The only question was what temperature the mate would raise in the course of the voyage.

In the fo'c's'le, the talk was in English, but with such an abundance of distorting accents that unless he concentrated, Benjamin thought he could be listening to a dozen different languages: Cork English; Liverpool English; cockney, Geordie and Glasgow English; scowegian and squarehead English; dago English and several Yankee variants. He had to admit, however, that none of it sounded any more exotic than his own Ulster English, with its northern Gaelic intonation, complex dipthongs and throttled consonants.

The men had not yet been assigned to watches—that would take place once they were underway—and they were camped at will on both sides of the fo'c's'le partition. After twelve hours of pushing coal around the hold, no one stayed up late. As he lay on his "donkey's breakfast" straw mattress in his lower bunk in the full daylight of the early-spring evening, which nevertheless barely filtered through the fo'c's'le ports, Benjamin reflected on the day. He had been grossly insulted, driven up the mast in fear and trembling, worked to exhaustion in the filthy innards of the ship, provided the nascent conditions for miner's lung and fed inedible food (except for the bread). He wondered what in the name of God he was doing there. Maybe tomorrow he'd just hop over the rail and walk away; he hadn't yet signed the articles, so technically it wouldn't be desertion. He still had a little money and could be back in Carrick in a couple of days and back on board a coaster in a week. Or perhaps he'd just work ashore for a while—he was a good carpenter. Maybe he'd forget about those exulting images he'd had of himself: at ease in the high rigging of a wind ship as it bowled along in the heel of the trades; or facing down a Cape Horn snorter and coming ashore in some American or eastern port; or back home, walking down Green Street in the Irish Quarter, with the contained, confident roll of a deep-sea veteran, a man for all the world to see, no questions asked.

The next day, however, he was still there, shoving coal around, sweating, choking on dust.

Loading and trimming took two more twelve-hour days, each one a copy of Benjamin's first. As the coal came in and was levelled off, its surface rose closer to the deck overhead. The men began to crouch, then to crawl, as they heaved the stuff out to either side of the hatch, their motions an unconscious mimicry of the miners who sweated it out of the low coal-face tunnels a hundred miles away.

A tight stow and lots of shifting boards to prevent dislodgement meant less chance of the cargo moving around when things got rough in the North Atlantic or the Southern Ocean. When a square-rigger disappeared at sea, it might have been one simple thing that did it in—an iceberg or a full-blown, unstoppable fire. Or many things might have happened in a disastrous cascade: a cargo shift that caused a list, the unmanageable ship then overwhelmed by storm waves it could otherwise have weathered; or the bulk of the cargo shifting all at once, capsizing the vessel immediately. Getting the coal loaded properly at least reduced the chance of that kind of catastrophe. Loading was the mate's sole responsibility, and under his hectoring direction, the men slithered back and forth until the coal was trimmed and the ship lay more or less even on its marks—the load lines painted on the hull, recently legislated to stop owners sending overloaded "coffin ships" to sea for the insurance money, and damn the men aboard.

On the second day, after dinner (the midday meal aboard ship), the mate shouted: "All hands, lay aft to sign articles. And keep your goddamn filthy hands off the captain's furniture."

He herded the seamen into the great cabin under the poop deck: an elegant, wood-panelled space with a polished table for the officers' meals, a navigation desk, neat bookshelves, a large bronze skylight and varnished louvred doors leading off to the small cabins occupied by the mates and the captain's steward. It was not yet secured for sea, and on the table, Benjamin noted with surprise a rum decanter, glasses and a vase filled with fresh-cut flowers. Rolled charts littered the narrow, plush settees. The contrast between the near opulence of this cabin and the bare iron structure of the fo'c's'le merely reflected the traditional divisions of comfort and deprivation aboard ship, like those of class and privilege ashore. Not unexpected on a British vessel, certainly, but things would have been the same on any nation's ships, including American "downeasters" and clippers. Rough-and-ready democracy was possible ashore, but not afloat.

The signing ritual was normally done in the owners' shipping office, but the captain and the mate had decided they didn't want to lose time getting the crew ashore. They also feared that the strumpets and booze of Sailortown might well provoke some early desertions. The *Beara Head* was still several short of its full complement, which would have to be filled at the last minute by crimps, maybe using shanghaied men. Captain McMillan didn't want to add to his problems. The crewmen already on board would stay there until the ship sailed; their voyage had already begun. The grimy sailors crammed into the great cabin and stood uncomfortable and silent.

The articles—Form Eng. (for engagement) 1—supposedly framed the contract between shipowner and seaman. British crews had been obliged to sign articles of agreement since the early 1700s. By 1877, the articles had become a thirty-four-page document that embodied the Orwellian dictum that everything not mandatory is prohibited. A clerk from the shipping office read the document, or rather, garbled it at high speed. It would have been an unintelligible speech even to a native English-speaker with some knowledge of the vagaries of legal and bureaucratic vocabulary and sentence structure. To these men, the clerk might as well have been talking in Bantu. They heard some brief mention of food. Benjamin thought he heard that everyone was obliged to wash, shave and dress in clean clothes on Sundays, and something about the number of able seamen required and penalties for desertion, numerous but incomprehensible. The clerk skidded to a halt with the words "no spirits allowed." The seamen signed with their names or, more often, their marks, coal dust sifting off them and onto the page as they moved. And that was that. They were legally bound to the ship for the next three years, for £3 or £3.15s a month, leave to be granted at the master's discretion, on a voyage whose "limits" were stated to be between seventy-five degrees north and sixty-five degrees south, which meant almost anywhere on earth with enough water for the ship to float in.

Not one seaman in ten thousand had the chance to read the articles himself. Not that any of them cared, or that it really mattered. Every man knew what to expect before the mast on board a square-rigger: at best, a benevolent dictator and officers who were hard but fair; at worst, a red-hot ship on which men were killed in the name of discipline; most of the time, one of the variations between these extremes. The arcane contents of the articles, whatever they were, had no bearing on anything.

With one exception: food. The first page, a list of provisions, set out a minimum amount and variety of food—the infamous "whack" mandated by the government's Board of Trade—which was then subject to negotiation. However, owners treated the scale as a set maximum, and thus condemned crews to a diet almost completely lacking in fresh food and vitamins. In fact, the seaman's whack (also known as the Liverpool Scale) was a recipe for near starvation, and sometimes even scurvy, the disease of vitamin-C deficiency. A man who often expended an athlete's calories was entitled, per day, to one pound of bread or biscuit (the traditional hard tack), one and a half pounds of salt beef or one and a quarter pounds of salt pork (both given the legendary nicknames of salt horse or salt junk), a half-pound of flour or one and one-eighth pounds of peas, one-eighth of an ounce of tea, a half-ounce of coffee, two ounces of sugar and three quarts of water for all purposes. Cocoa could be substituted for tea, molasses for sugar and rice for flour. Rice, peas, beef, pork and flour could be substituted for each other. The freshest part of this diet was probably the insect life it invariably contained: cockroaches in the tea and coffee, and weevils, which the men often couldn't avoid eating, in the hard biscuit. You could always tell an old sailor by the way he rapped his hard tack on a convenient solid surface before each bite to shake out the worst of the weevils. The biscuit might be so hard that it had to be soaked in tea or coffee before human teeth could deal with it, drowning the weevils and producing a less nutritious, but more palatable, mouthful. The same old sailors often professed to like this food, and if they'd been beachcombing for a while, they couldn't wait to get back aboard a deep-sea ship for a good feed of pea soup you could stand a spoon upright in, with salt horse and hard tack. This might well have been food that, over a long period of time, ruined a man's taste for anything else. For a newcomer, it was difficult to get used to, and Benjamin struggled to down the gristly salt meat, wondering how he would survive at all on grub that in real life was even worse than its description.

Of the daily three quarts of water per man, two went to the doctor, as sailors called the cook, for his dubious purposes, which at least included tea or coffee. A seaman could drink, or wash with, the remaining quart; it was up to him. In any event, most ships left with water tanks half or two-thirds empty, counting on the plentiful rain of the horse latitudes and the doldrums to top them up. It was at least pure water, unlike the questionable stuff available ashore. The crew often did their heaviest work in the hot, windless region of the doldrums, doing a lot of what sailors called pulley-hauley, bracing the heavy yards one way and the other, hour after hour, for days on end, to take advantage of every vague whisper of wind and work their way through.

When it rained, they could drink what they wanted; when it didn't, they went thirsty.

❋

Signing the articles pitched Benjamin headlong into a peculiar domain: the social community of a ship near the end of the nineteenth century, an isolated and specialized society with a number of unusual characteristics.

Shipboard life was marked, for example, by a mixture of strict hierarchy—the divisions between officers and men—and democracy— the brotherhood of seamen before the mast. Discipline on a sailing merchantman wasn't as rigid as aboard a warship, but it was, nevertheless, exacting and harsh. Tight regulation was necessary simply because the ship went to sea, an enterprise whose hazards and uncertainties required a decidedly top-down decision-making process. There was often time for only one person to decide what to do and order it carried out. Discipline induces people to put their lives in danger. Because sailing ships were such successful widow-makers, their hierarchical discipline was particularly strict. The captain was absolute king, unconcerned with constitutional niceties; the mates enforced his will with uncompromising orders—and with physical force, if necessary.

The routines of the seaman—the same day in, day out, except for the disruption of bad weather or landfall—had a cadence to them, a comforting predictability in the face of the unknown future at sea. It also calmed down the men themselves, many of them turbulent wanderers without homes, roistering their way around the world from whoring drunkards in port into sober and precise workers at sea. It usually took only a few days, Conrad observed, "for the soothing deep-water ship routine to establish its beneficent sway." The sailor's life aboard ship ran according to deeply conservative, ritualistic rhythms. There was caste and subordination, and a kind of moral law that resembled the most traditional of societies. Even as life ashore was being transformed in the nineteenth century, iron necessity at sea maintained the archaic forms of regulation, compliance and punishment.

When the deep-sea sailing ship got clear of the land and began its voyage, one of its master's most important obligations was to "keep the sea": to stay out there; sidestep the entanglements of shoals, capes and rocks along the way; avoid putting into harbour and sail the ship, unless it suffered catastrophic damage from storm, fire or collision. The only port the owners of the ship and cargo wanted to hear about was the one the vessel was bound for.

The result of such single-mindedness was that Benjamin and his crewmates were, literally, confined for months in the small space of the vessel. There was always a paradox in the sailor's life at sea. In the midst of the great wilderness of the ocean, the untrammelled freedom it represented and seemed to promise to him, the seaman was, nevertheless, an inmate in a kind of "total institution" for the duration. He could have been in a jail (as Samuel Johnson famously remarked, a somewhat better proposition than being on board a ship) or an asylum. In this isolated, sometimes violent, all-male society, which as a matter of course operated under the stress of perils of the sea, the sailor submitted himself to be ruled, with or without benevolence, by the captain and the mates. As for prisoners, the normal cares and uncertainties of the working man ashore were removed. Each man had food and shelter (of sorts) and the comfort of a sure routine without needing to make personal decisions, or having the opportunity to make them. Life was simple: tend the needs of the ship.

This was what Conrad missed the most when he finally came ashore to a liberalizing and increasingly rambunctious society. He looked back with nostalgia, and with some of its faulty recollections, to "that untempted life presenting no disquieting problems, invested with an elementary moral beauty by the absolute straightforwardness of its appeal and by the singleness of its purpose."

Conrad was a seaman for twenty years, and like a retiring athlete or a veteran home from war, he had trouble giving up his life of danger and adventure aboard ship for the writer's sedentary and solitary one ashore. More than a year after his last berth afloat, after the publication of *Almayer's Folly*, he still hankered for the life and cast about for a sea job. Even three years later, after he was married and had a baby, he thought of going back to sea with his family. After *Youth* was published, he travelled to Glasgow looking for a ship. He was "almost frantic with the longing to get away," he wrote. Perhaps "get away" are the key words: escape from the uncertainties of writing for a living, a venture, for him, more nerve-racking than the risks of the sea, which could at least be provided for by the sailor's rage for method and vigilance.

He was still a seaman, he wrote, one of the rootless men whose "home is always with them—the ship; and so is their country—the sea." The assumed nationality of the wind ship had been so necessary for him, the thrice-exiled man, son of a political prisoner, mother dead when he was seven, father four years later, an unreliable uncle his guardian, maybe a suicide attempt in Marseilles, a gunrunner to Colombian and Spanish rebels, shipwrecked, an unrequited love for "the woman of all time," a near-fatal duel—all before he went to sea on a British square-rigger for the first time when he was twenty-one. No wonder Conrad sought out a life of "no disquieting problems" aboard ship. No mystery that it was hard to give up. But sail was dying its slow death; there were no commands—he needed a captain's job to take his family along—and his hope withered away. He reconciled himself as best he could to life on the unkempt and irreverent shore.

Conrad saw life aboard ship as contained and orderly, but all around it was the perpetual fetch of the sea, forever threatening and unpredictable. Its huge emptiness, the inability of humans to control it, or even to mark it with their presence, was appalling, and still is. The brawn of the sea in a storm and its latent power when calm impress the sailor with his own fragility because his existence, his very self, is in constant jeopardy. No wonder it's difficult for those on land to understand the ways in which the lonesomeness of such surroundings leaves its mark. "The intense concentration of self in the middle of such a heartless immensity, my God! who can tell it?" wrote Melville.

"Ports are no good—ships rot, men go to the devil," complained one of Conrad's first mates, who was returning dead drunk and self-loathed to his ship yet again. Ships should be out on the sea where they belonged, not shut up in noise and filth, chained to the corrupting land. Ports were necessary to get, or to get rid of, cargo, and to replenish supplies and crew, but ships in dock were literally out of their element, which was the infinite, cleansing sea. He has seen ships come out of some docks, writes Conrad, "like half-dead prisoners from a dungeon, bedraggled, overcome, wholly disguised in dirt." You'd think that a ship would die from the ill-usage of ports, "like a wild bird in a dirty cage." But because they are faithful to their men, ships survive their incarceration and live to go to sea another day.

CHAPTER TWO

Un-homing, Up-anchoring

Fare ye well ye Liverpool Molls,
Painted, powdered, "pretty" dolls,
Ladies! laugh at my downfall,
Fare ye well and to hell with you all.

SHANTY

On May 5, 1885, the *Beara Head* finished loading and was ready for its liberation. Paddy Fearnaught, one of the most notorious Liverpool crimps, had delivered the last-minute seamen, if that was what they turned out to be. They came on board in a drunken herd of five men, most without any belongings, who collapsed in unison on the main hatch cover and fell asleep. The mate merely glanced at them; he had lots of experience with riff-raff. They weren't worth expending any effort on now, but when they woke up, sore-headed and confused, then his time would come. Benjamin already felt sorry for the late arrivals. The mate would bring down his anger on them like the wrath of God. Their voyage would not get off to a good start.

The captain had received all his necessities: clearance from customs and the port authorities, permission from the Bramley-Moore's all-powerful berthing master to undock and a letter containing last-minute instructions from the owners. It took him some time to read it.

He had their full confidence, they wrote, on this his first command for them. Nevertheless, they understood very well the risks involved in any voyage round the Horn, and they chivvied the captain like maiden aunts will the butler before a summer lawn party. He was to conduct his business as if the ship belonged to him and was uninsured. If he did that, no matter what happened to her, he could never be in the wrong. In particular, he must ensure that a good lookout was kept, especially when approaching land, when the leadline should also be constantly employed. He was to maintain the ship to a high standard and should use every opportunity to check the frames and inspect the floors (parts of the iron hull). When he removed the mast wedges for any purpose, he was to make sure the masts were scraped and painted where the wedges had been. When the holds were painted, every inch of metal had to be thoroughly coated. He was especially urged to keep all moveable ironwork in order so that it didn't become "froze," to look after the boats and to keep the topsides in good order. He had to be careful always to have enough ballast, and too much was better than too little. The owners were sure that he had already discussed with his predecessor the amount of ballast required for the ship to stand in dock—250 tons was the minimum—and to move in a river or tideway. He had to keep track of the ship's draft both fore and aft whenever it changed; he had to report his position to shore stations or other vessels at every opportunity; in case of an accident, he would on no account give up command of the ship to anyone. He had to be as strict and rigid with disbursements for food, gear or repairs as was consistent with safety and efficiency. He would not buy anything from other ships at sea. Good officers were hard to find, the owners admitted, and so he should take his time finding them. Their sobriety was a *sine qua non*. He had four green apprentices, although two had spent time on a school ship, and he should do all he could to turn them into smart young officers.

As for the crew, the number was up to him. They preferred British seamen, the owners wrote, but if he had to take foreigners, Scandinavians would do. He was to give the sailors as little advance pay as possible to reduce the chances of desertion. The crew's well-being was of great concern to them, and they recommended cleanliness as the mainspring of health. The seamen should be fed larger amounts of food in cold weather than in hot. The captain should keep a very strict account of all stores and inspect the steward's books weekly. Keep the beef and pork as cool as possible, they wrote, and take the bungs out of their barrels occasionally to make sure the meat's pickle was all right.

On and on went the list of exhortations, page after page. The captain might have been lord of the ship at sea, but in a home port, he was painfully subject to the worries and whims of the owners of his command. And vulnerable to extortion by chandlers and tradesmen as well, not to mention the crooked crimps, who had probably, once again, given him a gaggle of malingerers, drunks and farm boys in place of the able seamen they had promised. No wonder captains mostly looked forward to getting clear of land; they went to sea like monarchs entering at last, and after endless trials, into their rightful dominion.

Benjamin saw one other person come aboard the *Beara Head* in the hours before the ship sailed. He caught a brief look at a young, pale, dark-haired woman accompanied by several wooden trunks; she went below immediately into the aft quarters. The captain had a wife, and it seemed that she was on board for the duration.

At slack tide, the crew warped the ship into the lock that lowered the vessel down to river height, literally manhandled it away from the wall with long lines taken from the lock bollards to the ship's capstan (the latter usually used to haul up the anchor or help trim the braces on the big lower yards). Near high or low tide, the Mersey ran fast—vessels were shot out to sea at ten or more knots over the ground on a full ebb, and that made it difficult to negotiate the channels through the bar at the river mouth. The *Beara Head* had to pick its time to leave when the current was relatively slight; the waiting steam tug could more easily get a towline on board and begin hauling the square-rigger down the river and clear of the land.

There was no ceremony on departure. As many as 150 vessels left or arrived at Liverpool on any day, fifty thousand of them every year. And there was nothing unusual about the *Beara Head*, not even the fact that it was a four-masted barque—square-rigged on the forward three masts and with the yacht-like fore-and-aft sails on the aftermost mast. (A barque, by definition, was a three-masted vessel unless otherwise described.) These four-masted square-riggers had begun to appear in the 1870s. Bigger ships were needed for bulk cargoes, and the new vessels were huge by traditional sailing-ship standards. Adding a fourth mast, fore-and-aft rigged for more efficient handling, was a good way of keeping the sails a manageable size.

By 1885, however, the new rig had become commonplace. A few longshoremen or crewmen on other vessels paused a moment to watch, but that was all. Benjamin was too busy keeping up with the mate's bellowed orders—hauling on warps and beginning the long job of scrubbing down the ship's decks and superstructure to rid them of the hateful, greasy, staining coal dust—to see what was happening and where the ship was, the transition from land to water sliding by unnoticed. By the time he had a chance to look around him, they were well down the Mersey, the city's docks had faded astern, and Liverpool Bay and the familiar Irish Sea were opening up ahead of them.

The beginning of Benjamin's first deep-water voyage was an unremarkable event in part because the *Beara Head* was leaving a busy port on a routine voyage, as just another freighter under sail hauling a dirty cargo round the world by the usual routes. But leaving the land lacked drama anyway, compared with the tense anticipations of landfall and arrival. That was when a ship was most vulnerable. After thirty or eighty or 120 days at sea, a tired captain found his navigation truly put to the test: his measurements and calculations were all directed towards finding a particular headland, one light, a familiar configuration of shoreline. Only the solace of actually seeing the right piece of land would relieve the ship of its anxiety. In rain, low cloud or fog, especially if the vessel was running hard before a strong wind, closing fast with the shore, the problems of making a safe landfall were compounded. "To see! to see!—this is the craving of the sailor," said Conrad. ("As of the rest of blind humanity," he added.) The haven of land close-aboard was transformed into a menace if the sailor couldn't see it. If the captain's science and guesswork were even a little wrong, it might be only a matter of short time before the ship found the piercing rocks with their strange passive power to rip apart iron and steel.

None of this applied to leaving land. "The departure is distinctly a ceremony of navigation," said Conrad. The seaman took back- or cross-bearings by compass on the known, familiar landmarks and noted them on his chart. With open water ahead, not being able to see didn't matter much. The crew, occupied with getting shore dirt off the ship and securing it for sea, usually didn't notice the passing of the real departure point and the true beginning of the voyage—the moment when the land disappeared. In the narrow confines of the Irish Sea, however, it would be a prolonged leave-taking for the *Beara Head*, as the ship's course took it close enough to the coasts of Ireland, Wales and England that they would be visible for days in good weather.

*

Before the barque had cleared the estuary of the Mersey, still under tow by the puffing little tug, the mate called the whole crew aft to the break of the poop deck for the first important ritual of any voyage: choosing the watches. The officers would divide the crew into two groups: the port watch, under the command of the first mate, and the starboard watch, under the second mate.

There were twenty-two seamen on board the *Beara Head*, a number that included the five thrown on board at the last minute by Paddy Fearnaught, but not the four apprentices or the so-called idlers—the cook (or doctor), carpenter, sailmaker and captain's steward. The idlers usually worked only by day, twelve hours on, twelve off, unless all hands were needed for large-scale sail changes or ship manoeuvres; then they turned to like everyone else.

The captain stood no watches, although he was often obliged to keep an eye on the second mate, who was often considerably less experienced than the first mate, to make sure that he did the right things when the wind began to kick up, and that he was able to deal with the malingerers, sea lawyers and hard men in his watch. As a matter of course, the captain spent more time on duty than anyone else on board. If he was supposed to treat the ship, as the owners had instructed, as if it was his, and uninsured, then he could never really consider himself off-duty. Before making landfall or approaching the narrow channels en route—for example, the Strait of Le Maire between Staten Island and Tierra del Fuego as the ship approached the Horn—the captain was always on deck, navigating or worrying. In bad weather, he might hold to his position on the weather side of the poop, braced against the jigger-mast shrouds, more or less continuously, day and night, until conditions eased. In the traditional strict routine of the ship, the mate on duty was then relegated to the lee side of the deck unless called across by the captain for instructions or the comfort of brief conversation. Just like the lines and gear of the sailing ship, each man had his neat, allotted place in the order of things. On a black night, when there wasn't enough ambient illumination to see anything except the phosphorescent slashes of wave crests, it was crucial on these lightless vessels to know the precise location of each element of its arsenal: the mate on the lee side of the poop, the watch clustered on the main deck at the break of the poop, or—some esoteric examples—the fore-upper-topgallant halyard belayed to the pin third from aft on the starboard side foremast pin rail, the main braces belayed port and starboard on the second pin aft in the four-pin cluster just forward of the mizzen shrouds.

This arrangement of things and men was always the same, with very few exceptions, from ship to ship, whether British, American, German or French, so that even a hungover, shanghaied seaman dumped onto the *Beara Head* without ever having seen the vessel before in his life would know exactly where everything and everyone was to be found.

The captain, in his dress jacket with brass buttons, stood on the poop overhang, the edge of his exclusive domain, and spoke down to his men like the absolute ruler he had become when the ship cleared the lock. His dog, the small brown-and-white terrier, which had already shown itself, like all captain's dogs, to be a biter of seamen, squatted at his feet. The mate scowled beside him on one side, and the second mate, a tall, dark-haired, olive-skinned twenty-year-old from Bristol, fidgeted on the other. Before the mates chose their men, the captain made a speech.

"Now, you men listen to me. Do your work and be lively about it, and we'll do fine. Any man who malingers, Mr. MacNeill here'll barge 'im." He nodded towards the mate, who made no acknowledgment but continued glaring at the men, as if he could barely contain himself from leaping over the rail to begin barging them right away. None of the crew, except for the Ulstermen, knew that barging was merely a kind of vehement scolding, but each drew his own conclusions; if it was something the mate was responsible for, they'd better beware. The captain's speech was rife with the part Scots-English, part Gaelic vocabulary of Ulster. He was from Islandmagee, a small community of a few hundred on the Antrim coast, which, in an odd confluence of tradition and family expectations, turned out ships' captains by the score. It was one of the reasons why his dancing career had not been merely disappointing; it had been scandalous, a suspect repudiation of his family and of the usages of the island itself—not to mention the fact that jigging was a Fenian perversion. But he had come around.

"I want no gurning about the whack or the work. You boys and green hands, you're not at your mammy's house now. No farles or buttermilk, just salt horse and hard tack—sailors' food.

"Now, you know that we're bound round the Horn to Valparaiso. That's a hard go, but I expect each man to stand his watch without grummiein'. Any men who doesn't will be all catched-on."

Benjamin heard a slight shuddering murmur run through the men. He realized he too had made a noise, a kind of hum or groan deep in his throat, not in any response to the vague threat of punishment, but because the captain had spoken the name of the place sailors feared the most. For Benjamin, it was almost as if the captain had broken a taboo by saying it aloud. Perhaps he had given the predatory rock warning; now it knew they were coming. Just superstition. Still, it was as if some die had been cast, the path he had entered on now made clear in all its hazard. Even at a distance, the malevolence of Cape Stiff seemed to loom over him. He recalled the song he'd heard one December 25, drifting in with the wind to the Carrick shore from a square-rigger towing out of the loch:

> A *hell of a Christmas Day, boys,*
> A *hell of a Christmas Day,*
> *For we are bound for the bloody Horn*
> *Ten thousand miles away.*

The captain again, in his harsh Antrim accent: "Remember: each and every one of you belongs to me and myself, body an' soul, while you are on this ship. Now the mates'll pick the watches."

So ended the captain's speech of welcome and initiation.

"Russell!" shouted the mate immediately. A tall, burly, black-bearded Yankee able seaman in his mid-thirties moved over to the port side of the main deck. Benjamin had heard that James Russell was the best seaman in the fo'c's'le, a veteran of a dozen American and British ships. A droll man with a quiet voice and a talent for mimicking almost anyone with tart accuracy, he had the mate down to a fault already, that officer's short-legged strut, his mad glare and, at a much-reduced volume to avoid detection, his bullhorn bellow. Russell had been resigned to being chosen first by the mate and took his place without expression.

The second mate began with another of the certified able seamen. There were nine of these men, the most qualified sailors on board, and they were distinguished from the less-skilled ordinary seamen and, at the bottom of the heap, the green hands, like Benjamin, and the boy apprentices. All the able seamen went first, no questions or hesitation, and the mate chose several more of them besides Russell: Jack Grey, a middle-aged Londoner who combined the accent of the Stepney streets he'd come from years ago with deep anger at the grievances he'd accumulated since then; Michael Heany from Skibbereen, County Cork, also in his forties, known as Paddy; a morose, monosyllabic Finn called Jan, who, in an earlier conversation with Benjamin, had asserted that the world was flat. An odd conviction for a sailor to hold, in Benjamin's view, given the horizon's curve and the progressive materialization from the top down of other ships as they approached— clear evidence, he thought, of roundness. The Finn looked even sadder when he was separated from the other Finnish able seaman on board and two Swedes, who were called into the starboard watch by the second mate. The ship's owners had stipulated a British crew or, failing that, Scandinavians, and the Finns and Swedes represented the captain's efforts to oblige. The mate's last skilled seaman was a German, Helmuth, who was the crew's "Dutchman" representative. (The word "Dutchman" came from *Deutsch*; real Dutchmen from the Netherlands were lumped in with Scandinavians as "scowegians" or "squareheads.")

The mate looked at the rest of the candidates for his watch with dour disapproval. He chose the tough-looking, muscular Anderson, who had at least distinguished himself under the coal shutes with his steady work and willingness to crawl into the smaller underdeck spaces to trim the cargo. Then, to Benjamin's surprise, he was next; happy to find himself with Anderson (although not pleased to be under the mate), he joined the men on the port side. MacNeill picked two of the four apprentices, apparently at random—they were fifteen- or sixteen-year-old, gangly, pimply boys and would be liabilities on the watch for a month or more—and four other ordinary seamen, including, at the end, three of the last-minute Liverpool arrivals.

The mate had pummelled awake the crimp-delivered men soon after the ship had cleared the lock into the Mersey, kicking and punching them into consciousness. These ragged, still-drunk men now stood haphazardly by the poop.

"Are there any seamen amongst you?" yelled the mate.

There were a couple. One was a small, thin, delicate man with huge ears, a high voice and a thick Liverpool accent; he was still tipsy but could think and talk. He had been a seaman aboard the ship *Dawpool,* to Calcutta and San Francisco and back to Queenstown, in Ireland.

"Port watch," said the mate without insults or oaths. These voyages represented several hundred days at sea. He was pleased to have an unexpected experienced hand, even if he resembled an elf.

Another of the Liverpool men had been a fisherman for five or six years, and he went into the second mate's watch, but the rest of them were just what MacNeill had feared: the usual assortment of Sailor-town tavern dross with the bad luck to have been drunk enough that the crimp's offers of more booze, free women and a little cash in their pockets sounded like just the ticket, the finest prospect imaginable. They hadn't quite caught the part about having to go to sea on a hard-case square-rigger as payment for the crimp's generosity.

One emaciated man, dressed in a navvie's rags, was still deep in his alcoholic fug and couldn't yet speak. The mate was running out of prospects and had no choice but to take him into his own watch. The mute drunk had no gear or possessions of any sort, and he would have to outfit himself from the slop chest, a store of goods of all kinds administered by the captain, who was free to mark up the prices for his own profit, one of the perquisites of his office. A seaman who arrived aboard without oilskins or a knife or clothes was obliged to buy them from the slop chest. The amounts were deducted from the pay owing to him at the end of the voyage, which became little, or nothing at all, if the sailor had been especially needy. Tobacco was the real money-maker. Towards the end of a long passage, seamen would have handed over their immortal souls for a quid of chew, never mind paying the captain's astonishing prices. Benjamin had been warned by old sailors about the dangers of slop-chest purchases, and as an abstainer from tobacco, he believed he was prepared to escape the captain's extortion.

Another of Paddy Fearnaught's shanghaiees was a strange sight on the deck of the *Beara Head*. Dressed in a cheap, worn suit and a collar no longer stiff and long-ago white, he had the pale, limp look of a man who spent all his time indoors, like a clerk, and in fact, that was what he turned out to be. He worked in a coal exporter's office, lived alone, had stopped into the Salmon tavern, opposite St. Nicholas's Church, for a quick drink—something he seldom did. There he allowed an affable fellow-drinker to buy him a whiskey, and then a few more, and then to introduce him to a woman whom he vaguely remembered as beautiful beyond compare, and now he found himself staggering in the waist of a square-rigged ship, trying to brace himself against a slight Liverpool Bay swell while a short, very loud man with a badge on his cap abused him.

"I should not be here," he told the mate with woeful indignation. "I've been kidnapped. I demand you take me back to shore, or I can go back on the tug. You can't keep me here. I'll report this to the peelers. My job, rent, my sisters in Birkenhead . . ."

Standing by the port bulwark, Benjamin watched as the mate demonstrated how hot a ship this would be. His wind-blasted face reddening, teeth gritted, fists clenched, he stepped close to the clerk and, without a word, punched him hard on the side of the head. The clerk collapsed and began to sob. Benjamin saw the captain, who had just come up the companionway from his cabin, stop and look down from the poop. He saw the blow, Benjamin was sure of that. The mate had no authority to carry out physical discipline without the captain's approval, but the latter's only reaction was that pause and a thoughtful look. Then he turned away and walked across to inspect the compass. The mate grabbed the man's collar, stood him up and kicked him, then drove him with more kicks forward along the main deck and into the seamen's deck house, shouting oaths and threats all the time. The clerk was well and truly shanghaied.

Returning aft, MacNeill spoke to the second mate: "Mr. Jagger, I'll have that man in my watch, keep an eye on the skunk."

"Aye aye, Mr. MacNeill," the second mate replied. With a slight twinge of distaste, Benjamin thought, and something else: perhaps fear.

Daniel Jagger had just won his second mate's ticket, and this was his first time as an officer. He had been chosen over many more qualified men because his father, a well-off Bristol merchant and alderman, knew the *Beara Head*'s owners and had arranged things. Jagger was indeed afraid—that his inexperience would be exposed, that he would not be able to command the men in his watch, and now that this loud and violent mate might turn on him sometime. He felt as if he was fighting on all fronts.

Following tradition, the cook belonged to the port watch. Benjamin had talked already to the garrulous doctor. He was another Liverpool man, born in the city to famine refugees but, like so many of its people, still as Irish as paddy's pig. Unlike most cooks, who were notorious for their slovenliness, he was a clean man, fastidious about his galley and himself. He was proud of his effort, seldom successful, to make edible the inedible salt meat and hard tack, and of his skills with the sailors' "delicacies": the baked hard tack-and-molasses dandyfunk, and lob-scouse, a mixture of salt meat and hard tack, with potatoes and onions if there were any (there seldom were), minced and stewed together. If all hands were called, the cook's job was to handle the fore sheet—one of the lines controlling each lower corner of the foresail.

"God sent the food, and the devil sent the cooks," went the old saying. No matter what actual duties the cook pulled on top of cooking, he always had to deal with his particular outsider status aboard ship. Food and how passable or execrable it was (it was never better than passable), how much was available, whether it was hot, the quality of the coffee, the degree of infestation by weevils and cockroaches—all were matters of obsessive interest on any ship. The cook was the front man for the sins of others; if provisions were delivered spoiled or underweight, and they often were, and the food was short or tasted foul, the crew more often than not held the doctor responsible (even though the captain's steward was in charge of the ship's food stores, and it was often his doing). On the other hand, no one wanted to offend the cook. If he was affronted or slighted, maybe he would sulk and withdraw the small favours—a dab of jam or a pinch of sugar reserved for the captain's table and begged from the steward, a palatable dandyfunk—that helped convince men in the fo'c's'le that they weren't in a hard-labour penal colony after all. As a result, the crew treated the cook with gingerly deference, which disguised resentment at having to do so, and which could change quickly to venom if things on board went generally sour. Seamen had little to do with stewards, whom they considered irredeemable supercilious tightwads.

The carpenter and the sailmaker were the other idlers. The carpenter was a privileged person aboard ship. He worked solely under the captain's command. The mate had no authority over him, at least when he was working his trade. Most carpenters were also qualified seamen; in that capacity, like the other idlers, they had to pull and haul and, when necessary, go aloft, and then the mate's word ruled. Alexander Johnston, the *Beara Head*'s carpenter, had come out of the navy after ten years in a sailing man-of-war and then in a steam corvette. He had a shaved head, a full beard and the quiet, straight-backed self-containment and trim neatness of the long-service man. As tradition demanded, he was assigned to the port watch. The sailmaker's duties as a seaman were the same as the carpenter's, and while he was sometimes assigned to either watch as necessary, the *Beara Head*'s sailmaker, John Page, a reserved, elderly Yankee, was permanently attached to the second mate's starboard watch.

As the mates made their choices, Benjamin looked out at the land falling away on either side of the ship. The *Beara Head* was beginning its long, easy pitch and roll to the scend of swells swinging down from the northwest, its masts already moving in looping parabolas against the hazy sky, the motion exaggerated in proportion to their height. Conrad wrote about the "preposterous tallness of a ship's spars," and they still looked that way to Benjamin. He peered at the opaque and enigmatic aspect of the sea ahead of the ship. In the haze, the horizon was only three or four miles away, sea and sky fading into each other, a grey anonymity. Although he smelled the clean, astringent spice of salt water, it was only the narrow Irish Sea. The real ocean, the great North Atlantic, out on the other side of Ireland, was still some days away. Nevertheless, looking past the *Beara Head*'s pitching bowsprit, he felt taut and elated, trembling as if he was up aloft again, clinging to the royal mast.

Benjamin's previous voyages had been short round trips. Starting out each time, he knew that soon he would turn around and come home, running or tacking back over the same ground. This deep-water wind ship was different: a one-way passage out to the vast sea that was already opening up to him as he stood on the deck, with the mates' voices tolling out the names of the men he might go through hell with, binding himself to them. It was the beginning of his great adventure, and he thought that things would never be the same now, that anything was conceivable; there were no limits on what he might do or become. He would prove himself on this ship carrying him on and out into the Atlantic, and beyond that, to the Americas and the East and all the lands of the world empire. He might have been one of its despised subjects in Ireland, but it was, nevertheless, open to him everywhere else, his origin and accent no impediment. Its ships travelled the earth's seas in their tens of thousands. They were his ticket anywhere.

The watches assembled around him. They looked like a crew of privateers, Benjamin thought: one man bare-chested with tattooes that spiralled and tumbled down his hard torso and arms; others bearded and ear-ringed; several in bright shirts, red, purple and yellow; some in plain, stained dungarees; most of them various shades of tan, and brown, lined faces of a hard life in the sun and a harder one in taverns, with scars showing on every skin surface; most barefoot in the warm spring sun; all wearing knives and spikes on their belts.

In their proximity, Benjamin felt as if he were already swelling up into a man. These weren't the handful of stay-at-home half-lubbers on a coast-hopping smack, complaining about missing Sunday with their families, dodging in and out of port in fear of yachtsmen's gales and the short Irish Sea chop. The *Beara Head*'s crew were men of the world—*from* the world, for that matter, as the babel of accents confirmed. Above all, these were men who had crossed the open ocean, or most of them were anyway. They had been round the Horn, and many of them the other stormy capes as well. They knew what it was like to throw themselves out beyond the safe pale of land into the immensity of water, where anything could happen at any moment. Benjamin had never admired anyone more than these ragged, tattooed, wild-looking men. He hungered to be like them and to be accepted by them.

As the little tug towed the *Beara Head* out into the waters of Liverpool Bay, Benjamin could see scores of sailing vessels: coasters ghosting along in the feeble northerly breeze; big square-riggers under tow in and out of the river; and farther out, ships, brigs and barques under full sail, trying to make headway towards or away from the land, barely moving their large bulks in the capricious conditions. There were steamers as well, ploughing straight lines in the sea, except where they had to wind through the bar channel, the wind irrelevant to their progress and the Mersey's tidal current just an inconvenience. But Benjamin's main impression on this spring day in 1885 was of ships under sail, or those that had just furled sails or were about to unfurl them. Even the puffing tugs, dwarfed by their overbearing tows, were almost invisible in the swarm of wind-driven vessels.

When he decided to go to sea, Benjamin never considered anything other than a sailing ship. He spent a few years in school; he worked as a butcher's delivery boy and later spent time as an apprentice carpenter. He liked working wood, and he had an easy coordination between hand and eye that made the saw, chisel and plane ride smoothly in his hands as he fashioned a dovetail joint, scarfed a repair or squinted and planed fair the curve of a bannister. All the time, however, he thought about going to sea. It was in the family and had been for generations. The little, walled harbour at Carrickfergus often drew him down to look at the trading and fishing smacks floating there or dried out at low tide, the dour Norman castle, with its English troops, shadowing him. On a clear day, he could see the Down shore across the loch and, in the foreground, the square-rigged ships towing in and out. Sometimes, he saw a wind ship run out under sail before a fair westerly towards the sea, courses and topsails set, men aloft loosing topgallants and royals or on deck hoisting staysails. His family couldn't afford to pay to apprentice him aboard ship, but when he was old enough to go as a hand, he took up the old family trade.

Benjamin was comforted by the abundance of sailing vessels in this small sea-precinct of Liverpool Bay. Their numbers made them appear so necessary. He could imagine the coasters being displaced by little steamers; that process was well underway. But the deep-sea ships— how could their range and independence ever be equalled by steam engines, with their unending lust for coal? Surely the big square-riggers, the real sailing ships, were safe. Harland & Wolff was still turning them out in Belfast; so were yards on the Mersey, the Clyde and the Tyne, not to mention the hundreds of wooden ships built every year in America and Canada. Benjamin couldn't believe that things so great and glorious could ever disappear.

Three hours after leaving dock, the *Beara Head* passed the Bar

Light Vessel and was clear of the silted shoals of the river mouth. The tug began a slow, wide turn to the west and a course that would parallel the north coast of Wales before turning south down the Irish Sea towards the St. George's Channel, marked by The Smalls and the Tuskar Rock lights, and out to the open Atlantic. Because of the persistent light air, the tug had been hired to tow the barque as far as Great Orm Head, another four hours away, in the hope of finding wind there. Meanwhile, the light northerly wind blowing across the beam—at right angles to the ship's direction—augmented by the tug's speed, was enough to fill the canvas and aid their progress. The *Beara Head* could set its sails.

The arrangement of sails and running rigging on a nineteenth-century square-rigger looks very complicated, but this impression is really the result of repetition rather than genuine complexity. There are only two species of sails—square sails and fore-and-aft sails—and each kind, together with its controlling lines, is a simple mechanism by itself. The apparent complexity arises only because each sail is repeated many times over. A four-masted barque like the *Beara Head* carried thirty-two sails, of which eighteen were square and fourteen fore-and-aft.

The upper side of a square sail is attached to its yard, the position of which is controlled by the braces. The yard itself is partly supported by lifts, which are lines leading from the yard to a point higher up on the mast. Most of the yards are stationary, but some—the upper topsail and upper topgallant on each of the square-rigged masts—are hoisted and lowered bodily by halyards attached to the yards themselves. Each lower corner of the square sail is controlled by a sheet (and, in the case of the courses, a tack), which is adjusted to form the sail into its most efficient aerodynamic shape in the variety of wind directions. Finally, each sail is fitted with two types of lines that are used to haul the body of the sail up to the yard before the crew goes aloft to furl and secure it. The clewlines, which are fitted to each lower corner of the sail (alongside the sheet and the tack), haul up the corners of the sail to the yard; the buntlines run vertically over the main body of the sail and are used to haul that part up as close as possible to the yard. When clewlines and buntlines have been hauled up tight, the square sail hangs down in looping folds, half under control and more manageable for the men who have to complete the job by furling it. All these lines are led directly down to the deck—or through blocks aloft, to give a fair lead—and then down. Each is secured to its designated belaying-pin, always the same one.

Fore-and-aft sails on a square-rigger are mostly the familiar triangular sails found on modern yachts. Each is controlled by a halyard to hoist it, a downhaul, a tack to secure one corner and a sheet to control the other, as well as set the sail in its proper orientation to the wind. On four-masted barques, the spanker, the large lower sail on the jigger mast, was either triangular or gaff-rigged, a four-sided sail that was just a variant of the fore-and-aft model.

The crew of a short-handed merchantman like the *Beara Head* could handle only a few sails at a time. In the old days—in the early 1800s, for example—on much smaller ships with fewer sails, a complement two to three times that of the *Beara Head* was able to work most of the sails simultaneously. That was especially so on naval vessels, with their many hundreds of seamen. The number of men making up the *Beara Head*'s entire complement might be assigned to the weather side of the main yard of a warship. When the order to make sail was given, the men swarmed the vessel and forty sails were set and trimmed in two minutes. The paucity of crew on these new giant merchant barques complicated things for the captains and mates. Every decision about which sails to set or take in, and when to do it, had to allow for the shortage of muscle power. A captain always had to gauge the collective strength and endurance of his raw material, his units of production.

"Now she's ready, Mr. MacNeill," the captain told the mate. "We will make all sail, if you please. Brace the yards round."

"Aye aye, sir."

The mate did a quick seaman's drop down the poop steps, hands sliding down the rails, feet not touching, and headed forward along the main deck, bellowing as he went: "Port watch to the fo'c's'le head. Get a wiggle on, you goddamn scum. Time to do some sailors' work, if you can remember how."

He turned to the second mate, who was following behind, his voice losing only a touch of its volume. "Mr. Jagger, you'll take the mizzen and the maintops'ls and to'ga'nts."

"Aye aye, Mr. MacNeill," the second mate replied with a touch of resentment in his voice. There was no need for this damned noisy mate to tell him that. The first mate's port watch took the foremast, the second mate's starboard watch the mizzen, and they divided up the mainmast between them, with the starboard watch or the apprentices handling the jigger as opportunity arose—the division of labour on all four-masted barques, each mate and his watch with their own inherited domain. As Benjamin, on his way forward, passed Jagger, he saw the man staring hard at the squat mate's wide back before turning to his own men, who were gathering around him. Maybe a hot ship in more ways than one, Benjamin thought.

In a minute, the crew was divided into three groups and clustered at the foot of each mast, the mate at the foremast, the second mate at the mizzen, the experienced idlers joining their respective watches without being told, an ordinary seaman at the wheel. (No need for any steering skill in these light conditions.) Then silence, as every man turned back towards the poop, where the captain, hands clamped to the rail, had watched the fluid deployment.

In the traditional division of place and labour on square-riggers, the captain stayed on the poop unless some urgent necessity brought him down to the main deck to emphasize an order face to face or to add his weight to a line or help repair damage, his sudden democratic presence a sign of how desperate things were. The mate was the only officer to go forward right to the foremast and the bow, traditionally the territory of seamen "before the mast," when general sail handling or anchor work was going on. Although the mate was the executive officer for the whole ship, the forepart of the vessel belonged to him. "He is the satrap of that province in the autocratic realm of the ship," wrote Conrad. The mate did not go aloft during normal sail handling; he was more valuable on deck with his overall view of the forepart of the ship, a perspective distant from that of the captain near the stern, and more important than ever on a four-poster more than three hundred feet long.

When big things happened on board—getting underway, coming to anchor, setting or taking in all sail or a substantial amount of sail in changing weather, altering the vessel's course by tacking or wearing ship—the captain took direct charge. Men jumped at his command then, and his orders were minute and detailed: when to haul halyards, let go buntlines and clewlines, trim braces; the all-important timing of "mainsail haul" (hauling the yards round to the other side of the vessel) during a tack. When the ship was sailing, the mate in charge of the watch could trim sails and take in or set more canvas without consulting the captain. The freedom accorded to the mates varied from ship to ship, and it depended as well on the personalities involved and the degree of amity and understanding among the officers. However, bigger sail changes meant the weather was changing, and the mate on duty had to tell the captain about that. He came on deck then and assumed direct command again. In practice, it was seldom necessary to call the captain to get up when conditions altered. All sailors become attuned to their vessels. A different pitch to the wind's blare, a new shudder in the ship's roll with a cross-swell or building seas—that was all it took to get the captain up onto the poop.

The captain had responsibility for navigating and working the ship—the word "working," in this case, meaning all the sail-handling and manoeuvring necessary to keep the vessel going in the right direction. It was one of the two great divisions of labour aboard ship. The other was the daily work of the crew: chipping iron and painting it (as endless a job as it had once been to prevent wooden ships rotting away or being devoured by teredo worm), maintaining standing and running rigging, splicing, serving (wrapping "small stuff," or very light line, around bigger lines to protect them). All this was the precise, timely and detailed work required to keep the machinery of the ship in good working order, no surprises, everything provided for—the sailor's obsessive concern with every scrap of line, wood and metal on board. The wind ship's genius lay in the intricate interconnectedness of all its gear, but so did its vulnerability. One small break or failure could mean the injury or death of a man, or many men, or the unstoppable slide into disaster of the whole construction. On a ship, the want of a nail could always lose the kingdom. For all this essential fussing with things, the captain spoke his general desires and concerns—about masts or rigging or iron work—quietly to the mate, whose duty it then was to see them done.

The second mate's position was summed up in the seaman's saying "Becoming second mate doesn't get your hands out of the tar bucket."

This officer had control of the starboard watch when the captain wasn't on deck, and he had the same responsibility and power as the mate to look after the ship, to trim or change sail. But he led by example as much as by orders. He was supposed to be the equivalent of the factory lead hand or the non-commissioned officer who goes over the top first—in other words, the most energetic and skilled man among the seamen. He was first aloft, where he was expected to take the most dangerous jobs—at the weather end of the yard, for example, or furling the bunt of the sail (its bulky centre part). When there was hauling to be done, he took hold of the line first, expended the most energy. As a matter of course, the captain spent a lot of time on deck, and then the second mate worked with the seamen, carrying out the mate's instructions for the day's jobs.

In the silence, the *Beara Head*'s crew, eyes turned aft, waited for their captain's order to make sail. Behind him, for the first time since their departure, the men saw the captain's wife, standing alone by the great-cabin skylight, a slight, still, dark-haired figure, looking out of place in the midst of this rowdy mobilization.

"Ready, sir." The mate's voice, Benjamin thought, could probably be heard back in Liverpool.

"Man the lee braces," shouted the captain, no mean long-distance communicator himself.

Some of the experienced crew were already at the brace belaying-pins on both sides of the ship, and they cast off the big coils of line, carefully capsizing them onto the deck so that they would run free without kinks. Heading west with a northerly zephyr, the barque would be sailing close-hauled on the starboard tack—that is, with the breeze blowing from the starboard side. It was easier to brace the yards into position—hard up against the wires of the standing rigging on the port, or lee, side—before the sails were set. Even in these light conditions, it took less effort to haul round the heavy spars without the added weight of wind resistance on the sails. In a real breeze, bracing the yards first saved a great deal of hard labour.

Now half a watch tailed onto each brace; in heavy weather, with all the weight of the wind in the sails, an entire watch would be hard-pressed to sweat a main yard round. Benjamin hauled with other men on the foresail lee brace in the stamp-'n'-go tramp down the deck. For a few seconds, they pulled at cross-purposes. Then the shanty began, the singer momentarily stopping the pull as the men adjusted themselves to the rhythm, and they began hauling away in unison. "Way, hay an' up she rises! Patent blocks o' different sizes . . ."

The shantyman—Paddy, the southern Irishman—sounded out strong and clear in a smooth, practised baritone. It was one of the "walkaway," or "runaway," shanties, used for heavy deck work like hauling up topsail yards or this present business with the braces.

> *What shall we do with a drunken sailor?*
> *What shall we do with a drunken sailor?*
> *What shall we do with a drunken sailor?*
> *Ear-lye in the morning!*

The men's voices joined in the chorus, loud and breathless as they stomped to the song's quick time—it was one of the few shanties sung that way—to the tune of an old Irish dance:

> *Put him in the long-boat till he gets sober . . .*

And:

> *Trice him up in a runnin' bowline . . .*

The chorus again. And:

> *Give him a taste o' the bosun's rope end . . .*

And on and on went the verses—Paddy knew them all—through the various punishments, and then they branched out into the usual obscene and blasphemous variants:

> *What shall we do with the Queen of Sheba? . . .*
> *What shall we do with the Virgin Mary? . . .*

Until the lower yards were hauled into position, and then the topsail and topgallant yards, all the braces tramped down the deck to the ring of the shanty, the "Way, hay" of the chorus a loud, savage yell.

This sudden burst of vigorous song astonished Benjamin, ambushed him with its joyful sound; he laughed with delight as the singer sang his solo verses. The choir of men's voices sounded out of place on the deck of this strict-regime barque, with its shouting, abusive mate and the malevolent Horn lying in its future. Yet it struck him like a song of celebration too: of the romance and beauty of the ship.

With the yards in close-hauled position, it was time to set the sails. At the foot of the foremast, Benjamin thought, This is it. It was bad enough to contemplate that spider-like crawl angling out and up onto the platform of the top. Now he'd have to go out along the yards on the footropes. And do something useful too, not just hang on and pretend he was somewhere else. At least on this benign day, he wouldn't have to worry about being bucked off into the sea or down onto the deck, an ignominious death his first time working aloft. In the long swell, the motion up there wouldn't be too bad. He realized that his worse fear, more intense than that of the waiting heights, was that he would disgrace himself in front of these other men. He felt hunger for their goodwill, almost a frenzy of desire for their approval. He would risk dying to try to win it.

"Lay aloft, you men," the mate hollered, indicating half a dozen members of the watch. "Gaskets off. Russell, you tell these new men the drill. Give 'em the topsail yards. You Liverpool rats"—he addressed the shanghaied clerk and the scrawny drunk, now more sober—"cast off clewlines and buntlines when Mr. Johnston here"—he nodded at the carpenter, his honorific part courtesy for the carpenter's privileged status, part sarcasm—"shows you where the hell they are."

The drunk stared hopelessly at the dozens of belaying-pins around him; the clerk cringed, keeping an eye on the mate, ready to avoid another punch, or worse.

By tacit agreement, the Yankee able seaman, Russell, was the leading hand of the port watch. He went up the ratlines followed by Cavers, the elf-like Liverpool sailor. Experienced men, they jogged up while Benjamin, Anderson and the two apprentice boys climbed slowly but steadily behind them. The apprentices had been sent aloft by the mate a few times when the ship was still in dock; Anderson, like Benjamin, had gone up only once before. It was still an ordeal for each man and boy, but Benjamin found it unexpectedly easy to heave himself up onto the foretop platform and to continue, without a pause, mounting the topmast ratlines. As long as the others above him and beside him continued to climb, he could too. He felt their movement, like a web of energy around him, carrying him up.

Cavers the Elf, deceptively fragile, scrambled out along the main yard—whose length, Benjamin realized as he climbed past it, was twice the height of the mainmast on his little coasting smacks—and began casting off gaskets. These were the short lines that were passed around the yards and the furled sails to secure them; releasing gaskets was the first step in loosing the sails. Russell led the green hands up to the topsail yards, which were still close together before the upper yard was hoisted up and away from the fixed lower one.

"Okay, boys—half of you out each side. Just pull 'em free and coil 'em," he said in deadly imitation of the mate's loud rasp, and Benjamin, even in his funk, grinned in appreciation.

"You'll get used to the footropes," Russell added in his own, more kindly, voice.

"In a pig's eye!" said Anderson, who was clinging to the ratlines and breathing heavily. "*You* get used to the fuckin' footropes."

Russell laughed and slipped the knot on the gasket, then coiled it and finished it off, passing a loop through the coil and stuffing the coil itself back through the loop. This produced a neat bundle that wouldn't chafe the sail. Benjamin swung off the ratlines onto the top of the yard, amazed again by its size. The scale of things on these four-posters! He stepped down carefully onto the footrope, which abruptly plunged eighteen inches under his weight. He slid farther out, trying to get a grip on the furled sail for support. The unyielding stiffness and weight of the canvas surprised him; it was entirely different from the light material of a coaster's sails, supple with wear. This was the *Beara Head*'s number-one canvas, bent on in anticipation of the North Atlantic gales they would surely meet. When Anderson stepped onto the footrope, it jerked up under Benjamin's feet. He yelled and clutched the sail, trying to wrap his arms around its bulk. The footrope evened out, but now it snapped back and forth, and his legs began to shake with fear and the strain of trying to stabilize the menacing line. Anderson was silent, but Benjamin could see that he was locked into the same frozen position. Looking down sixty-five feet, Benjamin heard the mate shouting at them. He was inquiring whether they were going to get the goddamn gaskets off or were they going to wait for nightfall or doomsday.

Dear Jesus, Benjamin thought. This is a calm day. How in the name of God will I do this when it's rough? It seemed unimaginable.

Slowly, much too slowly for the mate's liking—did he have to climb aloft himself to cast off a few fuckin' lines and then knock some useless jackasses into Liverpool Bay?—the new men got the gaskets off and coiled. At least coiling was easy for Benjamin, no different from what he'd do on any vessel.

"Main and tops'l gaskets away," Russell yelled down to the deck.

And all hell appeared to break loose there, a bedlam of orders from the captain and the mate:

"Slack away clewl'ns and buntl'ns!"

"Haul away main and tops'l sheets!"

"Slack away jib downhauls! Haul away all jibs!"

"Belay fore and main course tacks!"

"Belay mizzen tack."

No shanties now: too many individual seamen and groups of them hauling and pulling many lines to different rhythms simultaneously. The work produced only the "unnameable and unearthly howls" of sailors on a line: the belly-deep yelling grunts represented, inadequately, in writing as "Yeo heave-ho."

> *Your maryners shal synge arowe*
> *Hey how and rumby lowe.*

The sails opened up, unfolding like horizontal theatre curtains, and hung down, barely filled into shape in the scanty wind.

"To'ga'nt and royal gaskets off!"

Benjamin joined the others climbing farther up and inched out along the higher yards, shaking off and coiling the light lines. The fear he felt at the greater height was balanced by the unfurled sails below him; they obscured the view down, creating an illusion of safety.

Back on deck, the port watch reassembled for the heavy work of hoisting the upper topsail and topgallant yards. These weren't as brutish as the main yards on each mast, but they were hard enough to move by muscle power alone. The iron upper topsail yards were sixteen inches in diameter and sixty-six feet long; the upper topgallants were wood and somewhat lighter, but still a foot in diameter and fifty feet long.

The halyards were stand-and-pull jobs, and as the men jerked the yards up inch by inch, Paddy the shantyman sang out one of the "hooraw chorus" songs, this one popular on Yankee ships with Irish crews, and on Liverpool ships as well:

> *Sometimes we're bound for Liverpool, sometimes we're*
> *bound for France,*
> *But now we're bound to Dublin Town to give the gals a*
> *chance.*

And the traitorous chorus:

> *Hurrah! Hurrah! for the gals o' Dub-a-lin Town*
> *Hurrah for the bonny green flag an' the Harp without the*
> *Crown!*

Benjamin was a Protestant and didn't know the song. Neither did Jan, the taciturn Finn, nor Dutchman Helmuth. But the Liverpool-Irish seamen on this British ship knew all about Parnell, Ireland's "uncrowned king," and his long, implacable shove against the ancient weight of England; they chorused with pointed gusto.

With the royals sheeted home, all staysails hoisted and trimmed, and the spanker and its topsail in place, the *Beara Head* was under full sail, although still not sailing anywhere; almost all its progress was thanks to the steam tug, which bulled along into the lazy northern spring evening. No sail-handling job of any sort was complete until the scores of braces, sheets, halyards, clewlines and buntlines that lay on deck after the work of setting all sail were carefully coiled, or flaked down on deck, and then hung with a loop on the belaying-pin. Each had to be ready for the next time it was needed, when a seaman could dump the coil on deck with confidence that the line would run fast and free. It was a matter of keeping everything aboard shipshape—a term that had become the standard for orderliness ashore but, aboard ship, was a necessary gauge of the vessel's ability to work and survive. The superlative was "shipshape and Bristol fashion," or just "Bristol"—vessels from that city having a reputation for fanatical neatness. Coiling was a job for the apprentices and green hands when a few lines were involved. Setting all sail at once used almost every piece of running rigging aboard, and then able and ordinary seamen helped to square it away.

The mate called the watches aft to the break of the poop; there was one more tradition of departure to go through. The carpenter had set up a small anvil, and one by one, the crew stepped up and presented their seamen's knives. The carpenter laid them on the anvil and hammered off the knife points, leaving a sharp edge but a blunt tip. Some of the older hands had pointless knives already and were waved away by the mate. Benjamin's good Sheffield blade, for which he'd paid several weeks' wages from his last coaster berth, was mutilated unceremoniously.

"Why are they doing this?" he asked Russell.

"So we can't stick 'em each other, m' boy," replied the Yankee.

Later, Benjamin discovered that the old hands carried their blunted blades in their everyday sheaths, but in sea chests among clothing or under straw mattresses, they had cached other pointed blades. They often needed them ashore; maybe they would have use for them on board as well. Who could tell what kind of men they would be obliged to share a fo'c's'le with, many months and miles away from safety?

Off Great Orm Head, the tug's contract was complete, and it signalled to drop the tow. The line was cast off, and the tug swung around and steamed by close to the *Beara Head*, its side paddles thrashing the water. Its captain bellowed over his wishes for a good trip, not mere politeness, but the usual shorthand for asking for the gift of a bottle of booze for his trouble.

"Go to hell!" shouted Captain McMillan, the parsimonious Ulster Protestant.

Nevertheless, the tug gave a few good-humoured whistle blasts, its crew waved and it headed back towards the Mersey, the shoreline invisible in the twilight haze.

"Full and by," was the captain's order to the helmsman.

Left finally to its own devices in the light breeze, the *Beara Head* was able to lay a course close-hauled for Carmel Head, on the island of Anglesey, close off the northwest coast of Wales, a day away in these conditions. No sail handling was necessary in the wind, which was steady in velocity and direction. The barque ghosted along, silent except for the soft creaks and groans of blocks and the faint wash and slap of the bow wave.

Soon, Benjamin heard the ship's bell ring out eight times in the traditional rhythm: two chimes, a pause, then two more. It signalled the beginning of the evening watch, from eight to midnight. The captain ordered the second mate's starboard watch to stay on duty and sent the mate's men, including Benjamin, below. The old tradition: the captain took her out (he being ultimately responsible for the starboard watch) and the mate brought her home—that is, the mate's port watch would stand the first night watch on the first night homeward bound.

Sending a watch off duty at the start of a passage also signalled the official beginning of the watches; from then on, they would alternate four hours on and four hours off—except for the two dogwatches, which were each two hours long, from four to six in the afternoon and from six to eight in the evening. Their purpose was to allow the daily alternation of night watches, so that no one was stuck permanently on the same schedule. As a practical matter, both watches sometimes worked on deck during the day in good weather, doing maintenance: the endless rearguard action against salt, sun and rust.

The times of watches were strictly observed and enforced, to be fair to everyone on board, but also because time itself—knowing the time—is crucial aboard ship, a hedge against chaos. The accurate record of the hours, minutes and seconds of Greenwich Mean Time by the captain's chronometer (now refined to the size of a pocket watch) helped the sailor to establish his vessel's latitude and was the practical means of finding its longitude. Without this mechanical division of the natural day, the sailor is lost on the boundless sea.

The seaman's interest in fine time divisions is evident in the breakdown of watches into smaller intervals. In an age when few people carried timepieces, the rhythm of life and work was based on the solar day or the gross intervals of the mine or factory: start, meal break, finish. No one needed to know more than that. Unlike other workers of the 1880s, however, sailors aboard ship lived in half-hour parcels of time, rung out by the ship's bell. In the afternoon watch, from noon to four, for example, one bell meant 12:30, two bells 1:00, three bells 1:30 and so on, until eight bells signalling four o'clock, the end of that watch and the start of the next, when the bell sequence would start over again. (During the two-hour-long dogwatches, four bells ended the first dog-watch and began the second.) The times of meals, the lookout's watch, the helmsman's trick—all were determined by the bell striking each half-hour section of the two- or four-hour watches.

Seamen did the never-ending work of setting, trimming and taking in sail, and of maintaining the thousands of interrelated parts of the wind ship's complex machinery. Often, however, the crew's main task was, literally, to watch—to make constant observation of sails and how they were set, of the lead and wear of lines and gear, of sea and sky for signs of how the inevitable coming change in wind and waves might actually unfold. Patient watchfulness and an abundance of time are the stuff of normal life at sea. That's why thinking about things came naturally to seamen under sail—an unavoidable introspection. "Meditation and water are wedded forever," said Melville's Ishmael.

Benjamin spent the four hours of his off-watch, from eight to midnight, down below. He was too excited to sleep, but even if he had been disposed to, it was impossible in the fo'c's'le racket. With so many men, the deck house felt narrow and constricted, in part because of the placement of the double-decker bunks, which were close together along the bare iron walls. He was used to the seaman's cramped quarters, although the *Beara Head*'s deck house was far roomier than the tiny cabin aboard his last coaster. Even so, when Benjamin lay down, he was only inches away from another man's snoring mouth or stinking feet.

It was the noise of everyone's individual and eccentric activities that took getting used to. In this first watch below, there was a card game, and not a quiet one either. Various discussions proceeded about the ship (consensus: not a bad one so far, but it was a hen frigate with the woman on board, and we'll wait and see what damned interference she gets up to), the mate (equal consensus: about as bad as these bad sons of bitches get), the second mate (a wet-behind-the-ears babby who got the job because his da knew somebody important) and the captain (spooning the wife at this moment, the lucky bastard), whose speech was so full of the Irish that nobody (except for the blamed paddies) could understand what the hell he was talking about.

Any man among them who slept, snored; it seemed to be a prerequisite for the job. Virtually every man awake smoked—some pipes, some cigarillos. The deck-house air quickly assumed the consistency of a dense fug, which would persist throughout the voyage. It improved only when the ports and door were opened in fine weather, or when a strong wind swept some of it away as the watch was coming in or going out. When men began to run out of tobacco near the end of the passage, and had no more credit with the captain's slop chest, foul tempers would replace bad air. The oil lamp smoked despite continual maintenance, the result of cheap oil foisted on them, said Paddy, by the goddamn penny-pinching Prod Scotchmen who owned this bucket.

The clerk lay in his bunk in his suit and former white shirt. He had sobbed for a while but now lay quiet and staring. No one knew his name, and no one asked; everyone called him Clerk well into the voyage. By the time someone found out that he was Sean McCarthy, the eighth child of Liverpool escapees from the Irish famine, his shanghai name had stuck, and Clerk he remained.

Jan the Finn and Helmuth the Dutchman played checkers in non-Anglo solidarity and with loud, mutually unintelligible exclamations in Finnish and German.

One of the ordinary seamen, Spyros Kapellas, stroked a five-string banjo. He was a Greek who had jumped ship in Liverpool, running from a Yankee hell-ship, a cross-Atlantic packet whose mate had it in for him because of his accent, or his black hair or smooth skin or missing left index finger—who ever knew the sources of a bucko-mate's rage. Now Kapellas wondered if he hadn't wound up merely exchanging one brute for another. Out of the frying pan.

The very thin Liverpool drunk, one of the other shanghaiees, was called Philip Maguire. Sobered up, he was a well-behaved, quiet-spoken man, curly-haired with a shy, slightly twisted smile, a labourer with no experience of water except to wash in it occasionally. He lay in his bunk cursing quietly, methodically and without pause the *Beara Head*, its officers, Paddy Fearnaught, the crimp who had shanghaied him, and gin—in grim combination, the agents of this radical change in his life. The last thing he remembered ashore was agreeing to accompany a damn fine-looking woman to the Salmon for a dram. He had a wife and three small children, and he didn't know how they would live now.

There was a second Yankee in the port watch besides Russell, its de facto lead hand. John Urbanski, the son of a Polish immigrant and labourer, had wanted nothing else but to escape his own version of his father's life sentence in a Pittsburgh iron foundry. At eighteen, instead of heading west like any other American yearning for escape and adventure, he had taken a train east over the mountains to New York and shipped aboard a three-masted schooner to the Caribbean. He had been dazzled by the bright orb of warm, sunlit blue sea; a horizon always distant, unobscured by trees or mountains; the green, floating sargassum; little fish that flew; the high islands with their easy ways and women. Now some years later, doing hard labour on square-riggers was still far better than shovelling foundry sand twelve hours a day.

Russell himself was from Baltimore. He had started out in a bay bugeye, hauling up the swimming Chesapeake crabs. One day, in Norfolk, he had seen a clipper, one of the old Yankee tea traders that had kept going after the collapse of American ships during the Civil War. Every element of its beauty entranced him: the fine run of the lines; the sweet sheer and tumblehome; clipper bow and transom in perfect equipoise; the delicate masts and yards tapering gracefully at their ends; brightwork and scrubbed teak. He was a man who had gone to sea for love and beauty. Now, in the spare iron fo'c's'le of a slab-sided limejuicer, he was a veteran, but not a bitter or weary one. No infatuation, his love had been true. Russell's devotion had become a clear-eyed force of habit, a wary yet interested participation in the drama of his own sea-play. Now, in the off-watch, the two Yankees yarned, in the easy American way, about ships and men they had known up and down the East Coast and in the ports and roadsteads of the world.

Benjamin gave up trying to sleep. Later in the passage, none of this would keep him awake. After the first couple of days, every deep-sea sailor became sleep-deprived, weary enough to slumber through any hubbub. The four-on, four-off routine meant that no one got more than three hours or so at any one time, and usually less than that. Everything else in the sailor's life had to be done during his off-watch: eating, keeping his clothes in repair—sewing or smearing linseed oil on oilskins—smoking enough to keep the addicted body happy, hauling his water whack from the deck tank, getting in and out of oilskins, and making and unmaking soul-and-body lashings to keep them in place in heavy weather (although he turned in all-standing—fully dressed and ready to tumble out—when things got very rough). All this took time away from sleeping. The off-watch could be curtailed or eliminated, and often was, by the need for all hands on deck—a frequent necessity on these undermanned vessels with their heavy gear.

None of the usual passage weariness had settled in yet, however, and Benjamin got up and chatted with Anderson. They had become friends, swapping backgrounds in between the deafening dumps from the tipping buckets as they shovelled coal in the hold, or in the fo'c's'le before they collapsed in exhausted sleep. It helped that they were both right-footers—that is, they had a shared Protestant view of the turbulent Irish world. It would have made a difference if one of them had been Catholic. Ulster was already an apartheid society, divided along the fracture lines of religion. There had been vicious riots in Belfast in the 1850s and 1860s, and during their own lifetimes too. Anderson was the son of a shopkeeper on the Ormeau Road, the descendant of Scots settlers who had been part of the seventeenth-century plantation. His many brothers had followed the empire's Irish pattern: one each to the army and navy, one each to Canada and South Africa, one who would take over the shop, one in Liverpool, and Anderson himself, first an apprentice in the shipyard and then, tiring of the routine, off to sea as a green merchant seaman.

Benjamin's lineage was more complicated, his surname an ambiguous signifier in the text of Irish history. An ancestor became famous during the Williamite wars of the seventeenth century, when he tried to hand over a besieged city to the Catholic enemy. Amongst Protestants, the name Lundy became a synonym for "traitor." And there are many Catholic Lundys in Ireland. Benjamin's family was part of the tradition of Protestants who supported the idea of home rule for Ireland, although they probably weren't prepared to fight anyone for it.

What Anderson and Benjamin shared, as they did with Catholic Paddy and the Liverpool Irish, was a guarded and ambivalent reservation about England. It had been a refuge for famine Irish and was still a place to go for work and a lower level of impoverishment. It was impossible not to feel some pride in being part of the astonishing ascendancy of this small island's world-empire. But England was also the ancient oppressor and occupier, the overbearing imperial power and destroyer of the old language, the enduring alien arbiter of Irish life. It had been a long and harsh embrace:

> *In peacetime, live in your little coops;*
> *In war, put up with quartered troops.*

Benjamin himself had never cared enough one way or the other to resolve his equivocal feelings about England and Ireland. Now, in the fo'c's'le of a wind ship bound round the Horn for the Americas, he cared even less. That old world was behind him, or soon would be; the new lay ahead.

At midnight, eight bells—the end of the evening watch and the start of the morning watch, or gravy-eye (sleepy men's eyes become thick and viscous)—and the mate called up his men from below to relieve the starboard watch. Benjamin and the others came on deck to find the wind stronger but so constant in direction that they didn't have to touch a line. No other work was possible in the moonless night, and the port watch took its station below the poop. The ship's rules did not permit them to sit down or to smoke. Benjamin had begun to notice the effect of the long day and his postponed sleep, and he felt a weariness and slight disorientation familiar from his night watches on the coasting smacks. However, there was no cozy port in the offing where he could recover; this time, the watches would go on day and night for months. He hung over the rail, staring at the dark sea, the gentle, shimmering curl of phosphorescence. He shivered with the cold and, in spite of his fatigue, the exhilaration of this first night of his first blue-water passage.

Other seamen walked up and down the weather side of the deck in twos and threes, talking quietly. The mate was visible in shadowy profile in his accustomed place by the jigger shrouds. As the helmsmen struck the half-hour bells, the sailors rotated through their turns at lookout and wheel; they were allowed to decide among themselves the order in which they changed. But tonight, the mate had specified only experienced hands on lookout, men who could make sense of the maze of lights of other ships coming and going around them, all weaving silent tracks in the sea. They had to bear off fast several times to avoid running down smaller vessels, fishing boats and smacks ghosting along. The *Beara Head* was on the starboard tack and, therefore, had right of way over ships inbound on the port tack. But sometimes, it was impossible to be sure what another vessel was doing, and some—the fishermen especially, in their time-honoured oblivion to surrounding traffic—were not keeping much of a lookout at all.

A few steamers passed in both directions, moving fast. They were so noisy in the gentle night, such large intrusions into the hushed neighbourhood of the wind ship, that tracking them was easy. They were required by law to give way to sail, but they often behaved like the "ram you, damn you" liners that scared the hell out of sailing captains and, in these cramped sea areas, all too often ran down their ships. Steamers were almost always at fault in collisions with sailing vessels. It was odd, because all steamer officers spent their apprenticeships in sail; they should have remembered something about how cumbersome the big sailing ships were to manoeuvre, yet how deceptively fast they moved, and that it was necessary to give them a wide berth. But it seemed that something happened to a man on the bridge of a steamer. He became a driver-conductor, obsessed with schedules and passage times and the continued comforting thump of well-fed engines, which, in any event, his engineers saw to—they were the essential men aboard. It had to be that the machinery induced a kind of amnesia— the lumbering vulnerability of a big square-rigger in close quarters slipped his mind—and contempt as well (those obsolete ships on their way out, their officers either too stubborn or too stupid to make the leap to the future, to steam, as he had done). Perhaps envy and loss too: the man on the steamer's bridge knew that he had given up the old skills and the chance to carry on with the sheer wonderful seamanship that only sail demanded. A man could be glad to be dry inside a steamer's wheelhouse rather than clinging, cold and soaked, to the weather shrouds of a square-rigger, while at the same time hating his comfort, regretting that he had given up the hard, calculating battle with the sea that those men were fighting as their ship crossed his bow, getting in his way. He knew where the better men were.

One of the *Beara Head*'s experienced seamen stood by the wheel with the new hands to educate them in the vagaries of steering, one man's muscles directing the big barque's five-and-a-half-thousand-ton bulk. When Benjamin's turn came, the Elf instructed him: "Don't oversteer. When she starts to swing, that's when you haul the wheel back the other way to stop her goin' too far. She ain't a pig like some of these big wagons. Maybe even better 'n the *Dawpool*. No sea tonight, so it's like drivin' your cart down the road. When there's a sea runnin', though, that's when you have to fuckin' sweat. In the Southern Ocean, you'll see; sometimes three men can't hold the goddamned helm. But those breakers down there can give you a famous shove when they're runnin' up your arse."

In fact, in the calm water, with only a slight swell on the beam, Benjamin was surprised by how easy the steering was. Looking forward, he saw the barque fade into the dark distance, only the gleam of the oil running lights showing faintly ahead. When the Elf stopped talking, in the silence of the ship's smooth progress, Benjamin had the sensation that he was riding a huge machine down towards the dark centre of the earth.

Above, he glimpsed the suggestion of the shapes of sails out of the corner of his eye, the way one caught sight of a faint star. In a swell, the barque would roll this wind right out of its canvas, but now there was just enough to put it to sleep. He had steered the little coasters at night, but they were nothing like this. It was hard to believe that his own arms turning the wheel had any connection with the movement of this monumental ship, which, in the dark, seemed to occupy his entire horizon. The barque's slow response to his efforts, a ponderous, reluctant swing to port or starboard, reinforced the feeling. He felt excited, too, and proud of himself. Here he was, his first night out, driving the barque; for an hour, he was responsible for its proper course, its well-being.

With a glimmer of the early spring dawn in the sky astern, Benjamin and the rest of the port watch went below at four o'clock—eight bells, the end of the gravy-eye, the start of the morning watch. This time he slept until shaken awake at seven bells, 7:30 in the morning and breakfast—salt horse, hard tack free of weevils (for now), coffee. To everyone's continuing surprise, the cook gave them more soft white bread, an unforseen delicacy now they were at sea. Feelings towards the doctor were warm; the fo'c's'le believed that the bread was all his doing. The bastard old man and his equally illegitimate steward—stewards always took on the characters of their captains—were patently too mean to do anything good for men before the mast.

On deck again for the start of the forenoon watch at eight o'clock, Benjamin twitched with anticipation. The morning was cool, the sun bright, the horizon clear, the sea pale blue and redolent with its pure, sharp savour of salt. It was the end of his first twenty-four hours at sea, the start of his second. He had a full gut and enough sleep for the present. Later, the work and the routine might grind him down, but now he felt strong, equal to things, eager for the ship's demands.

The *Beara Head* had passed Carmel Head. If the north wind held, they would have an easy run down the Irish Sea and out through the Western Approaches into the open Atlantic, a lucky start to the passage.

The north wind did not hold. The captain was about to order the course changed and the yards squared away for a run south when the wind died into a mild, sunny calm. The barque drifted all day, the men wearing themselves out bracing the yards one way or the other to catch a few puffs of moving air, mostly imaginary. The captain's irritability in the morning became rage by the evening. The lull was like mid-summer weather; in May, you expected strong wind, maybe gales, but at least wind you could work with, not this damned doldrums frustration. He drove the crew without rest, watch by watch. When his anger became unbearable, he stamped below and left it to the mate to ramrod. For a few hours, the men worked the yards to shanties, but the sail handling was constant, and no shantyman's voice could hold up. By afternoon, the watches were hauling the braces in a sullen and weary tramp-and-go, sweating and thirsty in the unaccustomed heat of the spring sun.

Above them, wind signs appeared, slowly taking shape: high, wispy white cirrus and gauzy-patched cirrostratus. Something was coming, but not from the north, not a fair wind. When it arrived, it would be southwest at best, maybe south. It was always the same in these narrow waters, thought Benjamin. He remembered many hops up and down the Irish coast beating into contrary wind, but not more than one or two with a fair one. A law of nature for coastal sailors: the wind always blows along the shore, and always from the direction you want to go in.

Around seven in the evening—two bells of the second dogwatch— they saw the first cat's-paws on the sea's surface and felt the faint breeze as it filled in from due south.

Work for all hands: they braced the yards round, hauled on sheets and tacks and got the *Beara Head* sailing close-hauled, on the port tack this time. One of Benjamin's fore-and-aft-rigged coasting ketches could have made reasonable progress to windward under these conditions—a full-sail breeze with no sea running yet. But a square-rigger was a much different proposition: designed to range the open oceans, looking for the fair-wind quadrants of the world's immense wind systems; arcing its courses to keep the wind as much astern, and for as long, as possible. With the wind ahead, the square-rigger was unhandy and unweatherly. It couldn't point high—that is, sail a course close to the wind—to begin with; the shrouds of the standing rigging prevented the crew from bracing the yards round far enough to give a good angle, and the square sails were not an efficient windward aerodynamic shape. There was the effect of leeway as the ship drifted downwind under the sideways pressure of the wind on hull and rigging, no deep yacht-like keel to hold it up. When waves increased, their constant slap and push on the windward bow had a cumulative effect, increasing the vessel's tendency to drift away to leeward. With all these forces accounted for, the barque could lay a course about sixty-five or seventy degrees off the wind, and with leeway, her course made good wasn't much better than eighty degrees. In other words, the ship could do little better than hold its own; its modest goal was to avoid losing ground while it waited for a favourable wind shift.

The south wind grew stronger until, by nightfall, it was blowing at around thirty-five knots, close to gale force but still a full-sail breeze for the laden ship. At the end of its second day at sea, the *Beara Head* pounded into the steep-sided waves of the shallow Irish Sea, throwing spray back down the deck as far as the poop and aloft almost to the foretop. The occasional wave began to break aboard and sweep across the low-slung main deck. Its heading seemed to take the ship west-southwest towards Wicklow Head, on the Irish coast, and some modest progress. But the tide running north and the effects of leeway were shoving the vessel back, so that Dublin, less than fifty miles away almost due west, was the de facto destination.

CHAPTER THREE

In Narrow Waters

. . . the nearer the Land the greater the Danger,
therefore your care ought to be the more.
CAPT. GREENVILLE COLLINS,
<u>Great Britain's Coasting Pilot</u>, 1693

Dublin was a good town for sailors with its whores and drink, although the dank slums and desperate, starving shanty Irish had turned it into a place even more dangerous than Liverpool, where, as in other Sailortowns, the brotherhood of the bottle could turn fast and with little warning into the rage of the knife. Dublin's attacks were more elemental: for a piece of a sailor's biscuit, a worn jacket or, with the latest upwelling of the old nationalist grievances, maybe just the ironic pleasure of sticking an Englishman with his own bright Sheffield steel.

The barque had a different itinerary, however: a load of soft coal, some of it a little damp, and a long, unrestful ocean voyage to the bottom of the world and back up the far side of the Americas. No interest in any ports, now or for the next four or five months. It had made the irrevocable transition from land to sea: cast off the tow, chosen the watches, blunted the knives, made all plain sail. Now all it had to do was get out of these narrow waters with their irritating shallow-sea slop that, nevertheless, impeded the ship's headway like an apprentice thug, unimposing but an obstacle all the same. The Atlantic was what it needed and sought—the wide-open deep sea and the strong, sure westerlies, which would batter the vessel, no doubt, but drive it south too, towards the trades, the doldrums, the trades again, the southern westerlies and the Horn—in a way, their real destination.

The land with which they were closing fast wasn't the only concern aboard the ship. Nightfall and thirty or more knots of wind on the nose hadn't reduced the traffic around them. Like the first, tranquil night under sail, ghosting along the Welsh coast, this second, blustery night was full of other vessels. The watch on deck could see lights everywhere, a surround of red, green and white, intermittent in the swell. Oil running lights were visible for two or three miles at most, and they were often obscured by waves; the captain had flares handy in case another vessel blundered too close. The lights moved without pattern as ships on the same course tacked, others running downwind came on steady and fast, and smaller coasters under reduced sail headed in all directions.

The *Beara Head*'s captain doubled the foredeck lookouts and never took his own eyes off the lights, mentally calculating their trajectories in a constant geometrical plot of tracks and crossing angles. Sometimes he sent the second mate up the foremast rigging to get a better look at some ambiguous signs. The mate was on the poop too, silent, braced by the jigger lee rigging, watching the circus around them.

During his watch on deck, Benjamin, in wool and oilskins, sheltered with the other seamen at the break of the poop, trying to keep out of the wind and stay dry for as long as possible. With the sun gone and the sea still winter-cold, the night was freezing. He paid no attention to the running lights around the ship. He regarded colliding with another vessel as a trivial event, much less important than the way he felt. He was seasick—not to the point of vomiting, but the headache, lassitude, hot rushes and queasiness of the pre-disgorging stage. He was familiar with this; his stomach always took a day or two to settle down—but as often as not, the little coasters had reached port again before he had a chance to adapt. In the meantime, the *Beara Head*, its bucko, stumpy mate, its Islandmagee peasant of a captain, with his jigging queeralities, and all the other irritating living flotsam on board could go to hell—he was concentrating on his innards.

Soon, in spite of his guts, Benjamin had to go to work.

The barque was closing fast with the Irish coast. The south wind had blown away the haze, and the lookouts could see lights on shore. It was time to tack: to change course from the port tack—that is, the wind blowing across the port side of the vessel—to the starboard tack by swinging the bow through the direction of the wind. The ship would head back out into the Irish Sea, towards the Welsh coast they had left five hours before. The *Beara Head* was forced to do the thing square-riggers were worst at: beat, tack and tack about into a head sea and a wind dead on the nose. And wait for things to get better—at least, a wind shift to give the barque a favourable lift on one leg so that it could make some southing before tacking again; at best, the wind hauling right round to the side, or astern, so that the vessel could make a comfortable beam reach or a run right out to the open sea. They needed just ten or twelve hours of fair wind to get clear of the St. George's Channel; then they would have sufficient sea room to make progress no matter what the wind direction.

With a green crew and enough of a head sea to check the ship's progress as it swung around, the captain might have thought about wearing ship instead—that is, changing course to swing the stern, instead of the bow, through the direction of the wind, bringing the vessel up close to the wind again on the new course. This manoeuvre had the vessel running off before the wind for some minutes, losing valuable and hard-won ground. But wearing ship was the only way to get onto a new tack in heavy weather, when the force of seas stopped the bow from making its arc through the wind's eye.

Tacking, on the other hand, was tricky and required close coordination of the seamen's work. The crew had to do many things at just the right moment. It was even more difficult on a moonless night; they had to do everything by feel: of wind on the captain's face as he judged the right moment to give the crucial orders, of hands on lines as the seamen found blind the ones they needed and hauled away fast and furious in the dark.

❋

The watch could see the shore lights now; it didn't take an experienced man to know that the barque, thundering along with a phosphorescent bone in its teeth, throwing spray up to the tops, making maybe ten knots or so, would have to tack or wear ship very soon. If Benjamin had been in a speculative mood—and paying attention, instead of trying with all his might not to throw a brash (an Ulster upchuck)—he might have been struck by the ship's imperious inertia. He might have believed that altering the wilful course of this vessel—its immensity stretching upwards and forwards from him into the dark, its five and a half thousand tons of momentum—was impossible. It felt unstoppable in its noisy, urgent rush towards the nearing shore. Could one man at a small wheel turn this runaway leviathan around? It seemed much more likely that the ship would keep going until it drove ashore in full stride, ploughing up some stony Irish beach or gutting itself on rocks until the cataclysmic friction finally dragged it to a stop, just another wreck landsmen would gawk at the next day and loot the following night. From his purely theoretical perspective, Benjamin might have seen the old and commonplace routine of tacking ship as a seaman's truly ingenious manipulation of the wind's force to turn the great ship to his will.

The captain of the *Beara Head* had his first opportunity to show what sort of master he was. There were many possibilities. He might be a cautious man who stood off and played safe, wearing the ship around, losing ground but making sure no mistakes could be made that might bring down a mast or destroy sails. At the other extreme, he might be a driver, like the tea-clipper officers who were almost willing to sail their vessels under in the race to get the sacks of pekoe or oolong first to the London market. In that case, he would tack away from the Irish coast. Or the captain could be anything in between. As long as he wasn't a drunkard or a coward.

"Keep her full for stays," he ordered the helmsman.

He wanted the ship steered for maximum speed for a few minutes so that it would be more likely to complete its swing through the eye of the wind. He was not a cautious captain. He was going to tack.

Benjamin's first square-rigger manoeuvre was a blur of nausea and confusion.

"Ready about!"—the captain.

"All hands on deck to tack ship! Ready about!"—the mate, doing his hands-only slide down the lee poop steps.

The mate's whistle; pounding on the fo'c's'le door; rush of the starboard watch from the deck house; the mate telling off men to sheets, tacks, braces; second mate joining in the welter of orders; able seamen chivvying the green hands; captain calling two apprentices onto the poop to handle sails there; a tumult of bodies; Benjamin sweating with nausea, stumbling along the deck to the mainmast—all this impetuous movement in deep shadow. Sails and lines in barest outline against a dark sky, some stars, a faint line of sun-glow in the southeast; the ship plunging along to the southwest, steep waves detonating against the hull, wind a loud, moaning hum in the web of rigging. Mate and carpenter onto the spray-drenched fo'c's'le deck to tend jib sheets, idlers standing by their sheets, two able seamen to handle the main tack, second mate clearing the lee main and mizzen braces, men to the mainsail and crojack braces. Then the silent, disciplined pause, most heads turned aft, Benjamin numb, panting, not vomiting, but almost.

"Ready, sir!" the mate shouts from the bow.

"Lee-oh!" hollers the captain. "Ease the helm down," he says quietly to the helmsman.

"Helm down. Aye aye, sir."

"Helm's a-lee!" the captain bellows.

The *Beara Head* turning to windward—carpenter and mate easing the jib sheets a little, the doctor the foresheet, to spill some wind out of the sails. On the poop, the apprentices hauling the spanker and topsail in hard—they'll provide wind leverage to help turn the barque like a weathercock—sails beginning to shake as the wind comes ahead.

"Raise tacks and sheets!"—the captain.

Seamen letting go main and crojack lines and hauling the sails up by their clewlines so they'll swing free; the barque coming up closer to the wind, sails flogging and rattling, lines flailing, masts vibrating—the seamen can feel it in their feet through the deck. The turn slowing now as sails lose the wind's force, waves slapping and pounding on the weather side, resisting the vessel's pivot. Now the ship is almost head to wind, its momentum lost; the captain standing square, facing forward, straining to see, his head swaying from side to side to catch the precise direction of the wind in relation to his ship's bow.

He can't see but senses the bowsprit within a point or so of head to wind. He yells the crucial order: "Mainsail haul!"

At once, the seamen haul hard and fast on the main and mizzen braces—Benjamin too hauling frantically, with all this diversion feeling less like dying, his nausea in abeyance. Yards swing round at the run,

just enough momentum to bring the bow through, foremast sails un-
touched, now aback, wind blowing into their forward side, helping the
ship's turn; men getting the yards round and trimmed as close as possible on the new tack before the wind fills the sails—to avoid hauling the
wind as well—sheeting the fore-and-aft sails onto the new tack, hauling round topgallant and royal braces as the main and mizzen sails begin to draw. The ship pushes ahead, waves helping now as they butt
the bow to leeward, men belaying lines—even the captain's little dog
feeling it, barking and snapping at the men on the poop.

"Let go and haul!" the mate hollers.

Men who know what they're doing—not Benjamin or the other
green hands—run forward, letting go the fore braces and hauling in on
the new side, cook gathering in the fore sheet's slack, the carpenter
and some able seamen taking down the fore tack, other men trimming
the sheets and braces, slacking off the mainsail and crojack clewlines,
hauling in their sheets and tacks; the barque gathering way on the starboard tack—picking up the pace, shouldering the waves, kicking up
spray—while experienced seamen climb up in the dark to overhaul
buntlines (the boys' job when they learn it); all hands, and Benjamin,
coiling and flaking down lines, the *Beara Head* making nine or ten
knots again, heading southeast for Wales but making a course close to
due east, back to the tip of Anglesey, which they left a handful of
hours ago.

"That'll do the watch."

The starboard watch, with half an hour left before its duty begins,
going below. Benjamin with the port watch gathering again by the
break of the poop. The captain posting two lookouts and taking his position by the weather shrouds. Another seaman relieving the helm.
The mate moving to stand beside the lee jigger rigging, and bracing
himself there against the ship's pitch and roll.

The tack took eight minutes, but another thirty were needed to
trim and secure and make everything Bristol, ready for the next one. In
four or five hours, they would have to do it again.

Benjamin stayed on deck awhile after the port watch went below. He
needed more time to look off towards the ragged horizon and try to pla-
cate his roiling stomach. He could see someone else vomiting over the

lee rail, although he couldn't tell who it was, nor did he care. The tack had been a shambles for him. He had had no idea what he was doing, what the line he was hauling on was supposed to do, even where he was on deck—by mainmast or mizzen? He couldn't remember. He had planned to watch carefully the first time the ship changed course, to get a rough idea at least of the complicated choreography their dancing captain put them through. All he had been able to do was stay on his feet and avoid throwing up or provoking the mate's rage, more the result of darkness than anything else. It would be easier in daylight, but when the next tack came, he would be as gormless and useless as he had been the first time. And then, the mate—he glimpsed his form on the poop, planted firm and weaving with the barque's motion—would see him.

Benjamin went below to a noisy argument.

"Thinks he's on the poop of the fuckin' *Thermopylae!*" Jack Grey, the cockney, was shouting. "Thinks he's a fuckin' driver!"

His short, greying hair, salted and stiff, stood on end; spittal had dribbled into his beard.

"Jackie, it went all right, didn't it? She went round without a peep," said Russell.

"Blind damned luck. If he's drivin' her now, two days out, by heavens, he'll have us pulley-haulin' day and fuckin' night around the line. And he'll push this bloody bucket like a clipper when we get south."

"He knows what he's doing. He mainsail-hauled in the dark just right."

"Goddamn it! He's a fuckin' driver. And you saw the way he just looked away when the mate rode down the Clerk? Did nothin' to stop him. A fuckin' bucko mate and a drivin' old sod. And you've got his favourin' wind up your arse, Russell. Lookin' for a bosun's ticket, are ya?"

Russell, the Yankee, red in the face, muscles vibrating with suppressed rage, glared at the cockney.

Benjamin was appalled; two days out, and two men—able seamen they'd need the most, who were supposed to know what all this was about—were at each other's throats already. He felt sicker than ever. He hated fighting, although he'd had to do it often enough in the miniature streets and alleys of the Irish Quarter. "Lundy's gang's the worst gang out. Hit 'em on the head with a bottle of stout." He had been a little Wellington for a while, directing his brigade of boy-Anglicans and Presbyterians in various sallies, reconnaisances and all-out frontal assaults on the Catholic gang two streets over. They threw and catapaulted stones and whaled away at each other with stick swords. But Benjamin had not been thrilled by the combat like the other boys. He had never been able to bring himself to punch another boy in the face; it seemed like breaking some sort of taboo. He felt that if he crossed that line, anything was possible. First you bloodied a nose, then you maimed someone or killed him. He avoided taking part in the hand-to-hand donnybrooks by planning and directing them. He watched from behind or off to one side as his battle plans, good ones, unfolded. His intelligence and tactical skill kept him above suspicion, and he was never challenged with the coward's blow, which meant fight or be branded a feardie. But he was afraid now, standing in the doorway to the *Beara Head*'s deck house, watching two men decide whether it was time to try to kill each other. It was hard enough bearing with the novelties of this huge barque and the prospect of the Horn without having to deal with a murderous fo'c's'le as well. Surely with the deep sea out there waiting for them with its careless enmity, they had to stay together. Like friends or brothers—a band of brothers.

"Come on now, boys, stand off. Belay all this." Paddy the shanty-man had decided to intervene.

"What does it matter, sure. Jack's right as rain"—talking to Russell—"the old sod did take a chance tacking, but"—turning to Grey—"Jimmy's right himself; she went round like a charm. Maybe she's one of them quick-about ships, and the old man knows that. Come on now, boys, knock off."

He kept talking for a minute or so, making peace, draining the tension away, and the moment for violence passed.

"Aye aye," said Russell quietly. He turned away and lay down on his bunk.

"Well," said Grey, standing down too but changing tack, "if that goddamn, four-legged, little son of a bitch of a dog bites at me once more, I'll scupper the little bastard!"

Most of the other men watched silently. A few, like the Greek, Kapellas, who had often seen this fo'c's'le head-butting, ignored it. Jan the Finn looked on with a smirk, a mixture of mild contempt and glee. This was typical of the roastbeefs: lots of shouting, like women. His people didn't waste time with that, just kept quiet and let the knife do the work. Helmuth the Dutchman lay in his bunk with his face to the bulkhead; he was alone in his language and had already decided to ignore any speech in English that wasn't spoken directly to him. After a few more seconds, Grey, staggering a little with the ship's motion, shoved past Benjamin and out on deck.

Jack Grey used to be William Ferguson. He changed his name after he killed a man in a bloody drunken mêlée in a tavern in Tiger Bay, London's Sailortown. It wasn't far from where he grew up: a hovel running with alcohol and rife with assaults and a confusing mélange of sexual encounters. His mother was a famine-Irish whore, his father any one of five hundred customers. The boy was a skilled liar, thief, street fighter—and a cocksman as well, not favouring either sex in particular—by the time he convinced the mate of an Atlantic packet to take him on as a steward's helper. If the warm, bright sea of the Caribbean had been an epiphany for the Pittsburgh Yankee, John Urbanski, the young cockney was gobsmacked in a different way by his first foray out of the noisome, narrow streets and alleys. For days, he couldn't bear to look out across the ship's rail at the dazzling sea; its emptiness, the absence of boundary or limit, terrified him. He kept his head down and, with uncharacteristic mildness, served the captain and the officers and the first-class passengers travelling to New York. When the salt horse and peasoup and scraps from the aft-cabin table—the best food he'd eaten in his life—encouraged his stunted body to grow a little, the mate moved him to the fo'c's'le, and he became a seaman. He worked the Atlantic packets and got his able seaman's rating.

Then, in one of the usual blowouts ashore, his knife, which he'd often wielded in an incompetent, drunken rage, found by accident a soft space between the other man's ribs and pierced his heart or kidney, or some other indispensable body part. After that, he shipped out on Cape Horners, far from England. On one passage, a man called Jack Grey died in a fall to the deck from the main topgallant yardarm. When his belongings were auctioned off, they fell to the cockney, Ferguson.

He jumped ship in San Francisco and kept the dead man's clothes, his knife and his tobacco pouch, and then took his name as well, severing his connection with the murder in Tiger Bay. No one missed Jack Grey, a homeless seaman, and no one missed William Ferguson either. Now in his thirtieth or fortieth fo'c's'le, the new Jack Grey had still not got used to the sea. He avoided looking out at it over the rail; it appalled and frightened him still. He hated its void, the oblivion it embodied and often threatened. In port, he was a steady cargo stower and iron chipper; ashore, a violent drunkard; on a passage, a good sailor. But he was always angry at the sea and his fear of it. On the *Beara Head*, he had kept his good knife, the one with the point intact, hidden from the carpenter's anvil. Now he had two reasons for thinking it might come in handy: the Yankee arse-licker and the dirty paddy.

The paddy in question, Michael Heany from County Cork—the shantyman of the *Beara Head*—was, like Grey and so many other men before the mast, a drunken, whoring brawler ashore; aboard ship, however, he calmed down—the ship's "beneficent routine" quieted him. He became a steady and cheerful seaman, and he was an inveterate peacemaker in the fo'c's'le. He had been twenty-five years at sea, sampled every Sailortown and learned every shanty there was. He was born in the first year of the famine, and he'd somehow outlived the great hunger and the diseases that followed. His numerous family died around him, one by one and in bunches, and when he was ten, he began to survive alone in the streets of Skibbereen and Baltimore, and along the coast to Queenstown. He found work on the docks there. He learnt to speak English.

One day, he was half shanghaied—it was all the same to him whether he went to sea or not—onto an Indian jute clipper, and he spent the next ten years running round the Cape to the Hooghly and back. He sailed on all varieties of ships all over the world: Welsh copper carriers, Yankee Atlantic packets, timber ships across the Tasman Sea, tea and wool clippers, Cape Horners of every kind. At sea, he had been wrecked, stranded, half foundered and dismasted. He was laid open by a pirate's sword in the South China Sea and shot in a mutiny on a Yankee hell-ship while trying to make peace between the desperate mutineers and the resolute afterguard.

He understood the fear and ignorance that drove his shipmates, who were often simple down-and-outers, men from the fringe of things, like him; he knew that seamen, if they stayed aboard long enough, became separated from life ashore. Where they came from no longer mattered; ships became their country. "We expatriate ourselves to nationalize with the universe," said Melville. Paddy knew that too. He tried to make peace in the fo'c's'le the way a wise elder of the tribe soothes the young men. They had to stand together because outside their small democracy were the hostile absolutists: the officers who drove them, the sea that would one day consume them, absent-mindedly or with malice. After his childhood hunger, his family's destruction and his solitary struggle to stay alive, Paddy thought of the square-rigger fo'c's'le as a place of milk and honey: a comfortable home with a reliable food supply, his own bed, his own sea chest with his own few belongings, and a life that comforted him with its constant, necessary rhythms and observances.

The *Beara Head* tacked back and forth between Wales and Ireland for the rest of the third day and night. The wind held steady at about thirty knots and from exactly the direction they wanted to sail in. Captain McMillan was as enraged by this contrariness as he had been by the preceding calm. He took it personally; it was an attempt by some goddamn party or other to make him look bad on his first command with these owners. However, they were making some progress. At times, the current was in their favour and the ship made twenty miles or so to the south before it had to tack again. In calm water, they could have worked their way out to the Atlantic in a couple of days. But the short, steep seas hindered their progress.

Each time the barque changed course, the captain tacked, rather than wore ship. He was damned if he was going to lose two or three miles every time he went around. That could be a fifth or even a third of the entire gain over a four- or five-hour tack. On a beat to windward, the square-rigger fought hard for every mile; he'd risk a little sail damage to hold the line. In the daylight, the tacking was easier and the new crewmen began to get the hang of things. On occasion, when the current was south-going, the wind blowing against it built up waves that were steeper and squarer than usual, like a series of fifteen- or eighteen-foot-high grey-green brick walls bearing down on them. Twice when they tacked into such seas, the ship wouldn't complete its turn; it was punched back by their force. The barque got into irons— head to wind—the helm couldn't turn it either way and it drifted back- wards. There was a lot of extra work to get things straightened out: seamen had to take in the spanker and the mizzen and jigger staysails, and then they had to square the main and crojack yards, heavy work at the braces, until the ship fell off onto the new tack, the sails filled and the vessel slowly gathered headway, when the spanker and staysails were hoisted again. Keeping it moving was the trick. Sailing vessels are under control only when they're making headway; the flow of water past the rudder allows it to exercise the necessary turning force.

During these extended manoeuvres, the mate bellowed without cease and often larruped the green hands with a rope end. Benjamin got the sting of it a few times. He felt outraged that he could actually be whipped with impunity, like a balky horse or an uppity dog, but he was also grateful that he was only one of a gang at the receiving end of it. He feared being singled out. When the mate concentrated and di- rected all his loud vituperation at one man, and threatened or dealt out blows, it was a hard assault to withstand.

By dark on the third day, the *Beara Head* had sailed more than two hundred miles and had made twenty miles southing. At the end of the fourth day, the ship had sailed an additional 250 miles to make another twenty-five miles to the south. But it managed to clear Bardsey Island, at the tip of the Gwynedd peninsula; this opened up a little more sea room, as the barque was able to make a longer tack to the east into Cardigan Bay, off the Welsh coast. By the morning of the fifth day out from Liverpool, the helmsman, still hugging the wind and sailing full and by, was able to lay a course for the Tuskar Rock light, off the south- eastern tip of Ireland, a marker of the narrows of the St. George's Channel and the path out to the Celtic Sea (between southwestern England and Ireland) and the Atlantic.

Finally, the wind changed. It veered to the west-southwest and fell off to ten knots or so. In the crew's tenacious, extended buck to wind- ward, this was the break they had been hoping for. In cloud and rain, with seas down to a few feet in height, the barque was able to squeeze clear of The Smalls light at the southwestern extremity of Wales and thus break out of the St. George's Channel. Making only three or four knots now, they carried on into the Bristol Channel until Lundy Island was in sight (Benjamin wondering again what his connections were with the bird-infested rock). Then they made a short, although dispiriting, tack back to the northwest, almost the direction they'd come, before the captain judged they had enough room to turn again and clear Wolf Rock at Land's End. They threaded their way between it and the Scilly Islands and the Bishop's Rock light, the last impediments to the barque's quest for the open ocean. In the afternoon watch, Benjamin took his last quick look at the empire's home islands; he would not see them again for thirty-two years. Eight days after dropping its tow, the *Beara Head* cleared English waters. The ship had sailed close to a thousand miles to make 240 miles to the south.

It could have been worse. If the wind had blown ten knots more, they would have been hard-pressed to make any gains. The captain would have anchored somewhere to wait things out and fume at the injustice of it all. Or they might have been forced to tough it out at sea.

One of the longest passages from Liverpool to Land's End would be made by the iron ship *Micronesia* in 1895. In a strong southwesterly gale, the vessel was under tow to take refuge in Holyhead, on Anglesey. The towing hawser parted, and the *Micronesia* had to make sail and get out to open water in a hurry. From then on, it was too rough to approach land. The ship clawed its way to windward into one head-on storm after another. It took twenty-two days to clear the Bishop Rock light.

It could have been very much worse for the *Beara Head*. Narrow, ship-infested seas are the scourge of all sailing vessels, and especially so the biggest, most unwieldy deep-sea square-riggers. Strong tidal currents, changeable winds and frequent gales, even in summer, mean that the home waters of the British Isles—the Irish, Celtic and North seas and the Channel—and the adjacent continental coast are crowded with all the varieties of dangers fatal to ships. A coastline with a limitless capacity to receive the ribs and bones of broken vessels and drowned men. It has done so for hundreds of years.

Bad weather completed the dispersal of the Spanish Armada begun by Francis Drake's fire ships off Calais. The galleons ran in all directions, and almost half of them were driven ashore. So many were wrecked on the Irish coast that the surviving sailors—enemies of England and therefore friends of Ireland—permanently altered the genetic composition of the shoreside population.

The wrecking of the Armada was merely one of the earliest large-scale slaughters of sailing vessels around the British coast. It happened every winter in the usual storms, and occasionally in the summer gales. Sometimes there were particularly destructive events. A random sample: in a storm in 1692, two hundred vessels and a thousand men were lost; in November 1703, a gale destroyed thirteen men-of-war and hundreds of men in the Downs, in the English Channel; there was the Big Wind of 1839 and the *Royal Charter* gale of 1859, in each of which scores of ships were lost in a day. In 1861, a reasonably typical year, 1,170 British vessels were wrecked, including thirty on one day. From 1874 to 1883, Britain's maritime losses totalled 699 ships and 8,475 men. In 1883–84, 121 British ships and 2,245 men were lost. During all these years, a great many more seamen were saved from burned or foundered ships—like Conrad was when the *Palestine* caught fire at sea in 1883 (the genesis of *Youth*). No one in Britain kept track of foreign losses.

The last notable storm in the age of sail was the Great Gale of 1894. Over two days, from December 21 to 23, eighty two British-flag sailing vessels were wrecked or foundered, including forty-nine that disappeared in the adjacent seas—everything from fishing smacks to four-poster barques. Five hundred men drowned or froze or were battered to death on rocks. The total does not include the fourteen steamships and trawlers lost, nor an undetermined number of small vessels under twenty tons. Again, there was no coherent count of foreign losses. Between 1900 and 1910, at least ten thousand seamen were lost on British sailing ships alone. In 1905, nearly 5 per cent of all British ships were lost: 501 vessels.

Getting out of British waters in one piece was something to be grateful for; taking eight days to do it was a minor inconvenience in view of some of the alternatives. Ships were lost in every week of every year around the coast, and many men with them:

1891: The four-masted ship *Bay of Panama*, a fast jute clipper built by Harland & Wolff, ran ashore near the Lizard in hurricane-force wind and snow. Fifteen men, including the captain and his wife, drowned or froze to death. Fourteen men survived until their rescue almost thirty-six hours after the wreck.

1900: The four-masted ship *Primrose Hill* went ashore near Holyhead after a towing hawser parted in a gale; the captain, his wife and twelve crewmen and apprentices were lost.

1904: The ship *Khyber*, with a cargo of grain from Melbourne to Falmouth, was swept by huge seas in a southwesterly gale off the Lizard. Crippled and blown close to the rocks, the vessel let go its anchors, which held for twelve hours. Then they dragged, and the ship was smashed to pieces. Three men survived.

1905: The four-masted ship *Eulomene*, the second of the name, went missing in the North Sea while sailing in ballast from Bremerhaven to the Tyne.

1912: The *Wendur*, one of the fastest four-masted ships ever built, set records for passages from the Cape to Melbourne and from Newcastle, New South Wales, to Valparaiso. While sailing from Plymouth to Swansea, the ship was wrecked on the Seven Stones and foundered in deep water. Three men were lost.

The *Beara Head* had spent more than a week in crowded coastal waters, often in sight of land as it beat its way out to the Atlantic. Nevertheless, the ship's sea-going routine had been set. The watches had settled into their debilitating rhythm, one made worse by the need for all hands each time the barque was tacked (interrupting sleep, wearing them out early in the passage). Benjamin had had more than sufficient chances to learn the elaborate interlocking tasks involved in bringing the ship around when the orders roared out from the poop. That had been one advantage of their struggle south. At least when they got out into the open ocean, they would have some sail drill under their belts. A fair-weather run out to sea was pleasant but often left a green crew unprepared for the Atlantic.

Benjamin had climbed the ratlines again only a few times—"Lay aloft, my Proddy paddy, and tie off that damned halyard." He climbed up to the foretop and out onto the yard, an easy job of a few minutes. The barque had kept under full sail in its drive to windward. There had been little need to send men into the rigging to do anything except overhaul buntlines. The mates usually sent the apprentices up for that purpose after each tack; it was one of their traditional jobs, but green hands were sometimes included. These lines, running down the forward face of each square sail, chafed if left hanging. They had to be pulled up a little, until they lay loosely on the sail, and tied off—stopped—with light rope yarn or a loop of the line through the block, which would give way immediately when the sail had to be taken in and the buntlines were hauled tight. The apprentices, young and agile and aloft on every watch, had adjusted quickly to the work and acquired already the easy swing and skip of the seaman at ease on his own high wire. Benjamin envied their opportunities to adapt there, and he worried each day about going back up himself to furl sails. That would mean that the barque was in heavy weather, its yards rolling and tipping, the footropes dancing their quick, twitchy rhythm.

The members of the port watch and the afterguard had mostly confirmed Benjamin's first impressions: Russell the Yankee, a superb seaman and a quiet leader, was the kind of man other men gladly worked for. Except Jack Grey, the cockney; his anger at Russell had intensified into open hostility.

Paddy the shantyman, jovial and gentle, sang from his unending stock of songs and watched the tense space between Russell and Grey. He'd often seen blood in fo'c's'les, and this one was turning into a prime risk.

Jan the Finn played checkers with the Dutchman and, in his few moments of unoccupied time, began to carve a model ship. He made one on each passage and sold it ashore afterwards; it was usually the only money he had to tide himself over to the next ship, the rest having gone to the slop chest or for the immediate pleasures of drink and whores.

Helmuth the Dutchman had become a little more sociable; he was beginning to use the small stock of his English words that didn't refer to sails or lines or what to do with them. He needed to acquaint himself with the English on board because his checkers companion scared him.

Finns were reputed to be as quick with the magic as with their knives; that was why they had no national nickname: to avoid occult retribution.

Two of the shanghaied Liverpool men were doing well. Cavers the Elf was wiry, willing and cheerful. With the pea soup, the gristly salt horse (already growing an odour) and the hard tack (not yet weevily), he was eating better at sea than he had ashore. That and temperance were good for him. He waited with impatience for the chance to go aloft; he always felt free up there, away from the hectoring officers. On deck, other men always overtopped him. Aloft, his slightness was an advantage; it gave him a slick agility on the footropes and ratlines.

By the end of the week, Maguire the landlubber labourer seemed resigned to becoming a sailor and seeing the world, or seeing the sea. So far, the work was no more strenuous than the digging and carrying he did ashore. It was just that you had to get up in the middle of the damn night to do it. He avoided thinking about going aloft. He missed his wife and children less than he expected.

The other Liverpool man press-ganged aboard, the Clerk, was profoundly depressed. He was struck in every watch by the mate, who was like an animal unable to stop harassing one of its own injured kind. The Clerk's debasement and misery inflamed the mate's rage; they were like torturer and victim, locked in a sordid embrace.

Urbanski, the other Yankee, was happy at sea and impatient to clear away from the surrounding land; he waited with eagerness for the warm sun and enfolding sea-and-sky blue of the trades, the hazy heave of the doldrums. Unlike most sailors, Urbanski wasn't bothered by calms. He enjoyed their peace—the living things seen in the water; the neighbourliness of the other ships, dozens or scores of which were within sight as the calm gathered them in, rolling and slatting in the swell. He would be happy for weeks in those undisturbed waters if it weren't for the constant pulley-hauley, dragging the yards this way and the other, and the bloody Horn always in the offing. For now, the only irritant was Kapellas. Urbanski couldn't say why, but the dago rubbed him the wrong way.

The Greek's attention was fixed on the *Beara Head*'s first officer— who, he saw now, was no different at all from the damned bucko Yankee mate he'd run from, with his knuckleduster and the bulge of the

revolver in his pea-jacket pocket. MacNeill didn't go around armed—
you couldn't do that on limejuicers—but he was as mean as the Yankee, maybe more so. The American often had a sardonic flash in his
eye when he rode down a man. This Bluenose bastard was all earnest
business; even his sarcasm was somehow without humour, and therefore more threatening than his loud barks. On top of all that, there was
the Yankee Urbanski, with the stares and irritation he'd seen so often
before, watching him when he stripped off. Every fo'c's'le had one or
two men like the Yankee.

Anderson, the strong, crooked-nosed Belfastman whose muscles
provoked even from the mate occasional respect, or a somewhat lower
level of insult, was a smart man, with his city's fast wit, like a cockney;
yet he was often unintelligible to anyone but Benjamin—or, presumably, the captain, if they'd ever had a conversation. Impatient to get to
sea, to learn this business of sailoring, he, like Benjamin, saw it as his
one-way ticket out; in his case, he wanted out of the cramped little city
hemmed in by its green hills, the famine terraces still visible on them,
with its prejudices and hatreds and the violence that promised every
time you rounded a corner. America and the East: San Francisco, with
its gold-rush sheen still intact in his eyes; Singapore; Calcutta; Rangoon. The fabulous cities of the empire—he would see them all. And
to do it, he'd jump from ship to ship when he damn well felt like it.

Benjamin himself, another paddy leaving home, was surprised to
feel sad at his last sight of Ireland, the green Wicklow coast, as they
tacked away from it; then, as he had expected, he felt happy to see the
last of England when the *Beara Head* passed Wolf Rock and headed
out into the Channel. In the fo'c's'le, where size and physical force determined status and safety, he was reassured by the presence of Anderson, his countryman and co-religionist; he thought he could put his
back up against Anderson's and stand off whatever might threaten.

The mate, the Bluenose, deserved every epithet in the seaman's
vocabulary. This hard man had shown no quarter, not a moment of in-
decision or vulnerability. His shiny skin, like a carapace, armoured him
against man and sea. His voice was already part of the raucous background noise aboard the barque, penetrating everywhere, dominating
and monopolizing. It was impossible to envision an officer like him
aboard a steamer; he was as dedicated a part of a wind ship, as non-

transferable to any other workplace, as a clewline or a Flemish horse. To every crewman, he was at the very least a goddamned son of a bitch; they winced when they caught sight of him, motionless at the lee rigging or bulling his way forward to assume command of the fo'c's'le deck during a tack. Even Russell and Paddy—and the Finn, who wouldn't hesitate to use his knife on anyone—were careful around the mate. He beat men; he would kill a man without a pause. The captain had turned away a score of times when MacNeill hit someone, and had thereby given his tacit approval to his bucko first officer's heavy hand.

No one could deny, however, that the mate was a seaman's seaman, "every hair a rope-yarn and every drop of his blood Stockholm tar." For the men who'd been round the Horn, the mate reassured as well as intimidated. He was the boy who'd see them through; he would be equal to anything. In the coming war with Cape Stiff, the mate was a man who would refuse to lose; they would ride around on his back, win with his fire.

The second mate, Jagger, the daddy's boy, sonnywhacks (a name usually reserved for the apprentices), seemed hard done by during this first week, lorded over by the mate, held in contempt by the seamen when word spread about his rich father and his queue-jumping, obeyed with that fatal pause between "Aye aye," and "sir." They thought that he was, at best, a seaman like any of them, and not necessarily an able one at that. Jagger suffered in this crossfire of disapproval. He had two consolations: the captain, who treated him correctly and left him alone on his watch; and the captain's wife. At mealtimes, the two mates, the captain and his wife ate together at the aft-cabin table; she smiled at Jagger and asked questions about Bristol and his family.

The captain's dog remained fierce and yappy; it snapped at the seamen every time they appeared on the poop. It chased the apprentices up the ratlines. Its shit appeared in odd corners and nooks, and men stepped in it barefoot as they were hauling on lines. The watches agreed that the dog might live to see Valparaiso, but maybe not.

Mary, the old man's old woman, was the hen in this hen frigate—her first voyage, first time at sea. Married a year, twenty-five years old, a piano-playing, tea-in-the-front-parlour upbringing in genteel Bangor, on the County Down side; her family was not happy with her inexplicable attraction to this Antrim hillside man, this sailor. But he had mesmerized her. When he told her what it was like to go up a mast and out along the yard in a blizzard—the closest solid objects the icebergs invisible around the ship, the wind howling and screaming, its force on his body like the hand of the Devil trying to pry him loose—she felt the exotic, erotic euphoria of danger wash over her. Her captain had just the right combination. He had lived rough at sea and around the world; it was as if he had been at war and had acquired the soldier's deep knowledge and understanding of things. But he had a civilized manner, and he had read books and could carry on a literate and witty conversation with her family and with the Bangor gentry who came to scrutinize this odd addition to their table. He was, or had been, wild, but he was safe too.

The mate was different; he appalled her. She thought he was like an animal that might go berserk and attack without warning. The captain had an alluring whiff of manly toughness; his authority over these fo'c's'le ruffians thrilled her. The mate was a pure, ferocious male, unadulterated and unsoftened by any sentiments. Her husband assured her that there was no alternative to eating at the same table with the man. She behaved as if he wasn't there. He didn't seem to notice.

The captain, the old man, the old bastard, the old son of a bitch, the bloke, etc., had done it right so far: tacked when it was just possible to do it (the cockney's complaints notwithstanding), had the sand to take her in close to get as much as possible out of each tack, knew exactly where the ship was. Always on deck, too, when they were near land—never left it to the mate, staying below sulking and depressed, like some captains the first days out.

Unlike the often-rootless seamen, most captains were family men. Starting a deep-sea voyage, a captain could go two years or more before he saw his wife and children again. If he was a gregarious man, the isolation of the aft cabin could be a terrible trial, especially if he was without the luxury of compatible officers. Captains could seldom pick and choose their mates, and they were stuck with each other, day in, day out, for the duration. They needed a great deal of self-control. There was also the matter of the captain reassuming his freight of responsibility for everything to do with the ship and the passage. That was what they often resisted for as long as possible. It took a while for them to accept that they had to assume the strain of it all one more time. However, this captain, happy with his first big-ship command, and with his wife along, was up to it from the first.

There was another difference between sail and steam. A steamship captain knew, down to most of the details, what to expect while his vessel was at sea. Even the ship's responses to the unpredictability of the weather—the only real unknown for the steamer—were limited and foreseeable: the vessel might have to alter course to take big waves at a safer, or less stressing, angle; at worst, it would have to abandon its course and dodge into especially large seas. Occasionally, its passage might take longer than scheduled; calms were irrelevent.

There couldn't be a greater contrast than a sailing ship's passage— its volatile singularity, its intimate dependence on wind strength and sea state, the large variety of tactics with which the vessel might have to respond. If the steamer's voyage is a single, simplified game of checkers, that of the sailing ship was a series of chess games, a long match. The wind-ship captain could be forgiven some anxiety at the beginning of the next elaborate, knotty contest. Sailing a ship was a true art, wrote Conrad, and therefore it required an artist of sorts to do it. Captains varied in their ability to meet this high standard. Some failed: they were incompetent or drunkards. At least they didn't last long in the constant proving ground of the sea. Some were modest in their craft: they understood their limitations, but because of that, they never did great things. The best, said Conrad, "combined a fierceness of conception with a certitude of execution upon the basis of just appreciation of means and ends which is the highest quality of the man of action."

It was never clear how good a captain was, however, until the ship's needs, its demands for devotion, became urgent. Then the sea, and the ship, would expose his weaknesses, momentary or chronic. Of all living creatures ashore or at sea, said Conrad, it is only ships "that cannot be taken in by barren pretences, that will not put up with bad art from their masters." It was early yet; the North Atlantic lay ahead, and the Horn, but the captain of the *Beara Head* had showed signs of artistic promise.

❋

The fo'c's'le had also been judging the ship's qualities as a sailer. Would she ghost in light wind? Punch through a chop? Run straight and true with a big following sea? Was she quick in stays, able to tack fast and easy? Above all, would she stand up to the Southern Ocean, put her shoulder into the battering waves, rise up over and over again, shake off the tons of rushing water encumbering her deck and fight on? You could never tell how a sailing vessel would perform. Even the sister ship of a quick, handy passage-maker, almost identical in its hull lines and rigging and built by the same yard, might turn out to be slower and mysteriously perverse.

Ships had a variety of faults: they could be slow, balky, unlucky. Some had a reputation as man-killers. Some were next to impossible to steer on certain points of sail. Even the best helmsman could not steer the iron ship *Soukar* when the wind was coming from dead astern. The vessel had to be tacked downwind, sailed with the wind on one quarter, then jibed over onto the other tack with the wind on the new quarter. Fore-and-aft-rigged vessels, schooners and the like, did this all the time, but to keep the sails full and drawing. Square-riggers were supposedly built to run dead before the wind.

Vessels like the *Soukar* were at least consistently unruly. Sometimes the crankiness was intermittent. The Yankee down-easter *William F. Babcock* did all right most of the time. But if its helmsman let the ship fall off the wind, beam-on into the troughs of big seas, it sometimes stayed there—"balked like a south Georgia mule," according to its captain, the rough-and-tough Shotgun Murphy. Once, off the Horn, the *Babcock* slid into a trough and moped there. It was one of those days when a regular procession of ships were in sight of each other, all rounding the cape. One by one, they sailed by the Yankee ship as it wallowed, with the crew trying everything known to seamen to get it sailing again: hauling the yards one way and the other, backing the jibs and the spanker, the captain and crew united in their hatred of the pig-like vessel and cursing in a dozen languages. It took hours to get the ship moving again.

Other ships were plain unlucky, some so much so that they acquired the man-killer tag. The *British Enterprise*, later the *Annesley*, lost men by drowning or accidents on virtually every voyage. And it seemed to be accident-prone as well. In 1883, it was rammed on a mooring by a steamship and sunk. Raised and recommissioned, it later fell over while in dock in Rotterdam—the top-heaviness of square-riggers—its keel rising out of the water and its masts crushing the wharf. One of its captains died on board. In 1910, another captain was lost overboard in terrible weather off the Horn. The ship had suffered storm damage to the masts and yards, and nothing could be done to rescue him. Like most who went overboard in heavy weather, he was a dead man when he hit the water. Later, near the end of the same passage, the ship ran onto the Tuskar South Rock, off the Irish coast, and foundered.

A ship could also become known as a widow-maker through no fault of its own, but because the men aboard failed in their art. Captains and mates made tactical decisions, but individual seamen, ordinary men with extraordinary skills, were often the crucial actors aboard: for example, the helmsmen when the ship was running before big storm waves. There were usually two men at the wheel in those conditions, and their feel for the wind, the size and direction of seas, and how and when to respond was a second-by-second test. A mistake could cripple or destroy the ship if it broached—that is, swung beam-on to the waves and was knocked down on its side. The shock might dismast the vessel or cause a catastrophic cargo shift, or simply allow the seas to overwhelm it. When a square-rigger disappeared from the face of the ocean, the disaster might well have begun with one ordinary man's small mistake.

These misjudgments could wipe out men too. One moonless night in 1895, the *Star of Russia*, a Belfast-built jute clipper, was sailing with the wind freshening to gale force and a big sea running. The mate and his men were on the bowsprit, hauling down the flying jib, doing everything by feel, beginning to snug down for heavy weather. The ship might have had a brief fit of obstinacy, or perhaps it was a moment of inattention by the helmsman. He allowed it to luff—that is, the vessel's head came up close to the wind—and the bowsprit plunged deep into a wave. It washed the entire watch overboard in a second. None survived.

A square-rigger was a hazardous workplace. Safety was the responsibility of each man alone, and sometimes the ship's situation was so parlous that the seaman had to take great personal risks on its behalf—its survival his own, and only, guarantee. Any ship could lose a man or two once in a while. Even the sweetest sailers with the best of captains had men washed overboard, pitched off yards into the sea and maimed or injured in various ways (crushed by a yard, slashed by a breaking line, swept into some unforgiving piece of iron by a big boarding wave, frost-bitten and gangrenous in the high southern latitudes). Some ships, how-ever, were too big for their rig—three-masted ships that should have been four-masted barques, with a greater number of smaller, more easily handled sails and lighter gear. Others were designed wrong: just little things, perhaps, but they made the difference in a tight situation. What-ever the reasons, some vessels lost men on every passage: *Gunford, British Isles, Serena, Kate Thomas, Hougomont, Blythswood, Zinita.*

Conrad envisioned quiet nooks in docklands, out-of-the-way back-waters, places of meditation "where wicked ships—the cranky, the lazy, the wet, the bad sea boats, the wild steerers, the capricious, the pig-headed, the generally ungovernable—would have full leisure to take count and repent of their sins." They might have had merely irri-tating faults, or, at the worst, been killers, but Conrad's intense nostal-gia for these beautiful ships made him generous and forgiving: among them, "there has never been one utterly unredeemable soul."

As for the *Beara Head*, its sins were venial so far. Its iron hull would eventually become foul, and therefore lethargic, with its underwater load of clinging weed and barnacles. They attached themselves to iron, happily taking both nourishment from the sea as it streamed past and knots off the vessel's speed; nothing could be done about that. For now, however, with a clean bottom, the ship slipped along pleasingly in light wind. It was quick in stays, tacking handily into thirty or thirty-five knots of wind and the awkward waves and current of the Irish Sea.

"She's a goer," said Paddy.

"We'll see what she does outside," said Russell. "At least she ain't hoodooed."

Benjamin was happy to hear it. He felt squeezed between the two kinds of malice, of the mate and Jack Grey, and he needed a bright side to look on. If the ship was quick and sound, that was something at least.

"She's loaded too deep," said the cockney. "She ain't got enough freeboard. Wait'll them greybeards come tumblin' down on her. She'll wallow like a wagon."

He half turned away, then paused.

"And don't forget the goddamned coal's wet."

CHAPTER FOUR

The Stateliest, Stiffest Frigate

Of all fabricks a ship is the most excellent,
requiring more art in building, rigging, sayling,
trimming, defending and mooring with such a
number of severall termes and names in continuall
motion, not understood of any landsman,
as more would thinke of, but some
few that know them.
CAPT. JOHN SMITH

O ne hundred and sixteen years after the start of Benjamin's voyage, I sailed round Cape Horn for the first time in my life, and afterwards, just as I stepped ashore, I saw the square-rigger. The coincidence or commingling of the two events was remarkable, in a way fateful.

We docked *Baltazar* in Ushuaia, in the Argentine portion of Tierra del Fuego. This southernmost city in the world is the base for the handful of charter boats—a few French and one American—that operate in these remote and precarious waters. It's only eighty miles north-northwest of the Horn as the brown-headed albatross flies. On this late austral summer day, the snow line on the surrounding mountains began 650 feet above the sea. In *Baltazar*'s unprotected cockpit, I wore every layer of clothing I had and still froze in the thirty- to forty-knot winds sweeping down the Beagle Channel. We had explored the icy fjords of the nearby Darwin Cordillera and, above all, we had sailed past the Horn, the mythic cape of my forefathers and of all the other vanished seamen. After that, the last thing I expected to see at the end of the world was a square-rigged wind ship.

It was ironic because I had gone to the Horn expressly to find out more about square-riggers by looking at the place they had feared the most. I expected to see the landscape of dread and death, but in a historical mode—old Cape Stiff, the nemesis of ships and sailors long gone. My own status confirmed this: I was a tourist (although the tour was a strenuous one and required some skill and knowledge on my part) on a voluntary outing, sampling danger for a few days with all the paraphernalia of contemporary technology to keep me comfortable and relatively unafraid. Cape Horn means something to me as a modern sailor because it is still a hazardous place to hang around in small boats. But to experience it as a nineteenth-century sailor did, as my ancestor Benjamin had done—as a place of most dire threat, the largest natural mass graveyard marker in the world—was impossible. I could come close only through a constant effort of historical imagination. The cape is still there, but the ships have disappeared forever.

Some square-riggers still operate. They fall into two categories: sail training vessels for the future naval and merchant-marine officers of various countries (it's still considered valuable for these people to experience the peculiar mixture of rigours and satisfactions of life under sail), and ships for tourists, whether sailing versions of full-fledged cruise ships, with buffets and drinks by the pool, or the more spartan "training" vessels in which passengers may or may not help out with sail handling and maintenance, as they prefer. But both categories of these vessels keep mainly to the warm seas and palmy islands of the tropics, or follow the artificial tourist agendas of the "tall ships" events around the world. The Horn has revived over the past few decades as a marker for sailors, but only for round-the-world racers or the more daring cruisers who are equal to sailing for pleasure in the latitudes of the "furious fifties." The old association between the most stormy of capes and square-rigged ships was broken long ago.

After we had secured *Baltazar* at Ushuaia's Club Nautico, I walked down the waterfront Avenida Maipu to the commercial dock to look at the anachronistic wind ship. It was the *Europa*, a Dutch barque—square-rigged on the forward two masts, fore-and-aft-rigged on the after mast. At about 150 feet on deck (180 feet counting the bowsprit), it was about half the size of the average iron or steel four-master of the 1880s and 1890s. Its great value for me was that it was rigged in close imitation of those old vessels. It was a revelation. I had studied many photographs, drawings and diagrams of the old wind ships. From these, I had tried to visualize the intricacies of their layout and gear: how the yards were attached; the run of braces and sheets; details of footropes, stirrups, shrouds; the universal organization of the lines and their dedicated belaying-pins. I had read accounts of how things worked—the sequence of line handling during tacking or wearing ship, hoisting sails, the order in which they were set and taken in. But it was, at best, difficult to wrap the mind around the actual procedures. What had to be done first? What could be done simultaneously? How differently were things done on various points of sail? In heavy weather? When ships were short-handed? The details were fuzzy, hieroglyphic. The square-rigger in Ushuaia was a personal Rosetta stone, the word and picture made flesh.

I went aboard and met the captain, the mate and some crewmen— tall, lean men from the Netherlands. They told me the ship had just docked after its third voyage to Antarctica with passengers (in this case, in the more adventurous and active trainee category). In a few days, it would sail for the Falkland Islands and then to San Salvador in Brazil, the Azores and back to its home port of Sheveningen, in Holland. A few years earlier, the *Europa* had crossed the Southern Ocean from Australia and rounded the Horn. It was built in Hamburg in 1911 as a light ship for the Elbe River estuary. When its active-duty days were over, a Dutch millionaire bought it and, for his own eccentric reasons, spent millions of dollars refurbishing the hull, and rigging the vessel as a replica nineteenth-century barque. The riveted iron plates I could see were original; the refitters had laid down new teak on the deck and built a steel deck house and a brightwork-finished wheelhouse above a traditional great aft cabin. They had added steel lower masts and yards, as well as wooden topgallant and royal masts and upper yards, just as on a nineteenth-century vessel.

I spent a couple of hours on board, making notes and looking at what I could see without going aloft; the captain wasn't about to assume liability for that little expedition. I realized I'd been confused about a few things: how the clewlines and buntlines were attached to the sails and yards, and how their lines were led; the arrangement of sheets, tacks and clewlines at the clews of the square sails; the lead of the halyards that hoisted the upper topsail and topgallant yards; and a dozen or so similar details. However, some things were just as I expected them to be: the overlapping iron hull plates riveted in place and streaked with rust; the layout of the belaying-pins, virtually identical to that of nineteenth-century barques like the *Beara Head* (except that the latter had a fourth mast). Benjamin and his shipmates could have come aboard the *Europa* and found any line they wanted almost straight away. The miniature barque was a most useful paradox for me: a living museum; a working, breathing artifact. In this scaled-down version, I could see the highest art and technique of wind-ship design and construction, the most sophisticated specimen of its kind. Conrad wrote that he had known the sailing ship in "her days of perfection." That's only a slight exaggeration, although it may lead to a misconception: that, like all highly evolved and specialized forms, the last barques and ships were fragile, brittle; that an altered environment could overwhelm them. In fact, even though their environment changed mightily in the second half of the nineteenth century, their extinction was a very slow one. It's surprising how long they were able to hang on.

In the customary view of the history of ships, the age of sail gave way to the age of steam, wooden construction to iron and steel. Historians used to see this narrow technological revolution as one of the main reasons for the ascendancy of the British Empire. It was a world-empire, went the argument, and depended on fast and efficient lines of communication and supply by sea. The rulers had to know what was going on in India or New Zealand or the Cape Colony, and they had to be able to tell their administrators and soldiers there what to do, both day by day and when subject people became refractory. Imperial troops and warships had to get to anywhere in the world without delay. The empire's trade depended on thousands of merchant ships to carry commodities and goods anywhere a ship could float. The sea was to Britain what its straight, ordered roads had been to Rome: the web of information and transport that bound the separated parts of the empire to each other and to the imperial centre. Britain outdistanced the rest of the world in the construction of iron and steel steamships, and that assured its political and military world-dominance. The small northern islands used to be protected by the "wooden walls" of its oak-hulled men-of-war; then cheap coal and an efficient iron industry made possible steamships, and the full manifestation of the haughty empire itself.

A corollary of this view was that change happened quickly and smoothly: a new superior technology displaced an older form. It was inevitable and linear, and it guaranteed the quick death of the sailing ship.

The other nail in the wind ship's coffin, said the traditional analysis, was a simple ditch, although the biggest one ever dug. The Suez Canal, begun in 1859, was completed in 1869, and it was a creature of the steamship—without it, there would have been no canal. Suez connected the Mediterranean and Red seas, and so provided all the things that steamers loved and required: a shortcut, a straight line, a depot to take on fresh bunker coal and the avoidance of stormy seas—in this case, the winds and strong currents around the Cape of Good Hope. (The Portuguese originally labelled it the Cape of Storms; it seemed like a very bad place until sailors discovered the even more formidable Cape Horn.) Steamships don't so much make use of the sea as exploit a highway, said Conrad. Suez was a necessary link in the steam sea-highway to the imperial jewel, India, and the other colonies and trade of the East.

There was no conceivable reason to dig the canal for sailing vessels. Theoretically, tugs could have towed square-riggers through it. But the winds round about were all wrong. In the Mediterranean, they were light and fluky or strong and contrary, depending on the season, and it was always impossible to sail north against the prevailing winds in the Red Sea. The rhumb line—the straight mechanical course from point to point on the earth's surface—meant nothing to ships under sail. Their choice of route was bound up with the direction of wind and current, a subtle adaptation to nature and to things that could not be changed, as always a matter of the sea-cunning of men. The sailing ship "seemed to lead mysteriously a sort of unearthly existence," wrote Conrad, "bordering upon the magic of the invisible forces, sustained by the inspiration of life-giving and death-dealing winds." A straight, dirty ditch through the desert sand was a mere road for mechanical machines; it was useless and irrelevant for the wide-ranging, opportunistic sailing ship endlessly seeking fair wind and sea room.

In fact, however, the opening of the Suez Canal was not a turning point for sailing ships. And steam went through a remarkably slow and uneven period of development and distribution. There was no one-two punch. The argument can be made that the greatest days of sail took place *after* the canal was begun and in spite of the growing use of steam—even that it was, in part, the spur of competition from steamships, as well as their hunger for coal, that was responsible for the efflorescence of sail in the second part of the nineteenth century. Sailing ships were not made obsolete by grandiose excavation projects in the desert or the mere existence of steamers. They were done in by tinkering, a slow and methodical accretion of engineering improvements that gradually found solutions to the serious and limiting technical problems of steam engines; only then did it become clear that the days of sail were irrefutably numbered. "The incredible defiance of the Industrial Revolution by sail during the second half of the nineteenth century," as Robert Foulke put it, finally ended.

The main problem with the steam engine was that it needed a boiler that would generate enough steam pressure to make all the noise and smoke worthwhile. A steamship had to operate efficiently enough that the coal it had to carry for its own use still left enough room for a decent amount of cargo. And a steamship had to be able to chug along a reasonable distance as well. If it couldn't cross oceans on the load of coal it could carry, it would never be a paying alternative to sail, except to scuttle around coasts from port to port. Until the 1850s, that was about all that a steamer could do, given the measly steam pressure its boilers could provide—about twenty pounds per square inch. They needed at least fifty pounds, and preferably sixty or seventy pounds, of pressure to become real deep-sea passage-makers.

The pace of technical improvement was slow, but it was steady. By 1860, Alfred Holt and John Elder of Liverpool had developed the so-called compound engine, and within a few more years, they had demonstrated that it was an efficient and reliable means of driving a ship. This was a two-cylinder engine, in which steam from the first cylinder was passed to a second one. There was less pressure in the second cylinder than in the first, but cumulatively, the new machine considerably increased the amount of power generated from a given amount of steam.

These engines produced close to sixty pounds of pressure per square inch, and the steamer was in serious business. This meant that a vessel of, say, 2,000 tons displacement and making nine knots, which previously would have burned close to 40 tons of coal a day to carry 1,400 tons of cargo, could now carry 2,000 tons of cargo while consuming a mere 14 tons of coal daily. In geographical terms, the longest route on which steamships could compete with sail in 1865 was about 3,000 miles from British or northern European ports—the Mediterranean fruit trade. Five years later, that distance had more than doubled: steamers could sail 6,200 miles to Bombay using the Suez Canal, or direct to West Africa.

It wasn't only a matter of boilers. Making the steamship go required two more things, and both took time to craft: effective paddles or screw-propellers (which presented their own design and structural problems), and above all, a brawny iron hull with the tensile strength to handle the weight and vibration of heavy engines and the forces created when the vessel hewed its direct path through the sea. Applying the technology of the Industrial Revolution to the open ocean was much more difficult than making things work on land or on the inland canals of Britain. Like all ships, steamers had to contend with the power of waves and the chaotic and unpredictable forces they brought to bear on ships' hulls. And steamers were isolated from ready aid from shore; once committed to the passage, they had to be as self-sufficient as a wind ship (although for a shorter time), despite the vicissitudes of a tossing, salty, damp environment that is hostile to all mechanical things. This required reliable hulls and machinery, and to the extent possible, redundancy.

The straight-line courses that were one of the main reasons for the steamer's fast passages also meant that it spent a lot of time butting its way directly to windward. The sailing ship always had to adopt devious means of making progress against the wind. Its course was a dogged zigzag affair of minor gains grudgingly ceded until the ship gained the necessary advantage. Its crew won through an understanding of the natural limits thrown up by wind and sea, by patience and wariness— the stuff of watches aboard ship—and by the will to keep going, to take advantage of every pause or waver in the forces mounted against them, which were far greater than those of puny men.

None of this for the steamer. It smashed its way to windward—no tactics, just trench warfare, brute force, head-on attrition. There was no attempt to deflect the sea's power in the graceful way of a sailing ship, by heaving to or by running off before wind and waves, always attempting to soften the blows. The advent of the steamship at sea meant many things, but most of all, it was part of a fundamental change in the relationship between the sea and sailors. Indeed, it gave notice that all the rules of the game between nature and humans had changed. Together with all the other machines of the revolutionary new age, the steamer said, We control, we disdain, we dispose, we destroy. We give the orders.

At first, the dirty ditch and the new machine threw a scare into sailing-ship men. They felt the first real chill of the future—the end of sailing days. The completion of the Suez Canal in 1869 and the perfection of the compound engine stopped large sailing-ship construction almost dead for the next several years; by contrast, in 1871, for example, the number of steamers built was double the year before. However, the collapse turned out to be short-lived. The hard-headed businessmen and daring traders of Victorian England and its expanding empire quickly recovered their confidence in sailing ships as they gained perspective on the limits of steam—even in conjunction with the new canal—and the continuing possibilities for sail.

In fact, in the 1870s, the two means of propulsion achieved a kind of equilibrium. The compound engine gave the steamship short-range preponderance and medium-range competitiveness. But the weight of coal it was necessary to burn to carry a given amount of cargo set a limit on both the steamship's range and its practicality. For the long hauls, to Australia, the Far East, or the West Coast of the Americas, for example, the square-rigger was an equally sensible means of moving cargo.

Most important, sailing ships were far better for carrying the bulk commodities that were the stuff of British industrial supremacy: iron, cotton and woollen goods, and coal. And of these, coal was king. Britain produced and exported more of the stuff than any other country. One writer has labelled Britain "the Persian Gulf of the period." Coal fed the developing industries of countries around the world, and of course, it also fuelled steamships. Again, the irony: the sailing ship became the most efficient replenisher of steamship coaling bases and depots everywhere.

Coal was also an outward-bound cargo. British square-riggers didn't have to sail out to Australian or American ports with a thousand tons of English stones in their holds to keep them upright. They could load up with coal instead. It was like carrying ballast that paid a low freight and made a nice profit. Emigrants, a living alternative to coal, could also be loaded as paying ballast for the outbound passage. It was then cheaper to bring back grain from San Francisco, wool from Australia or nitrates from South America. And until the Suez Canal Company lowered its rates in the 1880s, it was still more economical to carry rice by sailing ship from the Far East. The wind ships continued to rule in Bangkok, Saigon and Rangoon.

The sailing ship became the cheapest warehouse in the world. Coal and other commodities were often bought and sold many times over while they lay in the holds of square-riggers, enjoying almost free storage until they were off-loaded. That was another reason why cargo was everything, why keeping it safe, dry and unspoiled was the highest good: because it was a continuing object of exchange. Ragged, ill-fed seamen endured dangerous, exhausting toil for meagre pay because they were an expendable means to ensure profitable ends. Elsewhere, other men—clean, sleek, full-bellied—turned profits on the cargo even while sailors suffered and died. Coal (like oil later) was an essential commodity, and a very effective money-generating substance. To snuff out men's lives in the process of producing and transporting it was considered an acceptable price to pay, unavoidable collateral damage. There is a straight line from the casualty rates of coal mines and the decks of square-riggers to the trenches of the Great War, a casual acceptance of dead men for dubious ends. *It's not coal you're talking about; it's men's lives.*

Steamers had other disadvantages. One is obvious: coal costs money; wind is free. Even though the supply of wind, its direction and quantity, was always uncertain compared with coal, sailing vessels had become able to use it more and more efficiently. By the time steamships came along, captains in sail knew much more oceanography than they used to, and that helped them make faster passages. Perhaps of greater importance for business deals, the duration of passages became more consistent. Storms, especially the high-latitude ripsnorters, were dangerous for any ship, but they weren't the real problem where carrying cargo was concerned—calms, and the unpredictability they created, were.

Up until the mid-nineteenth century, sailors could find out the best way of avoiding regions of calm, or at least the worst of them, only by talking to other sailors. There was a body of traditional lore built up over the previous few hundred years. But this was unsystematic and anecdotal, and sporadic as well, since only survivors got to add to the pool of knowledge—perhaps the most valuable information went down with the lost ships. Relatively few vessels crossed oceans, so there wasn't a large body of knowledge to draw on. And it was difficult to perceive patterns because of the variableness of weather in different places and seasons, and from year to year. Sailors did come to recognize the existence of the belts of the northeast and southeast trade winds, and their benign reliability. They developed a full and bitter appreciation of the doldrums as they languished there, or crept through them, water going bad, scurvy cutting them down. They knew roughly where the calms would likely envelop them, but nothing more detailed than that. Far-ranging seamen knew about the terrors they could expect at high latitudes and around the stormy capes like Good Hope and the Horn, but that was mostly just the knowledge of almost-certain pain. There was no pattern or paradigm of the good and bad times, or of the bad and worse times, to make a passage.

The first systematic approach to oceanography began in about 1830, with the Englishman Maj. James Rennell. He published a collection of charts showing winds and currents, based on interviews with ship captains. But the American naval officer and circumnavigator Matthew Fontaine Maury was the real founder of scientific ocean routeing. He took Rennell's limited statistical approach and developed and perfected it.

Maury was invalided out of the United States Navy in 1839, and the government put him in charge of its depot of charts and instruments. He was passionate about wind and currents. He distributed special logbooks to naval captains and studied historical logs. From these sources, he charted the routes of hundreds of ships that had made the same voyages at different times of the year and in different years. He noted the winds and currents each ship met along the way, as well as ocean temperature and magnetic influences on compasses. He convinced ship captains from other countries to gather similar information, and he eventually integrated this into his compilations, the first of which was published in 1847. His work also inspired the first international conference on marine meteorology, held in Brussels three years later.

Maury told long-haul square-rigger captains that they had it all wrong in their courses through the Atlantic and Southern oceans (the latter being those portions of the South Atlantic, Indian and Pacific oceans below forty degrees south latitude). Ships heading to Australia from Europe or the East Coast of the United States, for example, always avoided the Brazilian coast as a potential lee shore. In the absence of other information, this was a seamanlike course for captains to follow; the unweatherly square-rigger had to stay well clear of the shore, whether rock-toothed or soft and sandy, whenever possible. Avoiding Brazil, the captains swung across the South Atlantic and usually passed close to, or even within sight of, Table Mountain, at the Cape of Good Hope, and then steered north out of the latitudes of the roaring forties to avoid the Southern Ocean storms.

Maury soothed these concerns. Steer for Cabo de São Roque, on the northeastern tip of Brazil, he advised the captains. Don't worry about the lee shore; currents and winds close to Brazil favour ships, tending to keep them off the coast, sweeping them south as the land falls away to the west from Cabo Branco and the port of Recife. From there, make a wide southeasterly arc through the South Atlantic to pass as much as ten degrees—six hundred miles—south of the Cape, and then stay there, down in the roaring forties, all the way to Australia. It was rougher at those latitudes, to be sure, but modern square-riggers were built to take that kind of weather.

Through his painstaking research, Maury had identified the westerlies, the great wind systems between forty and about fifty-five degrees north and south of the equator. For ships that wanted a fast passage, there was no alternative: their captains had to take them south, into the strong west wind, to "run their easting down."

Ships that took Maury's advice got quick and dramatic results. In 1854, *Red Jacket*, a crack White Star Line clipper ship on its maiden voyage from England to Australia, went so far south into the Southern Ocean that spray froze on its bow. Nevertheless, the vessel set a record of sixty-nine days out, and then it stayed well south across the southern Pacific to come home in seventy-five days. This wasn't the fastest circumnavigation (although it was close to it), but *Red Jacket* "tied the knot"—that is, crossed its own outward-bound track—in record time, just under sixty-three days. This was beaten only a few years ago by big, fast, state-of-the-art multi-hulled yachts.

These voyages were, admittedly, exceptions. *Red Jacket* was one of the nimblest sailing ships ever built, and it was lucky with its wind and its calms. Still, Maury's guidelines also produced big improvements in the average passage times made by average ships. The time to the equator dropped from forty-one to thirty-one days, and one American captain, scrupulously following Maury's directions, sailed from Baltimore to the equator in twenty-four days. The eleven-thousand-mile journey from England to Australia took about 125 days in 1850. Using Maury, the average dropped to ninety-two. This was a revolutionary improvement by any standard, an example of the ability of the scientific method, or its statistical branch, to make an immediate and startling impact on real life, even in the supremely contingent environment of the sea. (Maury's advice has held up. Modern circumnavigating racing yachts follow essentially the same route, although they are able to make constant adjustments to long-term variations or atypical local conditions through the use of detailed satellite weather information.)

Maury's inspired data collection and the conclusions he drew from it had important implications for sailing ships in the competition with steam. The distances to Australia via Suez and round the Cape of Good Hope were close enough that the reliable westerlies south of forty degrees, driving square-riggers at average speeds of ten to fourteen knots, enabled them to clock times about equal to those of steamers. Maury's pilot charts kept wind ships in the game and helped make possible the so-called golden age of sail from the 1860s to the 1880s, when the elegant, speedy, gracile clipper ships had their day and the big iron and steel barques came into general use.

Sail and steam continued a relatively comfortable co-existence for a little more than ten years. But the technology of steam machinery was quite unlike that of wind-driven canvas. The steamer severed an ancient kinship: the intimate, necessary embrace of the seaman and nature. "The machinery, the steel, the fire, the steam have stepped in between the man and the sea," said Conrad. Men no longer exercised the sailing ship's old skills in a hard, yet somehow elegant, duel with wind and sea; they merely broke their backs in heat and dust, their work a repetitive and impoverished mimicry of the machines they serviced.

Things moved to a completely different rhythm. Improvements in the design and operation of sailing ships and their gear took place slowly; it was a conservative world intent on minimizing losses and disasters by sticking as much as possible to what had worked in the past. Innovation could be dangerous; the price of a failed experiment might be the lives of many men, or of the ships themselves. The obvious dangers of crossing the oceans under sail encouraged caution. And in any event, by the 1870s, sailing vessels had arguably reached a state of sophistication in design and function that was difficult to improve upon. Maybe a little tinkering here and there to make the work easier and safer, or to carry a smaller crew or more cargo, or to go a little faster. Perhaps a refinement of Matthew Maury's advice on the best routes to follow or how to get round the Horn a little less painfully. But by and large, the wind ship had become a set piece, a static object of combined utility and beauty.

None of this applied to the onrushing Industrial Revolution, in which change was embedded. That was what made it a revolution. It was part of the nature of the new machines that their creators would make them better as fast as possible. And as a new technology at a primitive stage of development, the steamship provided constant need and opportunity for improvement. Anyone could see it: a device like the marine compound engine was just the beginning. Once this process of using steam to power machines had got going, it would follow an urgent imperative. The rush and flux of things would never cease.

The inevitable new engine came to pass, and it was the true ringer of the wind ship's knell.

On April 7, 1881, the steamer *Aberdeen* left Plymouth bound for Melbourne. The vessel arrived a mere forty-two days later with a cargo of four thousand tons, having made only one coaling stop along the way, at Cape Town. The *Aberdeen* was fitted with a triple-expansion engine: a third cylinder, even larger than the second, in which the steam was able to complete its expansion, was added to the first two cylinders. This process worked because the engine's boilers could handle steam pressure of 125 pounds per square inch. Higher pressure made the engine far more efficient, and that meant that it needed much less coal to run on (allowing, in turn, more crucial cargo space). The usual iron boilers would have exploded at these pressures. But the triple-expansion engine's cylindrical boilers stayed in one piece because they were made out of good-quality steel, which tolerated much greater pressures. By 1890, a steamer could run at nine knots using half an ounce of coal for each mile per ton of cargo, about one-tenth the requirements of the old compound engines.

The *Aberdeen*'s achievement startled and worried sailing-ship owners. The passage from England to Australia was one of the main demonstrations of the square-rigger's usefulness, a feature in its repertoire of long-distance voyages. But forty-two days! And obviously, that was just the beginning. Even the fastest sailing ships, like the tea and wool clippers *Thermopylae*, *Cutty Sark* and *James Baines*, took sixty to seventy days on their best runs; the *Thermopylae*'s sixty days was a record.

True, compared with stolid steamers, plugging along at nine or ten knots, the clippers were capable of far greater speed, exploding for magnificent bursts of downwind surfing while running their easting down in the roaring forties. The *Cutty Sark*, famously, did 363 miles in a day, an average speed of more than fifteen knots. Back in 1854, the Black Ball Line clipper *Lightning* rushed 436 miles in twenty-four hours on its maiden voyage to Liverpool. That same year, the *Champion of the*

Seas supposedly made 465 miles in one day while running in the Southern Ocean towards Australia, the fastest any sailing vessel had moved until the late twentieth century. The crew heaved the log several times during the day, measuring speeds of eighteen knots and over. (Heaving the log involved allowing a knotted line to run off the stern for a set period of time. The number of knots that ran out gave an approximate speed.) Measuring distance run using this method was often inaccurate, and a certain amount of wishful thinking sometimes entered into the calculations. Nevertheless, no steamship of the time could make speeds anything like those of a square-rigger in its joyous, downwind, sail-bellying glory.

The problem was consistency. The sailing ship was often the elegant hare; the patient steam tortoise always got there faster in the end. The square-riggers flashed their virtues for a few days at a time, or even for a whole, very lucky, passage. Once again, it was lack of wind that was always their downfall. Even passages planned on the latest scientific routeing principles, laid down by Matthew Maury and improved on ever since, could not avoid areas of light or no wind. A ship just had to get through the horse latitudes, the doldrums, the summer calms of any ocean. No steamship deckhand had to stick a knife in the mast or whistle to draw the wind gods' attention. Or climb into the ship's boats with his mates and pull at the oars for eighteen hours a day, dragging the mother ship out of the doldrums yard by blistering yard. Or drive screaming horses overboard from a long-becalmed ship to save water—long supposed the origin of the name of the horse latitudes.

The age of steam was also the age of iron. The energy and the material were the soul and heart of industrialization—the equivalent of the computer chip in our own technological revolution. (Iron had been around a long time, of course; the Industrial Revolution involved the discovery of how to produce it in large quantities. When rolling mills began to turn out iron plate, the material was ready for all its subsequent building possibilities.) They were the primary inventions that allowed the creation of all the subsequent variations and elaborations: the specialized machines and devices that clanked and huffed around the world, changing its shape and sound forever.

Steam had to be compressed to work as energy so that steam engines could make other machines move in whatever directions: a locomotive, a lathe. Containers capable of putting steam under pressure and maintaining it had to be iron (and later, steel). Steam engines and their boilers were massively heavy and vibrated like nothing in nature. Only iron had the strength, durability and hardness to withstand the punishment of steam pressure. Therefore, steamships also had to be iron ships.

This connection was immediately clear and compelling in the case of steamers. For sailing ships, however, the argument for iron construction was not as apparent. Iron was an unknown material, and wood was the way things had always been done. The conservatism of shipowners and seamen, who were innately reluctant to do anything differently, was neither surprising nor irrational, given the character of the sea and the seaman's utmost wariness of Conrad's "great autocrat," the most "destructive element" (so often, for the sailor, a watery killing field). The poet Mary Oliver:

> *The sea*
> *isn't a place*
> *but a fact, and*
> *a mystery . . .*

We deal with actual mysteries, and dangerous ones in particular, through obsessive behaviour: an anxious repetition of the actions that have preserved us before. The superiority of something new has to be incontrovertible. Wood certainly had its problems and disadvantages as a construction material for ships, but the problems had the great advantage of being familiar. Who knew what would happen with iron, its strange, hard blankness? If the sea was a mystery, iron was another one. Scant comfort to a shipowner who might lose his livelihood or a seaman his life, one mistake the last mistake.

The wooden ship was a masterpiece of independent mobility. Its seamen could turn up or improvise the materials of its hull, masts and yards, and its hemp rigging, even in the most remote of new-found lands and islands. They were as self-sufficient as spaceships. Once launched out onto distant and unknown seas, wooden ships were small, self-sustaining worlds that could keep the sea out and make do without port facilities almost indefinitely. You could give a wooden-ship seaman a knife and a forest, went the old saying, and he could build you a ship. And rig it too.

Before steam tugs came on the scene to tow square-riggers in and out of port, wooden ships had to do the job themselves, working the wind, doing a dozen difficult things at once (sounding the depth, watching the currents, tacking, taking in sail, getting warps ready)—everything done under the vessel's own wind-driven steam. The captain who could work a six-hundred-ton ship into a small tidal harbour, or manoeuvre against the fast-running stream of an estuary while coming up to a dock under sail, or warp in, or pick his way into a slot in a crowded anchorage, was a plain-and-simple genius under sail.

One maritime historian describes this body of seamen's knowledge as "perhaps the most complex and demanding pattern of skills ever acquired by ordinary men." A man could acquire and keep such sophisticated skills only if he devoted his life to the sea. The effort involved in becoming a wooden-ship seaman meant that a man had to cut himself off from shore life and any proficiency with its demands, simpler and safer though they were. This claim on his time, together with the actual separation from land, helped to form the distinct brotherhood of seamen, who were alienated in almost every way from society ashore.

When James Cook began to explore and chart the east coast of Van Diemen's Land, the great undiscovered (by Europeans) world of Australia, he quickly found himself tangled up in the 1,300-mile-long dead-end channels and coral heads of the Great Barrier Reef. Even the Great Navigator couldn't help running aground. Modern boats with global positioning systems and detailed charts run aground there regularly. It was what Cook did afterwards that was so remarkable. On that remote, barren shore, with no idea of what lay around him, hemmed in by the intricate reefs, in shifting winds, his vessel's hull ripped open, Cook oversaw the salvation of the *Endeavour.*

In twenty-four hours of desperate but organized labour, the crew pumped, jettisoned heavy gear and, on their second high-tide attempt, managed to pull the vessel off the coral with the ship's boats. In deeper water again, they were able to reduce the flow of sea water into the vessel by rigging a collision mat, an old sail, over the hole, allowing the pumps to keep up. Then they set sail. For five days, in fickle wind, they tacked and wore ship in the narrow coral-strewn waters, pumping for their lives, until at last Cook found a small estuary in which to run the *Endeavour* ashore. After that, repairing a gaping hole in the hull and dragging the ship off into deep water again was a piece of cake. Cook sailed on, wiser now in the traps and treacheries of the reef, stayed out of further trouble and finished charting the coast. This was merely a somewhat more robust example of the sorts of skills the wooden-ship seaman took for granted.

Having acquired such hard-won adroitness and expertise, seamen were reluctant to let these skills go. They resisted innovation; their conservatism was the self-assurance of the confident expert whose work has always stood up to inspection and trial. At the same time, a worldwide trading system, and a string of port facilities, had come into being based on the technology of the wooden ship. Sailors would accept changes that made all this obsolete only if they were convinced that the resulting economic gains would be overpowering.

In fact, iron proved itself relatively quickly as a superior building material for sailing ships because (issues of knowledge and skill aside) wood's deficiencies are serious ones. Wood is fungible, biodegradable, in many ways fragile. It rots. In salt water, the sharp-toothed shipworm, the *teredo*, bores into wood, munching endlessly until the ship's hull is honeycombed, its strength literally eaten away. Many little pieces of wood have to be joined together to form a hull. This means weakness at the attachment points, and the failure of one piece jeopardizes the whole assembly. As a result, wooden ships leaked all the time, and often in copious amounts. Crews manned the pumps every day, and in each watch in heavy weather. The hulls of wooden vessels distorted during loading or taking the ground in harbour at low tide. They grew convex or concave, like spavined horses clapped out by hard use.

Wood might have been universally available and aesthetically agreeable, but if something better came along, even hidebound wooden-ship shellbacks could be quickly persuaded. They might well consider it a good tradeoff: hanging up some of the old skills in exchange for a hull that wouldn't fall apart in a Southern Ocean buster, or that didn't need pumping for hours in each twenty-four. They perhaps really wouldn't mind not having an opportunity to sing the old bilge-pump shanties any more. As for shipowners, they had little or no emotional investment in the old ways and required only proof of economic advantage to make a change. It didn't take seamen or owners long to realize the clear superiority of ships made of iron.

Iron could be joined together with iron rivets to produce a single, tough, seamless, near-watertight whole. Iron had a great deal of longitudinal strength, so that an iron ship would never distort like a wooden vessel. Ships were already becoming longer and narrower before the advent of iron; they were faster and more seaworthy than the older, beamy, tub-like designs. But a wooden ship with a length six times its beam would almost certainly deform. Iron ships with lengths eight or nine times their beam, meanwhile, even if they were loaded with heavy bulk cargoes like coal or railway iron, remained serenely unwarped. Iron ships were as likely as wooden ones to be driven ashore—that was still a matter of seamanship and luck—but they were much more likely to survive the experience.

Like any new technology, however, iron had bugs. For one thing, it played hell with ships' compasses. When shipwrights built an iron ship, they often had to beat the iron into submission, to hammer the obstinate material into the shapes they needed. In the process, they created complicated and permanent magnetic fields that surrounded the ship's hull and deck; that this happens is just one of the natural qualities of iron. But the sailor's oldest and most valuable navigational aid, the mariner's compass, went haywire and could not be trusted. No shipbuilding material could survive if it had this effect on the compass. That tool had set early sailors free from reliance on landmarks and dead reckoning, or on the oral tradition of minute observation and feel-of-the-scrotum techniques later perfected by Polynesian seafarers (memorizing star patterns, observing birds, getting the feel of swell and wave-train patterns). Compass deviation was such a big difficulty that insurers refused to cover iron vessels on deep-sea voyages.

The problem was mostly solved when the Astronomer Royal, Sir George Airy, and Sir William Thomson (Lord Kelvin), members of the Admiralty Compass Committee, prescribed a series of magnets and soft iron correctors to be set up in and around the binnacle, the compass container. The solution was, in fact, flawed, and ships that relied on the distinguished astronomers' method sometimes found themselves on the rocks anyway. Nevertheless, the magnets were a great success during their first real test: aboard the Liverpool-built sailing ship *Ironside* launched in 1838, on its voyage to Rio de Janeiro and back. (Airy and Kelvin were essentially right: the modern solution is to place soft iron balls on either side of the compass and other correctors in the base; this neutralizes the magnetic fields around it.)

The most serious problem with iron hulls was fouling—their propensity to attract large communities of marine weed and shellfish— a problem especially in tropical waters and during times of repose in ports, where, it seemed, whole crops sprang out from ships' bottoms overnight. Hull growths triggered expenses for owners. Frequent docking to scrape iron hulls clean increased maintenance costs. Various toxic coatings and paints were used to poison the hardy weeds and barnacles, but none of them worked very well. Even moderate fouling involved tons of extra weight. The increased friction and drag meant that a ship might need 10 per cent more energy to drive it along, an inconvenience for a steamer (a little more money for greater coal consumption; passage times a little slower). For a sailing ship, the reduction in mobility could be alarming. Overall speed was reduced when there was wind, but in light conditions, a 10 per cent reduction in efficiency could make the difference between ghosting along, making some progress towards areas of wind, and staying put like a driftwood log, adding many days, or weeks, to a voyage. "An iron ship begins to lag as if she had grown tired too soon," wrote Conrad.

Marine growth could become bad enough to affect a wind ship's manoeuvrability. Embayed on a lee shore, a hard-pressed vessel (which could have clawed its way free with a clean bottom) might lose the hard and absolute struggle because of the encumbering weed, making it unable to save itself and its men. A ship off the Horn, trying to come up into a gale to heave to, might fail in the attempt with a foul hull; unable to carry out survival tactics, it might be swept and overwhelmed by Southern Ocean greybeards, then posted missing, its disappearance yet another mute sea mystery. A foul bottom could destroy a sailing ship.

For a while, some builders responded with composite construction. Wood planking over iron frames could be coppered in the traditional way (sheathed with yellow metal) to prevent fouling but still offer some of the virtues of iron (smaller frames allowing more cargo space, and cheaper building costs). However, the premise of composite construction was the undemonstrated superiority of iron. In the course of the 1870s, everyone saw that, except for the fouling problem, iron was the better shipbuilding material. The rationale for composites lost credibility, and construction ceased.

One of the last composite vessels built was the *Torrens*, a fast, sweet-lined passenger clipper on which Conrad served as first mate during two voyages to Australia and back, from November 1891 to July 1893. This was near the end of Conrad's twenty years at sea and the beginning of his writing career. In his time on the beach looking for another ship, and in his spare time in watches aboard or waiting for cargo in Amsterdam or Samurang, he had been writing a novel. But it seemed as if he would never finish *Almayer's Folly;* he had been dragging it around with him between ships and boarding houses for three years. Almayer and Nina were in suspended animation, he wrote later. Then he met a young consumptive passenger aboard the *Torrens.*

During a bookish conversation one stormy day east of Tristan da Cunha, with the wind rising and Conrad thinking he'd have to go on deck any minute to order the topgallants taken in, he suggested on an impulse that the passenger—Conrad refers to him only as "a young Cambridge man" and "Jacques"—read the manuscript. Later, Conrad asked this only other person on earth to have read it (he was William Henry Jacques) if it was worth finishing.

"'Distinctly,' he answered in his sedate veiled voice, and then coughed a little."

Was he interested? Conrad asked.

"'Very much.'"

That was all it took to release Almayer from his stasis. Jacques died soon after the ship returned to England; Conrad finished the novel. It was published two years later under Jozef Korzeniowski's pen name.

The *Torrens* was a writerly catalyst for Conrad in another way as well. John Galsworthy, not yet a novelist, was a passenger on one of its return trips from Australia. In the course of his conversations with this aberrant ship's officer with the Polish-French accent, Galsworthy listened to his "tales of ships and storms, of Polish revolution, of his youthful Carlist gun-running adventure, of the Malay seas, and the Congo; and of men and men." Galsworthy became one of Conrad's enduring friends and his literary champion.

The wind ship could not begin to match the passage times and scheduled voyages of steamers like the *Aberdeen*, with their new, efficient triple-expansion engines. By the mid-1880s, the steamship could make three passages for each one completed by a sailing ship. This made the steamer, even with all its extra capital costs and manpower and running expenses, as economical as the latest big sailing ships built.

Shipowners in sail resorted to building massive metal four-masted barques—like the *Beara Head*—in a last-ditch effort to compete, and they became the standard British-built square-riggers. (Wood still predominated in the United States.) Some owners fitted them out with the latest labour-saving gear, like steam donkey engines and brace winches; all owners cut crew numbers back to the bone. And they specialized in the commodity bulk cargoes of the late-nineteenth century. It was enough for a while. But after 1881 and the voyage of the steamship *Aberdeen*, no one had any doubts about the square-rigger's obsolescence and mortality.

A few remnants of sail persisted for longer than anyone in the early twentieth century could have predicted. There was a brief revival of all kinds of sailing vessels during the First World War, when freight rates shot up and it became worthwhile to dig some of the remaining old, decrepit wind ships out of mothballs for a few more years. A handful of the big iron and steel four-posters (and a few five-masted ships too) carried grain and nitrates for German and Scandinavian owners (not British; their transference to steam was abrupt and irreversible) until the Second World War. Other cargo-carrying sailing vessels hung on elsewhere: American, Canadian and Portuguese schooners; rice haulers around the coasts of India and Ceylon; fishing and cargo boats elsewhere in the developing world.

But the large metal sailing vessels of the close of the nineteenth century and the start of the twentieth were the end of the line, kept going by hard men who pinched pennies, and harder ones on board who dealt with the consequences. They were long, lean, tough giants—four or five times the length of the tubby, hemp-rigged, high-ended little wooden vessels worked so skilfully and courageously by Magellan or Drake, and far more seaworthy too. But they weren't bulletproof, an impossible standard for a seagoing sailing ship, no matter what its size or what it's made of. The sea abides unchanged—as Melville describes it, "The dark side of the earth, and two-thirds of the earth." Even for the last big iron and steel wind ships, it was the enemy, ancient and dangerous.

CHAPTER FIVE

That Unsounded Ocean

Ah! The good old time—the good old time. Youth and the sea.
Glamour and the sea! The good, strong sea, the salt, bitter sea, that could
whisper to you and roar at you and knock your breath out of you.
JOSEPH CONRAD, Youth

The *Beara Head* beat to windward out of the western approaches to the English Channel in a light wind that had veered slightly and was now west. If it held, the ship would clear Île d'Ouessant, at the western tip of the peninsula of Brittany, by a wide-enough margin. The island, and its vicinity, was a mess of rocks, tides with a range of thirty feet or more and the strong currents that went with them. It was so obvious a danger that no deep-sea square-rigger captain would ever risk getting close inshore. There was another good reason to stay clear: south of Brittany lay the Bay of Biscay, the French Golfe de Gascogne. Unless their destination was a port in a bay, square-riggers avoided these bodies of water between headlands. They were always potential traps if the ship got caught in one with an onshore wind, and Biscay was one of the worst. The west winds, usual at that latitude, drove ships towards the French coast, turning it into a treacherous lee shore. In a gale, waves that might have travelled a thousand miles or more across the Atlantic piled into the bay. As they crossed the lip of the continental shelf, the abrupt shallowing—from a thousand fathoms to sixty or seventy—bred the steep-sided, breaking waves that played havoc with a sailing ship's ability to claw to windward. The square-rigger wasn't built to beat clear of lee shores. Failure meant the end of the ship, and maybe the men too, in the chaos of sand, rocks and surf. The *Beara Head* had to get to the west, away from dangerous water.

The ship had a definite itinerary, which was set out in the Admiralty's volumes *Ocean Passages for the World*—a summary of detailed and precise oceanographic information and route instructions. To get to Cape Horn, and to get round it, the *Beara Head* would do it by the book: "On leaving the English Channel, at once make westing, as the prevailing winds are from that direction. With a fair wind from the Lizard [on the Cornish coast], steer a WSW'ly course to gain an offing in 10 or 12° W." From there, "shape course to pass Madeira at any convenient distance, giving a wide berth to Cabo Finisterre, in passing it, as the current from the Atlantic usually sets right on-shore there." In the winter months, the mariner should pass to the west of Madeira (an archipelago about 475 nautical miles southwest of Cape St. Vincent, at the tip of Portugal), says the Admiralty, because the frequent strong westerly gales funnelling down the sides of the high islands create violent squalls to the east of the group. With its tall top-hamper and acres of sails (which required a lot of time to take in), the square-rigger was especially vulnerable to sudden wind, like williwaws off mountains or *pamperos* off the River Plate, farther south on the Cape Horn route, or vicious white squalls in the tropics, which would dive down on the ship at a hundred miles an hour out of a blue and innocent sky. In a few seconds, a ship could be knocked down onto its beam ends, its canvas blown to shreds, top and topgallant masts collapsed, cargo shifted, men in the water.

Captains could expect different wind conditions depending on the time of year. From roughly October to March, a ship, beating hard, sailing either full and by or comfortably on a beam reach, depending on its luck with the precise direction of the westerlies, might get no closer than 450 or so miles to Madeira before slowing down in the horse latitudes at close to forty degrees north. In winter, Madeira is almost dead in the middle of this high-pressure zone of light and variable winds. It's not nearly as troublesome an area as the doldrums. The winds are more uncertain in strength and direction than they are absent, and ships could often work their way through without too much delay.

However, a captain could never be sure his vessel wouldn't get stuck in these latitudes before he could coax it along—or rather, run his crew ragged bracing and squaring the yards for days on end—to the fair and constant northeast tradewinds. In spring and summer, the belt of horse-latitude variables squeezed together at its eastern side.

At the time of the *Beara Head*'s passage in May and June, it was possible for the ship to work its way out of the high-pressure zone and into the northeast trades—they were called the Portuguese trades in that region—as far north as thirty-eight degrees latitude, or about 360 miles from Madeira. If it was lucky, it might romp its way south from there, past Madeira and on to, and beyond, its next waypoint south.

That was the Cape Verde Islands, lying a little more than a thousand miles farther to the south-southwest and about 350 miles off the African coast (of modern Senegal and Mauretania). According to *Ocean Passages for the World*, the navigator should aim to pass to the west of these islands, where the wind is stronger and steadier, and just within sight of them, so he can confirm his position.

The *Beara Head* had to buck the eastward-setting North Atlantic current at the beginning of its passage, as it beat out of the Channel approaches. Then, somewhere across the latitude of Biscay, the ship would pick up the favourable south- and southwest-setting Portuguese and Canary currents, which it would ride all the way to the doldrums.

Nothing to it. Except that anyone who has used sailing directions and pilot charts (updated versions of Maury's originals) while under sail knows the experience: dead calm when the chart assures you that thirty knots is your likely lot. Or the reverse: you're putting a third reef in the mainsail even though the chart shows five-knot wind. The Admiralty sailing directions improved the odds, and they certainly made a huge difference in passage times compared with the old days. However, day-to-day and year-to-year variations could nullify all of this instruction. It was, after all, based on averages over long periods of time. On any given day during a passage, anything could happen, or fail to happen.

"Full and by, sou'-sou'-west."

"Sou'-sou'-west. Aye aye, sir."

The helmsman kept the *Beara Head* as close to the wind as the ship would tolerate without pinching and shaking the sails, slowing it down. His job was to keep the barque beating hard yet footing along.

"Sou'-west by south. Keep a good full."

"Aye aye, sir. Sou'-west by south, a good full."

Day by day, the few orders had been much the same. The light wind held steady for three days, coming mainly from the west-southwest. The crew had little sail-handling work beyond some small tasks: overhauling an errant buntline, sweating up halyards, squaring or checking yards a little to accommodate the wind's slight shifts. And so, in the cool, overcast days, the seamen chipped iron, spliced line and wire, scraped wood, tarred rigging, layered on red lead and inspected sheets, braces, halyards, ratlines and footropes. The eternal burden of the ship: warding off the degradation of its parts. Even if it ain't yet broke, fix it. During the night watches, the men paced the main deck below the poop, chatting quietly or watching the flash of phosphorescent waves breaking away from the ship's iron side in endless, hypnotic kaleidoscope. The mate stood motionless by the weather jigger shrouds or beside the helmsman, staring up at the faint loom of white sails.

The helmsman alone sailed the ship, its passage almost silent as it slipped through the sea. By day, crimping and squeezing his neck until it ached, he looked almost straight up to watch the windward edge of the mizzen royal, that mast's highest sail. If it shivered a little, he was right on course. The square sails on a wind ship were trimmed with careful precision in a slight corkscrew pattern, so that each successively higher sail was braced a little tighter than the sail below it. Near the sea's surface, the wind was slightly deflected by friction as it blew over the water. The sails were adjusted in response to this vertical variation in the wind's angle. The mizzen royal was braced in the farthest, and so it became the bellwether: if it was just full, then all the other sails should be too, if they had been trimmed with skill. On moonless nights, the helmsman had to try to catch the suggestion of movement aloft out of the corner of the eye, or if it was too dark, he had to judge his course by the feel of the wind on his neck and face.

By noon on its eleventh day at sea, the *Beara Head* was where it didn't want to be: a little more than two-thirds of the way across the

Bay of Biscay, on a course that would not allow it to clear Cape Finisterre. *Ocean Passages for the World* was very specific about this troublesome cape at the northwestern tip of Spain, the southern boundary of
Biscay: give it a wide berth. It was a cape like any other, and ships were
to treat it with the usual exaggerated respect; but on top of that, the
current set right onshore there, and that multiplied its menace.

One thing was certain: if a captain wanted to use *Ocean Passages*, he'd
better know where in the world he was. The Admiralty's exacting
guidelines assumed that the mariner knew his latitude and longitude,
or close to it. In the 1880s, that was a good assumption, and had been
for some time. The deep-sea navigator's repertoire of aids and instruments was well established: the sextant for latitude and to aid in establishing local time; the chronometer for longitude; the logline for speed,
and therefore distance run; and the much older compass, an Arab or
Chinese invention, used as early as the eleventh or twelfth centuries
by Italian and English navigators.

The sextant measures the altitude of a celestial body—that is, the
angle between the body and the horizon (usually the sun, or selected
stars at dawn or dusk)—in order to get a position line. If this angle is
obtained at exactly the time of local noon, when the sun is highest in
the sky, the sextant sight produces an instant latitude line. The sextant's predecessors—the astrolabe, quadrant and octant—all did the
same thing with increasing accuracy, and finding a vessel's latitude was
relatively straightforward from early on in the period of European sea
exploration.

The problem was longitude. For several hundred years, deep-sea
navigation consisted of sailing a line of latitude until the lookouts saw
land or the ship ran into it. In other words, a navigator could place his
ship on a reasonably accurate line running in an east-west direction,
but he had to guess how far it had sailed along that line. The only way
to transform this hit-or-miss technique into accurate navigation was to
find the vessel's longitude by means of astronomical sighting and calculation: the navigator measured the difference between his local time
and the time at other locations (known by means of a nautical almanac)
by measuring the angular distance between the moon and other celes-

tial bodies. He then had to do some difficult mathematics to arrive at the ship's longitude, or an approximation of it. This procedure required skill and a facility with mathematics that was uncommon even in the navy, and relatively unknown among captains of merchantmen.

It was also unreliable. When the scholarly Frenchman Louis Antoine de Bougainville made his great voyage of exploration and scientific inquiry beginning in 1766, he carried on board an astronomer who took thousands of sights, measuring the angular distance among various stars, planets and the moon. There were always inconsistencies, which themselves followed no discernible pattern. Land that should have been there wasn't; land appeared when it shouldn't have. Even the charted positions of well-known places didn't agree with the calculations. The Azores, for example, were two hundred miles away from where de Bougainville's astronomer put them.

There was a second, potentially much simpler and surer method. If a navigator knew, simultaneously, his local time and the time in Greenwich (the Royal Greenwich Observatory marked the zero-degree longitude line), he could deduce his longitude. Therefore, he needed a reliable chronometer that would keep Greenwich Mean Time over the course of a long voyage and in spite of temperature change, dampness and the ship's motion and angles of heel, all of which could perturb the delicate mechanical components and balances of a timepiece. Finding a method of calculating longitude was the eighteenth century's philosopher's stone for navigators, astronomers, mathematicians and inventors. European traders and navies, imperial adventurers and freebooters all needed a solution to the longitude puzzle so that they could cross oceans more safely to do their work, or their damage. Errors in calculating longitude had caused an unending string of sea disasters for single ships and fleets. In one of the worst examples, in 1707, four men-of-war of a British squadron ran aground on the Scilly Islands, off Land's End, losing the ships and two thousand men.

In 1713, the British government, desperate for a working chronometer, offered rewards of £10,000, £15,000 and £20,000 to anyone who could solve the problem to within sixty, forty and thirty nautical miles, respectively. (Twenty thousand pounds is about one million pounds in modern currency.) The amounts were impressive enough to induce hundreds of admirable, and cranky, proposals, which were evaluated by a Board of Longitude, but it took half a century before anybody won the money.

In 1736, an English village carpenter turned mechanical genius, John Harrison, built a chronometer to a design that solved the main problem of inaccuracy caused by temperature change. He proved it by taking the device to Lisbon and back aboard a naval vessel. It worked well enough. However, his chronometer was big, heavy and expensive, and Harrison spent the next thirty years perfecting it. In 1759, his fourth effort looked like a pocket watch. Cook took one on his second voyage; it was "his never-failing guide," he said. Harrison's son carried this version on a return voyage to Jamaica in 1762, and over a period of more than five months, in heat and cold, and in spite of the North Atlantic's jolts, it varied by less than two minutes, a very manageable error by the standards of the time. Even so, it took another eleven years, an act of Parliament, and the intervention of the Astronomer Royal and, eventually, King George III himself for Harrison to pry the full amount of prize money out of the parsimonious and recalcitrant government.

The chronometer and the sextant became the navigator's dual means of finding out where he was. Each was an ingenious device, a clever contrivance for its specific purpose. But they were so much more than mere instruments. Each also represented a hard-won and sophisticated body of knowledge about the planet, space and time. Each was, in a way, the culmination of the long gathering of that knowledge: that time could be used to locate an object in space; that the geometry of stars was related to an observer's position on earth's surface; that in the void of space lay the means to find oneself on the blankness of the sea. What a breadth of view that implied! What subtle connections! "What wonder that mariners tended to be reflective," says Robert Foulke, "when they had to deal with abstract time and celestial space just to find out where they were in the watery world."

By the time of the *Beara Head*'s voyage, dependable and inexpensive chronometers had been standard equipment aboard ship for a hundred years. Captain McMillan would have no trouble making sure he followed the instructions in *Ocean Passages*—provided, that is, that he could see the sky. If the sun or appropriate stars were invisible because of cloud or fog, then sextant sights were impossible and all bets were off. In the roaring forties and the furious fifties, around the Horn itself, the sky was cloudy most of the time. Then the captain had to fall back on the navigator's old seat-of-the-pants method of figuring out where he was: dead reckoning.

Samuel Johnson thought dead reckoning was important enough to define with his magisterial dash (it was presumably a well-known term in sea-absorbed Britain): "That estimation or conjecture which the seamen make of the place where a ship is, by keeping an account of her way by the log, by knowing the course they have steered by the compass, and by rectifying all with an allowance for drift or lee-way; so that this reckoning is without any observation of the sun, moon and stars, and is to be rectified as often as good observation can be had."

This was really what separated the men from the captains. Most people could learn how to shoot the sun and do the sight-reduction mathematics to get a latitude line. The ability to do dead reckoning, however, was one of those complex human skills that included close attention to speed and course and their precise notation, but it also involved the inchoate churnings of both sides of the brain in nimble tandem as they analyzed and collated the other elements affecting the vessel's position (set of current, leeway, wave action, the helmsman's yawing, compass inaccuracies, the rough-and-ready speed computation of the knotted logline and sandglass). Experience and a talent for the feel of things—the ship and the sea as they moved in uneasy conjunction—were what nurtured good dead reckoning.

Many square-rigger captains were notorious for doing slipshod celestial navigation and relying too much on dead reckoning. There's no doubt that sometimes they had very good results with it. As the full-rigged ship *British Isles* approached the Strait of Le Maire, northeast of the Horn between Staten Island and the mainland, in the terrible winter of 1905, cloud with heavy sleet and rain squalls blanked out noon latitude observations for two days in a row. Nevertheless, its captain, James P. Barker, drove the ship on under full sail, relying entirely on his dead reckoning to thread the twelve-mile-wide, current-ridden needle of the strait. As they got close, land appeared and the captain himself went aloft (his doing so displayed considerable anxiety) to take bearings. The *British Isles* was exactly where he thought it was. According to Barker, he maintained a remarkably accurate dead-reckoning plot during the ship's subsequent fifty-two-day ordeal rounding Cape Horn. For six weeks, he had little or no confirmation of the ship's position during its mortal combat with the Horn's winter snorters; when he did get a bearing, it was almost the same as his dead-reckoning account. (However, there has been a lively revisionist discussion of some of the claims made by the flamboyant Barker—for example, by the Australian square-rigger seaman and writer Alan Villiers—and they may not be reliable.)

The problem is that it's only a matter of time before the accumulation of errors, big or small, makes dead reckoning perilously inaccurate, and eventually, it breaks down completely (Barker's claims notwithstanding: either he was very lucky or he exaggerated). The ample literature of wrecks is a record of captains who weren't where they guessed they were. Before chronometers, the problem was even worse since, in overcast weather, the navigator was in the dark about both latitude and longitude. During his voyage of circumnavigation from 1740 to 1744, George Anson struggled to round the Horn for more than a month. At one point, his dead reckoning put him ten degrees west of Tierra del Fuego. When he stood to the north and got a sight of land, he found that the east-setting current, as well as storms and wave action, had him stuck still on the wrong, or eastern, side of the cape. He was more than 350 miles out in his guessing (and wishful thinking). When James Cook sailed from the Society Islands to New Zealand in 1769, he was four degrees (or more than 150 miles) out on his longitude by the time he got to the New Zealand coast. The lesson? Even the best navigator could not trust his dead reckoning.

After the advent of good chronometers that provided continuous and accurate longitude, navigators only had to worry about keeping track of their latitude when they couldn't take sun or star sights. Nevertheless, a high proportion of ship casualties were the result of the defects of dead reckoning. Even steamships succumbed with regularity. The captain of the SS *Bovic* might simply have been dead lucky when he made a remarkable transatlantic passage from the Tuskar Rock, off Ireland, to Nantucket in eight days without a single latitude sight. Later, he piled up the SS *Suevic* on the Maenheere Rocks, near the

Lizard and then the SS *Highland Hope* on the Burlings. In both cases, he was still relying on dead reckoning. The SS *Swazi* steamed for several days on end across the Atlantic in the winter of 1916–17 and, when able to get a bearing, found that it was 150 miles out in its reckoning. That experience was the rule; the *Bovic* was the exception.

Dead reckoning was more complicated for sailing ships, with their indeterminate courses, frequent zigzag tacking and leeway. Whenever a wind ship went ashore or gutted itself on off-lying rocks, it was usually because the captain had not had visual confirmation of his position for some time—or had not taken advantage of opportunities to take sights or shore bearings or to sound the depth—and his dead reckoning was out. Official enquiry after enquiry blamed the captains.

Even though chronometers were usually accurate enough, they were, nevertheless, mechanical machines in an inimical environment, and they often failed—that is, their time errors became so large that they were no longer usable (if the captain was lucky enough to discover that before disaster occurred). Then the navigator had to fall back on sailing down his latitude and dead reckoning his longitude, just as in the old days. Once, the famous clipper *Cutty Sark* made for Java Head with its chronometers out by as much as eighty miles; the captain had to sail until he found the latitude of the Sunda Strait and then head down that line until he could get sight of land and take shore bearings.

Every square-rigger captain could take sun sights with a sextant, but many couldn't do stellar sights, which were more complicated (although more accurate if they were well done). The captain of the *Bidston Hill* told his enquiry that he did not know how to take star sights, that he had not been able to shoot the sun for two weeks, and that he was relying on dead reckoning when he attempted to beat through the Strait of Le Maire in light wind in 1905. When the wind dropped away, the ship was much closer to shore than the captain thought, and the current and swells set the vessel on the rocks of Staten Island. Half the crew was lost. The captain of another ship nearby was able to take star sights each morning and evening, and he knew exactly where he was.

Currents were most often the stealthy killers of ships, setting them onto rocks or into the breakers along the shore, the land looming up, sudden, shocking, lethal. Even when their directions and speeds were generally known, currents varied from place to place and time to time.

And many, with their side streams, tributaries and backwashes, were little known at best. The worst areas: the foggy coasts of Newfoundland and Nova Scotia, the shore of Chile and Peru, the southwest coast of Africa, the Channel and the Baltic approaches, scattered rocks rising up offshore such as the Casquets and Burlings, fishing grounds such as Iceland, and the entire sea area in the vicinity of Cape Horn. Over the years, square-riggers went ashore by the hundreds when current set played hell with dead reckoning.

In *The Mirror of the Sea*, Conrad wrote that he had been chief mate on a ship when it was stranded (driven ashore by tide or current or blundered there in fog, he doesn't say). It wasn't a fatal contact, and Conrad worked for ten hours directing the crew in laying out anchors. (They were able to heave the ship off at next high water.) His captain, who had stayed below all that time, sent word to have his mate come below for a bite to eat. When Conrad entered the cabin, he was struck by the strange immobility of the gimballed table, which had swung continuously for the previous seventy days. He was also brought up short by his captain, who sat at the head of the table "like a statue." Conrad saw that the man's bald head, usually ruddy with sun and wind, was dead white. And something else: the captain looked somehow untidy, and Conrad realized that it was because he had not shaved. He had managed to do so every day, in even the worst of gales and throughout the roaring forties, but he had not shaved that day. More than anything else, Conrad thought, this omission revealed the shock of stranding: "The fact must be that a commander cannot possibly shave himself when his ship is aground."

In fact, none of this ever happened to Conrad in his real sea career. But the book is a narrative of how he wanted things to have been (like many memoirs and a great many personal sea-voyage accounts). Or of how things should have been for the purposes of telling the story later. Stranding a ship was one of the most awful events that could happen to a seaman. Conrad created a one-page novel of his own supposed experience—for dramatic effect, but also to give authority to what he wanted to say about how it felt to run a ship ashore, to experience that sudden death of the living ship's motion. More than anything else that happened to him, stranding brought "to the sailor a sense of utter and dismal failure." If the first rule of seamanship is to keep the sea out of the ship, the second is to keep the ship off the ground. Conrad describes the surprising sensation when a vessel takes the ground (even though he never experienced it): "It is as if your feet had been caught in an imponderable snare; you feel the balance of your body threatened, and the steady poise of your mind is destroyed at once."

The nature of the stranding is important. It's somehow all right, Conrad writes, to be driven ashore by stress of weather—losing a battle to claw off a lee shore, for example. That's a catastrophe, to be sure, but it's an honourable defeat. However, "to be 'run ashore' has the littleness, poignancy, and bitterness of human error." Once stranded, the ship may be saved by the crew's quick and knowledgable work, but its rescue also requires that the captain (who is responsible for navigation—the mate too, but less so) is able to bear up "against the heavy weight of guilt and failure." Save the ship, maybe redeem the man.

The *Beara Head* was sailing deep within the bounds of the Bay of Biscay, a potential trap if heavy weather developed. Cape Finisterre lay about 160 miles ahead. Unless the wind changed, the ship would have to tack away from the coast to the northwest, almost the reverse of the direction it wanted to go in, losing ground and time. The captain had maintained his course into Biscay, rather than tacking to sail out of it, gambling that the wind would haul round to the west, or north of west, just enough to free up his ship and allow it to sail clear of the cape. Sea room: the sailor's term for that comfortable margin of ocean around the ship that allows it to absorb a gale or two and remain in safe, deep water. Sea room was what the *Beara Head* was about to run out of.

The wind did change. It veered slightly, so it blew straight out of the west, a better angle for the ship, but it began to strengthen as well: ten knots to twenty and then thirty within an hour. In the great cabin, the captain recorded in the log the barometer's slow, steady fall. Overhead were cirrus clouds, the "mare's tail and mackerel sky" of an advancing front; farther west, grey altostratus, the most reliable indicator of rain and wind, rolled towards the ship. The westerly swell was building. The signs of heavy weather, close and closing in, were plain for any sailor to see. With a storm sniffing down their trail, it was more important than ever to keep well clear of the meddlesome land. Captain McMillan demonstrated again that he was not a cautious man. When the wind shifted, he was able to steer a little farther to the west than before, but it was still not sufficient to clear Finisterre. Yet he held the ship to its course, hoping that the wind would go round even more, to the northwest, giving them enough of a lift to avoid tacking away. He ordered the royals taken in and furled, and hoped for the best.

The captain lost his wager. The wind increased, gusting close to gale force, and continued to blow steadily from due west. The *Beara Head* was running out of sea room in a hurry, closing with the Spanish coast and getting shoved deeper into Biscay by mounting seas and increasing leeway as it heeled to the wind's growing pressure. There was most likely a current too, stealthily setting the vessel inshore. Now there was no choice.

"Clew up upper to'g'ants! Haul down the flying jib! All hands stand by to wear ship!"

The captain roared from the poop, spilling out anger and frustration. No wind, then headwinds, then light winds playing a jink on him, keeping him back, driving him into goddamned Biscay when he'd had a right to expect the strong spring westerlies from the start. Now he had to turn back to the northwest to escape, and by the looks of what was coming, they'd have to plug away in that direction for at least a day, maybe two.

Benjamin, who had been chipping rust off the base of a deck capstan, gathered his tools together, getting ready for the coming manoeuvre. This time, they would not tack; the seas were already too big, the wind rising by the minute. The barque was over-canvassed. For the first time, they would wear ship, swinging away from the wind, bringing it behind them and then onto the other side as they trimmed the sails and yards onto the new course. He stowed his tools in the bosun's locker, under the fo'c's'le deck, amid the bustle of the starboard watch hauling up clewlines and buntlines, slacking away halyards, smothering the upper sails on each mast. Then up aloft they went, the mate offering his usual loud encouragement. Benjamin felt a mixture of fear and relief. Soon he would have to go aloft again, and as he'd bitterly predicted about his inevitable next hike up, it was blowing a bloody hooley. But at least he wouldn't have to go right up near the mastheads; the royals were in, and the starboard watch had the upper topgallants in hand. He knew it was an illusory comfort: 110 feet up instead of 160—it would make as much difference as bedamned.

On deck, men moved around with the sailor's jerky, lurching trot, adjusting to the rise and fall of the deck as the big barque punched its way forward. Seas were breaking across the low main deck already, though only knee-deep for now; the washports (openings in the bulwarks with hinged doors) clanged as they swung open and shut, draining the water off, too slow as always. The starboard watch finished furling the upper topgallants.

"Haul up mains'l and crojick! Brail in spanker and tops'l! Down mizzen stays'ls!"

The captain got the *Beara Head* ready to wear round onto the other tack by taking in the fore-and-aft sails on the jigger, together with the three staysails set between the mizzen and jigger masts, so that they wouldn't impede the ship's swing. He ordered the biggest, lowermost sails on the main and mizzen masts—the main course (or mainsail) and the crojack (pronounced "crojick")—hauled up to the yards by their clewlines and buntlines, both to reduce sail area and the number of lines the crew would have to handle and to ease the job of bracing the yards up to the wind on the new tack. The square sails would be left hanging in their gear, not furled, so they could be set again quickly when the ship had been turned.

"See all lines clear!"

As for tacking, all braces, sheets, tacks and so on had to be clear for running during the wear-ship manoeuvre. A brace that jammed and prevented a yard from swinging across or a sheet that didn't run free could cause a sail to rip, or injure men if it suddenly jerked clear under pressure. Seamen lifted the coils off the belaying-pins and capsized them on deck with care.

"Put the helm up," said the captain quietly to the man at the wheel, an experienced seaman. The ship needed a fine touch for this; too fast, and the men at the lines wouldn't be able to keep up with the turning ship.

"Up helm!" the captain shouted to signal the start of the manoeuvre.

In thirty-five knots of wind, the *Beara Head* bore off fast to leeward.

"Slack away main and mizzen tops'ls! Handsomely now! Let 'em lift a little!"

These topsails were allowed to slacken and shake slightly, to keep the main weight of wind out of them and to slow down the turn; with the aft sails taken in and the wind's pressure there removed, the vessel's head came fast off the wind.

"Square the main and crojick yards!"

The men hauled the yards round as fast as they could so they were square with the ship's beam, the wind coming round onto the quarter and then right aft, the yards following the wind through its arc as the helmsman made his slow, steady swing. Now the barque was sailing downwind, with the main and mizzen topsails and lower topgallants squared and pulling hard. This was where the square-rigger lost ground when it wore ship. By the time the crew had got the sails trimmed this far, the *Beara Head* had sailed a mile. It made another mile before the yards on the foremast could be hauled round, and as they were squared in their turn and filled with wind, the ship's pace picked up. It took more time to get the jibs and staysails over to the other side. Then the helmsman continued his turn. The main and mizzen yards were braced round again, to follow the wind as the ship came up close to it on the new tack. The seamen set the spanker once again, released the mainsail and crojack and braced the foreyards round, trimming them all with care, the square sails in their slight corkscrew pattern. With the royals and upper topgallants furled, the mizzen lower topgallant became the helmsman's guide. The tired men and boys, muscles burning, did this workout to a shanty or to the universal, drawn-out hauling shout "E-yah! Oh-ho! Ee-i-i-yo-ho!" hauling together on "yah" and "ho." They belayed and coiled the lines, ready for the next time.

A properly manned ship could wear round in a few minutes and lose only a mile or so. The *Beara Head*'s crew, half a full complement by the old standards, took almost fifteen minutes to do it, with every man aboard except the captain lending his weight, and lost more than three miles. That wasn't important this time because the barque had sufficient sea room, but it was a different story for a vessel embayed by a wind too strong to tack against. Then the ship could only sail hard to windward from headland to headland, limit its net loss as much as possible each time it wore, and hope for a wind shift. Without that, it could be only a matter of time until it struck shore.

On the port tack, the *Beara Head* beat hard, heading a little west of north-northwest, its course a safe trajectory away from the Spanish and French coasts and the shallow water of the continental shelf, where the waves steepen and break. But with leeway and the North Atlantic current setting it east, the ship was really making a course back to Ireland. If the wind held as it was and they did nothing, they'd run into the Fastnet Rock, off the Cork coast, in three days or so—not all that far from where they'd started two weeks earlier.

To windward, Benjamin could see what was coming: the low, grey-black clouds of a North Atlantic spring gale. That was what he'd expected from the beginning of this passage: fun and games. These first few days had been an uncharacteristic hiatus. Now the real Atlantic was introducing itself.

The *Beara Head* began its square-rigger dance with the storm, which would start as another hindrance but end by lifting the barque on its way. The wind had backed a little from west to the southwest, showing the leading edge of the gale's southern semicircle. The centre was north of the ship, so that when it passed by that longitude, the wind would veer rapidly to the west and then the northwest. The *Beara Head* could then wear onto the starboard tack and, finally, sail south with a strong fair wind.

The ship's tactics were those of any weaker, yet wily, antagonist: to strip down to essentials; try to hold its own for as long as possible, then give way stubbornly; to be patient; wait for the opportunity to gain ground; to keep in mind that, like the guerrilla always up against superior forces, the wind ship that doesn't lose, wins, but to be prepared— calm and centred—for catastrophe, because at sea this is more than possible (most likely in the form of a despoiling wave whose size and strength are too shattering to survive).

The gale was a run-of-the-mill strong North Atlantic depression, one in the string of constant low-pressure systems—more frequent and severe in winter but characteristic of all seasons except brief high summer—that form from fronts sweeping off the North American coast. It had gathered energy from the warm Gulf Stream and the un-obstructed stretch of ocean, and it had become a full-formed cyclonic depression, with wind spiralling into the centre in a counter-clockwise rotation. It brought heavy rain and hail and, even in spring, tempera-tures low enough to freeze sailors. The whole spinning mass of air moved forward at an average speed of twenty knots or so, the forward movement adding to the strength of the wind where its direction was the same as that of the whole system. Such lows could be two hundred miles across or two thousand; their winds blew at gale force at least, and sometimes in the big, strong variations, into hurricane strength (sixty-four knots and over). Storms blowing for days over the long fetch of sea produced waves that sometimes came close in height to the monsters of the Southern Ocean: more than one hundred feet. Wave height is a result of wind strength and fetch—that is, the dis-tance over which the wind blows. In the high southern latitudes, the systems had unlimited fetch as they rolled right around the planet, no continents to get in their way. The North Atantic wasn't as expansive as that, but a ship on its eastern side encountered storms that had been building up waves for one or two thousand miles, more than enough fetch to create mean wave heights of forty feet or so, and to breed rogues and killers twice as high.

As the gale approached, the crew continued their submission to the wind's building strength. The orders in quick succession:

"Clew up courses! Haul down outer jib!

"Haul down main and mizzen to'g'ant stays'ls!

"Brail in the spanker tops'l!

"Clew up the lower to'g'ants!"

Benjamin was with the port watch, at the foremast as usual. He tailed on to the fore course and topgallant clewlines and buntlines, hauling the big-bellied sails up to the yards. Paddy, the shantyman, sang a halyard song, good for this hard pulling too:

> *Growl ye may, but go ye must,*
> *Handy, me boys, so handy,*
> *Growl too much an' your head they'll bust,*
> *Handy, me boys, so handy.*

Haul away and show yer clew,
Oh, we're the bullies for to drive 'er through.

We're bound down south around Cape Horn,
Ye'll wish to hell ye'd never bin born.

The verses rolled on; there were always as many as needed to finish any job.

The mate ordered Grey, the cockney, and Kapellas, the Greek, to douse the outer jib. The bowsprit buried its end in a wave from time to time, but each sea sent some green water across the exposed spar. The mate might have been a killer when necessary, but he didn't fancy losing two experienced men. He yelled aft to the captain: if the helmsman could bear off (head away from the wind) a little more until the jib was secure, he'd be much obliged. It would give the two seamen an easier time.

The ship was making heavy going, a corkscrew pitch and roll, as it slammed into the seas, parting them, sending volleys of spray a hundred feet up and out to leeward, and taking tons of water over the windward rail. This boiled and tumbled down the sweep of the main deck like big-river rapids, burying the seamen up to their waists, sometimes their chests. Now the main deck showed its true nature: it was an artillery field of fire in which sailors were mowed down by each shipped wave.

In a few years, the obvious hazards of open decks on the new giant ships would prompt the building of Liverpool houses and other kinds of raised amidship decks with catwalks between them; all were designed to break the waves' force and keep men off the sea-swept range of an exposed deck while they worked.

The *Beara Head*'s crew were half submerged every half-minute; the lines they were hauling became temporary lifelines. Sea boots filled up with North Atlantic water, although in this sea area, part of a far-travelled northern Gulf Stream eddy, it was warmer than the air and bestowed on the feet, at least, an almost pleasant, wet glow. The rain that had been falling for hours—it began before the wind got up ("When the rain comes before the wind, / Your topsail halyards you must mind")—felt bitter cold. Soul-and-body lashings around wrists, boots and waists were no use on deck (where men swam as much as they walked), although they would help keep oilskins attached aloft, where the wind blew even harder. It was the wind that leached heat out of bodies.

The starboard watch was already contending with the main course. The port watch, except for the cockney and the Greek on the bowsprit, went up the foremast ratlines. Benjamin's much- and long-feared time had come. He exchanged a look of wry resignation with Anderson, whose face was drained pale beneath sunburn and the shine of salt water.

"Lay aloft, me boys. Lively now! The course first."

The mate's voice was as strong as ever, but his words were softer than usual. It seemed that as the gale built up, the mate stood down, his violent rage diminishing in some obscure, strange adverse ratio to the increase in the wind's velocity. He had ordered the barque off the wind to protect his men on the bowsprit. Sending his watch aloft, he was no more than gruff.

Except in his dealings with the Clerk. The mate kicked the man hard in the small of the back, forcing him in the direction of the weather rigging. His invective was reflexively intact for the greenhorn. A pathology had evolved between them. The mate's insults followed the Clerk as he trailed the others up the ratlines. Under his slop-chest oilskins, for which he owed the captain several months' pay, he still wore the remains of his formerly black suit and once-white shirt. He hadn't been able to pay for sea boots, and so went barefoot until Russell extorted donations from each member of the watch (except Grey) to buy footwear for him.

Now began the most difficult and exhausting few hours of Benjamin's entire life—so far. This time, climbing the ratlines meant hanging on hard to the shrouds, pausing now and then as the ship rolled more than usual or pitched into a trough with a stomach-roiling dive. He could feel the weight of the wind and hear its keening sound in the rigging. As he hauled himself out around the foretop and onto the platform, heavy spray pelted his back, sixty feet up. The half-muffled sail billowed up and back towards him in forty knots of wind. A flick of the canvas could knock a man off the yard. At least it was bloody daylight, he thought; he could see what he was doing. He could also see down and around himself to the awash deck and the sea, which was streaked with foam in the direction of the wind; the height of the seas, for some reason, did not appear diminished from this high up. He eased out onto the port-side footropes of the fore main yard.

"Ben, Anderson, lay along with me to the end!"

"Goddamn it! Keep aholt, ya bloody Dutchman!"

"Wait till they're at the end, Janny, an' we pull together."

"Farther along. Spread yourselves!"

"*Satan!* Maguire! Led go and ploody move!"

"My father has a milk white steed an' he is in his stall, / He is a clever circus horse, he can balance on a ball, / Singin' blow, ye winds, in the mornin', / Blow, ye winds, high-ho!"

"Now haul away, lads. Haul up!"

"Goddamn it, my nails!"

"Don't get blood on the hooker's sails, ya dirty bastard!"

"Who'd sell a farm an' go to sea?"

"Let go'a the jackstay, you greenies, and lend a hand!"

"Keep aholt, I said, you goddamn Dutchman bastard!"

"Who let go? Who let go o' the fuckin' sail? Goddamn it! Haul away and hold on!"

"Belay your jaw!"

"Now, boys, on my word, all together. Ready? Now pull!"

"Eee-yah!"

"You ain't the goddamn bosun yet, cock!"

"Let's do the job. Hang on! Get your bellies on it! Now, ready?"

"Ohh-ho!"

"And pull again! That's it!"

"Then it's goodbye, Mavourneen, / We're off to sea again, / Sailor Jack always comes back / To the girl he's left behind!"

"*Satan!*"

"Goddamn it!"

"Hold the fuckin' sail, ya bloody Dutchman!"

"Lay out here, Clerk! Lend a hand! I'll be buggered if I can think of a more useless lump I've ever seen aloft! You must'a been a bricklayer's clerk! Lay out here or when we get down I'll put you in bloody drydock!"

"More beef! Haul away!"

"Bring it in. Handsomely now!"

"Keep ahold!"

"Keep aholt, ya goddamn Dutchman!"

"By heavens, every time I get aloft, I feel like I have to pump ship."

"Piss away, Johnny. No one'll know in this breeze."

"Ride it down, boys, ride it down! Keep 'er footing! Hang on and ride it down! One more time! Battle the watch!"

"Goddamn it!"

"*Satan!*"

"Sonovabitch!"

"*Gott im Himmel!*"

"Keep ahold! Who let go? Can't you hang on to a bit of canvas?"

"Keep aholt, ya bloody Dutchman!"

When they were clawing in the course, Benjamin experienced for the first time that swoop down the face of the sail as the seamen lunged to get a grip on it. Their bellies against the yard acted as a kind of pivot, and the momentum of their bodies as they swung them downwards to grab the sail simultaneously flipped the footrope out and up until their legs and feet were braced up in the air, level with their bodies, sometimes higher. The first time it happened, Benjamin thought they were falling in some inexplicable, terrible accident, the whole watch about to plunge down onto the deck or into the sea. He restrained himself from shitting but pissed into his clothes, the sudden wet warmth. Next to him, he saw Anderson's strong wrists, his hands clamped to the jackstay like pincers.

The port watch furled and lashed down the fore course in a little more than half an hour. Under the conditions, it was quick work. The mate's voice boomed out over the gale's racket, carrying up to them with ease: half of them to furl the lower topgallant, the rest to come down and aft to the mizzen course, which the starboard watch hadn't got to. Benjamin found himself in the topgallant gang. Muscles trembling with strain, he climbed up past the two topsails, which were rockhard with wind—they would be the last sails to come in as the gale grew—and up the topgallant ratlines to the yard. Then he ventured out again on the treacherous footrope, which he feared now more than ever.

The topgallant was a smaller sail, but there were fewer men to muzzle it and the wind was rising all the time. Looking down from the yard, 110 feet above the deck, Benjamin once again had that feeling of disconnection from the familiar, a sense almost of puzzlement, as if his brain could not absorb the novel information it was receiving through his salt-stung eyes: the barque's slender hull seemed to flip from side to side beneath him, leaving him suspended over the water, which was striped with foam, half white with breaking seas. As the ship went through its simultaneous pitch and roll, he felt the strong confirmation of the phenomena of inertia and gravity; the mast's upside-down pendulum motion whipsawed his body through sixty degrees of arc in a few seconds. His fear paralyzed him for several minutes. By chance, he was the only green hand on the yard, and the others—Urbanski, the Finn, the Dutchman, the Elf—did the work around him, cursing him, the weather, the ship and each other, damning God, the gale o' wind, the mate and all the whores of Sailortown. It was a kind of high, cold, open-air hell, Benjamin thought: around them the black, boiling clouds of the storm that was on top of them now; the rain, with some hail in it, coming at them like little icy bullets; the abstract pattern of the sea below; the snapping canvas and the tipping yardarm; the men around him screaming in three languages; bloodstains spreading on the sail.

Once, part of the topgallant flicked back at Urbanski, catching him in the chest. He had no handhold, only his knee braced against the yard, and he lost his balance, beginning to teeter backwards into a fall as the footrope swayed out from under his centre of gravity. Beside him, the Dutchman reached out and grasped the Yankee's arm, pulling him back towards the yard and away from the hole opening up behind him. The two men glanced at each other and reached down again to grapple the sail. A strange, casual gift of life, Benjamin thought, and its mute acceptance. It must happen all the time up here in the high rigging.

After ten minutes or so, he was able to free one of his hands and began to pull at the sail with the other. Soon, he found he could haul with both, or at least trap with his body weight the canvas the others had hauled up onto the yard while they dived down for more. The wind aloft had increased to close to forty-five, maybe fifty, knots, and it was a hard skirmish for the five men. It took nearly an hour to get the sail secured. When they climbed back down to the deck, the mate sent them up the mizzen to do the same thing to its flogging lower topgallant. Another hour of sweating, cursing, fisting, hauling bloody labour before the last gasket was passed around and tied off.

Back on deck, Benjamin was so exhausted he could barely stand up. The ship still beat to windward, heading about northwest under six topsails, two jibs, a jigger staysail and the spanker. The gale grew more severe, however, and almost right away, they had to haul down another jib, brail in the spanker and, to Benjamin's horror, lower the upper topsail yards, then clew up and furl their sails. At nightfall on the barque's thirteenth day at sea, the worn-out, short-handed seamen of the *Beara Head* hauled on downhauls, clewlines and buntlines, and then went aloft to furl again.

Tired men take much longer to do things. It was close to two hours later and pitch dark, the wind howling at up to fifty or fifty-five knots, before all hands had brought the upper topsails under control. In the process, the ship suffered its first damage: a clewline parted. Suddenly freed of strain, the line whipped around in the dark and slashed one of the starboard-watch seamen across the arm and chest, shredding his clothes and opening up a long cut. He was a lucky man. The shock didn't quite knock him off the yard, and he managed to hang on and then climb back down, blood spattering the deck and blowing off to leeward in the wind. A few inches closer and the line could have cut into muscle and bone, lopped off a limb or even decapitated him. The sail flogged itself to tatters in a minute. The seamen gathered in every shred of canvas still attached; on the frugal wind ship, everything was saved.

Afterwards, in the faint light of a single oil lamp at the break of the poop, the doctor served a mug of hot coffee and hard tack to each man, their first sustenance in more than eight hours. The captain, who was also ship's physician and surgeon, called the injured man aft, gave him half a mug of rum, bathed his two-foot-long laceration, applied the stinging iodine and bandaged the wound, stitching the deepest parts of the cut; he sewed away, an experienced tailor of flesh. The sailor, an able seaman from Liverpool, would be on light duties for a week or two at least; one good man short. The mate sent the starboard watch below; Benjamin and the other beat, bone-cold, soaked men took what shelter they could find under the poop overhang, braced here and there against the ship's incessant, extravagant motion.

✻

No two gales are the same, Conrad says. Each has its own physiognomy. The seaman remembers gales by the way they stamp themselves on his emotions: one made him plain miserable; one came on fierce and weird, sucking his strength away. He might remember "the catastrophic splendour" of one storm, or another that was "draped and mysterious with an aspect of ominous menace." In each gale, however, there is a common moment that seems to contain its entire feeling.

Once, in the roaring forties of the southern Indian Ocean near the Kerguelen Islands, Conrad, then a first mate, came on deck at four in the morning to take charge of his watch in the confused roar of a Southern Ocean gale. "I received the instantaneous impression that the ship could not live for another hour in such a raging sea." The quick, although mistaken, assessment became his memento of that storm.

The defining mark of another gale was one man's silence. The storm hit the ship like a *pampero*. In seconds, it blew out every sail and laid the vessel over on its side in the hissing, boiling sea, the crew swimming or clinging on, depending which side of the ship they were on. The noise of wind and men shouting was terrible. Yet out of all the confused tumult, Conrad remembered "one small, not particularly impressive, sallow man without a cap and with a very still face." The gale's abrupt assault had caught the captain unawares. He gave a few orders and then no more, seemingly overwhelmed by his mistake. Conrad and the crew worked without pause for hours, doing what they could to save the ship and themselves. They succeeded, but all through their exertions, they were aware of "this silent little man at the break of the poop, perfectly motionless, soundless, and often hidden from us by the drift of sprays."

As always, Conrad disdained the steamer man's experience of the sea compared with that of the square-rigger seaman. A sailing ship disabled by a storm has suffered defeat in an honourable battle, a struggle that is the essence of "the inner drama of her life." To look at a wind ship with its spars gone "is to look upon a defeated but indomitable warrior." Once a mast is jury-rigged and a scrap of sail raised to keep the vessel head to wind, it faces the waves again with its customary, and unsubdued, courage. In contrast, the steamship shoulders its arrogant way along "in blind disdain of wind and sea." When its engine fails or its shaft snaps under the press of weather, the steamer at once becomes a passive, drifting log, "a ship sick with her own weakness."

In *Typhoon*, Captain MacWhirr of the steamship *Nan-Shan* remains disconnected from the appalling, apocalyptic storm building around his vessel. It has no reality for him as a force that might disrupt the mechanical routines of his machinery and the ship's course. His blindness is psychotic. Only when he goes below to discover that the vessel's violent motion has made a shambles of his cabin does he realize that something is happening outside. He begins, finally, to question his ship's isolation from the natural world surrounding it—which is now in furious rampage and, to the captain's surprise, actually threatening the steamer's existence.

None of this could have been remotely possible aboard a ship under sail. The captain and crew had to react, immediately and with supple imagination, to sea and wind around them, making it up as they went along, their survival a matter of determined improvisation. That was why gales had their own personalities—and those single defining moments—only for sailing ships. It was natural, Conrad said, because the sea for a sailor is an "intimate companion," not a mere "navigable element." The sea, friendly today, can become dangerous tomorrow; that's its nature. For the wind ship, gales are "adversaries whose wiles you must defeat, whose violence you must resist, and yet with whom you must live in the intimacies of nights and days."

The *Beara Head*'s first gale of this voyage was Benjamin's first gale ever. And it had that moment for him too, the thing he would remember about it most of all when he looked back from the calmer future. It was just after they had furled the upper topsails—when they had come down to coffee and biscuit, the captain having just called the casualty aft and the starboard watch had gone below—when Benjamin wasn't sure he had a future. Tired beyond exhaustion, shivering, wet to his innards, hands bleeding, in agony where a fingernail had been ripped clean off, clinging to the poop-steps railing, he looked forward down the length of the deck and barely restrained his fear, only just avoided breaking into tears. It was what was missing that terrified him. The barque was there only intermittently; most of the time, he could see only the outline of the mizzen and jigger masts and the staysail, and fainter still, the loom of the topsails and the midships deck house. Apart from that, he saw the sea: masses of breaking crests and rushing foam, little distinction between the main deck and the sea itself. He wouldn't have been at all surprised if someone had told him the ship was going down, driving its bow under. ("She's scuppered, mate! Down to Davy Jones!") He knew the boats were no refuge in this weather; if the *Beara Head* was ruined, so, in their intertwined mortality, was Benjamin Lundy. And the noise! He knew gales were loud; every shellback he'd ever talked to had mentioned their din. This one screamed and shrieked so that you had to put your lips against the ear of the man next to you and shout right into his brain. That was bad enough. Worse was the racket made by the ship itself. It creaked, groaned, slapped, cracked, ground, barked, squealed, wailed, popped in every conceivable key and tone. It was the protest and lament of a living thing under torture.

Then the moment came. Russell, the Yankee, was hanging onto the poop steps on the other side from Benjamin. He swung across the gap between them and grabbed the railing so that they were side by side. He put one arm around Benjamin's shoulders, his mouth up to his ear, and shouted: "As long as she creaks, she holds!"

Benjamin nodded. His fear receded. It was not only the confident, humorous tone of the words, but also the comforting arm, the touch of another body and Russell's face in the faint light—shadowed, gaunt, strained, but with a wry smile. Everything together revived Benjamin; he was warmed by the subtle brotherhood of men.

It was his initiation. The gale would end, and then there would be fine weather and more gales, and above all, the Horn, but it was the lot of a seaman. This was at the heart of life, the mouth's dry taste of danger and holding yourself together and coming through it. It could never happen ashore, except in war. After this gale, he would be a different man.

The wind stayed steady, but the seas ran higher and higher as they drove in from the centre of the big weather system. Half an hour later, the ship's speed into the big waves was down to a few knots. Worse, however, the seas battered the barque with such ferocity that the captain, who had stood for twelve hours without a break at the weather jigger shrouds, began to fear for its masts. He could see them move with the vessel's shuddering jerks as it topped one wave and plunged into the next, like hitting a stone wall. No wooden ship could have stood up to this for long, but even a nearly new iron barque, stiffened with a heavy cargo, was vulnerable in these conditions. It would take only one marauding wave to damage deck houses and hatch covers and tarpaulins—maybe peel them off—or sweep away the boats. The masts were stayed with new wire rope, but seas like these created enormous loads that could wrench the spars out of their steps or bring them down, perhaps the first mischance in the spiral down to ruination. Anyway, there was no need to keep pounding away to the northwest, back across goddamned Biscay towards bloody Ireland. The captain knew, without knowing why, that at some point in the storm's progress, the wind would veer quickly to the west and then the northwest and blow harder. Then they could square the topsail yards and run off to the south on a broad reach, scudding before the gale, which would then blow over their starboard quarter, on the right track at last. Until they could do that, it was prudent to heave to. The purpose of the manoeuvre was to give a ship a little rest and relaxation in its battle with a gale, and so the more quietly the vessel lay, the better—it could tuck its head under its wing and wait for better times.

There is no more succinct definition of heaving to than Dana's: "To stop the progress of a vessel at sea . . . by reducing sail so that she will make little or no headway, but will merely come to and fall off by the counteraction of the sails and helm." The vessel gives up trying to get somewhere and gives in to the sea, drifting slowly to leeward.

There were at least a dozen ways to balance a square-rigger so that it would properly heave to. The *Beara Head* was happiest under a jib, the main lower topsail and a jigger staysail or the reefed spanker: the symmetry of a sail at each end and one in the middle. That meant taking in the fore and mizzen lower topsails. Captain McMillan decided to wait an hour for the change of watch at midnight before giving the order. He was a driver, but he was a realist too. He knew how undermanned his barque was, and how far he could push these men. You had to husband the ship and its gear, but the men were valuable too. You had to gauge their reserves of strength and resolve, and conserve those. Another hour below for the starboard watch, then all hands to muzzle the topsails; afterwards, the port watch, on deck continuously for almost eighteen hours now, could go below for three hours' sleep. Not much, but enough to tip some life back into them.

The *Beara Head* was hove to by two bells of the gravy-eye watch (1 A.M.). Benjamin's main recollection of furling the fore topsail was a haze of pain where the fingernail on his left middle finger had once been, and of fatigue so profound that he sometimes closed his eyes and drifted off for a few seconds at a time, even as he perched on the footrope ninety feet in the air and fisted and scrabbled the hard canvas. To his astonishment, there was a shanty. For the first time in the gale, Paddy was up on the yard with the watch. The man could not *not* sing. The song's energy surged through the men, and Benjamin even felt it put some spunk back into him.

> *Timme way ay-ay-ay high* ya!
> *We'll pay Paddy Doyle for his* boots!

They hauled away, or tried to when they could get a grip, on the yelled-out last word.

> *Timme way ay-ay-ay high* ya!
> *We'll all throw shit at the* cook!

> *We'll all shave under the* chin!

> *We'll fuck her well and be* done!

> *The dirty old man's on the* poop!

In his exhaustion, Benjamin couldn't fathom how the old hands could sing. He couldn't do it, even if he'd known the words. This storm had worn him out, damn near killed him. He thought it was as bad as things would ever get on this ship; even rounding the Horn couldn't be worse. Before he went below, he urinated into his sopping clothes and oilskins, too tired to do anything else, welcoming again the brief warmth.

Then he stumbled into the deck house and onto his bunk all-standing. Sea water sloshed across the deck and up over him from time to time in his greenhorn's low-slung berth. It might rinse the piss out of his clothes, he thought as he fell asleep.

The ship continued its slow, controlled drift for the rest of the night and through the next day and night. The big storm took its time, moving more slowly than most, and the wind stubbornly continued to blow from just south of west. The crew stood watch and watch, with nothing much to do but try to keep their wet bodies out of the cold wind, eat their whack, and sleep. It was a paradox aboard ship that the most restful times might come during a gale that dictated heaving to, when it was too rough and wet to chip iron or overhaul gear. Grateful men and fretful captain alike could do nothing but wait with the sailor's patience. Another chance for the shantyman to sing—about ships:

> *A Yankee ship came down the river,*
> *Blow, boys, blow!*
> *Her masts and yards they shine like silver,*
> *Blow, me bully boys, blow!*

Or women:

> *To Chile's coast we are bound away,*
> *An' we'll dance an' all drink pisco!*
> *We are bound away at the break o' day,*
> *Where them little Spanish gals are so bright and gay,*
> *Timme heave-ho, hang 'er Hilo!*
> *Sing o-lay for them Dago gals!*

And drink:

> *Oh, whisky is the life of man,*
> *Whis-ky, John-ny!*
> *Oh, whisky from an old tin can,*
> *Whisky for me, Johnny!*
> *Whisky when I come ashore,*

So I can fuck that nigger whore.
Whisky gave me a big, fat head,
But I'll drink whisky till I'm dead.

And so on, from Paddy's endless stock of songs and their permutations—until the mate stumped up to the poop rail.

"Belay down there! There's a goddamn lady on board," he yelled, so that the lady couldn't help hearing. "Haven't you scum got any fuckin' manners?"

On the third day of the gale, the cold front drifted across the waiting barque's longitude. The wind became ice-cold and blew harder, but it veered to the northwest.

"All hands on deck to wear ship! Do ye hear the news there?"

The watch below coming out on deck, making their way aft hand over hand along the heavy-weather lifelines through the unceasing gush and spout of seas across the main deck, leaping like acrobats into the rigging to avoid the worst of them, mustering at the break of the poop. The sky had cleared here and there, and in the intermittent sun, the waves turned iridescent—ink blue, violet, aquamarine—their crests breaking white high above the deck, level with the maintop, wind-foam striping the sea. Benjamin looked off to windward: in the troughs, he saw steep walls of water hemming him in; from the crests, a heaving blue-white barren running to the horizon. He remembered the old song "I went to sea to see the world / And what did I see? / I saw the sea." He thought: "God-a-God! How bloody beautiful!"

"Stand by braces, tacks and sheets!"

Into the waist of the ship, wading through water to the knees, hips, chest. One cascade upended three men and bowled them along the deck for forty feet before they brought up against the starboard bulwarks, lucky to be bruised and not broken, even luckier to be aboard and not swimming. The wind maybe a steady fifty knots, gusting higher. Not a dry inch on any man; boots and pockets full of sea water, which felt even warmer now compared with the polar wind. All hands at their stations, the mate and carpenter, wetter than any seaman, on the plunging fo'c's'le deck to handle the jib. The captain watching the seas bearing down on his ship, waiting for a smoother spot. The ship is vulnerable as it makes its turn and comes beam-on to the seas. Especially these breakers, which are confused because of the wind shift, the northwesterly waves overlying a westerly swell. A knock-down might cause the coal to shift or pluck out the topgallant masts, or it could capsize the whole shebang. At night, he could only chance it; he could see now.

"Up helm!"

Around went the barque, spinning off to leeward, beam-on and one quick roll over to fifty degrees; the main yardarm disappeared into the back of a receding wave—the men struggling to stay on their feet and cast off sheets, haul on braces—and the ship turned downwind, running fast, riding the seas like shooting rapids or surfing in a three-hundred-foot boat. Then it turned up to windward a little on the starboard tack, the gale on the quarter, broad reaching, the square-rigger's best point of sail, and finally to the south-southwest, at last the direction they wanted to go in, two men on the wheel, main lower topsail yard squared and the fore-and-aft sails trimmed, all of this a fast job because the crew had to handle just three sails, only one of them a square sail. For once, there were enough men aboard to carry out a manoeuvre pronto.

"Loose fore and mizzen lower tops'ls! Set the spanker and inner jib! Set the main and mizzen topm'st stays'ls!"

The mate ordered Benjamin and Anderson aloft with the apprentices to cast off the gaskets. There was no less motion, but it was easier: a slower rolling and pitching as the quartering waves picked up the barque and carried it along, twisting it starboard quarter-up and port bow-down, then quarter-down and bow-up. As he coiled the gaskets on the fore lower topsail yard, Benjamin looked out at the wilderness of the sea and down at the barque's narrow hull driving hard along. It was as difficult as before to hang onto the yard and its footrope. Often, he could still look straight down into the sea beneath him, the ship having rolled out from under his high perch, the yardarm end pointing the steep way down. He was still afraid, wary of the height and the precarious attachment of his body to the ship. Then he heard it: the sound of a square-rigger picking up its heels off the wind, going full sail ahead; the great roaring from aloft, now all around him, as the wind alone, flowing fast through the tophamper, propelled the ship through the sea. "And oh, the marvel of it! That tiny men should live and breathe and work, and drive so frail a contrivance . . . through so tremendous an elemental strife," wrote Jack London.

Benjamin had found the work on a square-rigger hard and testing beyond anything he had imagined. Nevertheless, as the barque turned away from the gale to run fast to the south, not slogging into the eye of the wind or hove to, but for the first time truly *sailing*, he became aware of something else: fascination, and the rapture of a young man in glamorous jeopardy. He saw himself from outside his body, riding the high yardarm not yet with insouciance, but with a kind of ease he hadn't felt before; a small, oilskinned figure gleaming with the sheen of sea water, cut by the north wind; a dwarf man in the rigging of the giant ship, perched aloft and doing his job like a real deep-water seaman. If they could see me now, he thought—my mother, my grandmother, my friends going home to spuds and buttermilk—how astonished they would be; my father's grudging respect at last, man to man.

"I wanted to live deep and suck out all the marrow of life," Thoreau wrote. If that was possible beside tranquil Walden Pond, it could certainly be done on a topsail yard in a gale of wind. In weather like that, said Conrad, the silent machinery of the wind ship "would catch not only the power, but the wild and exulting voice of the world's soul." It was "a wild song, deep like a chant." No seaman could ever forget it. Benjamin heard it, and he understood that the chance to hear it again might well, all by itself, balance out the burdens of life on a wind ship; it might be enough to bring a man back.

The hands braced the yards round and then tailed on to halyards and sang the topsail-halyard song "Whisky, Johnny" to do real work this time. The apprentices went aloft to overhaul buntlines so that they wouldn't chafe. The seamen hoisted the fore-and-aft sails and trimmed them for the broad reach. The *Beara Head* plunged and surfed, in its element at last, doing twelve, fourteen knots.

In twenty-four hours, the ship ran nearly three hundred miles; at the end of its seventeenth day at sea, it had passed Cape Finisterre a healthy fifty miles off and reached the latitude of Vigo, in northern Spain. In one day, the *Beara Head* made more miles than it had in the previous week. Madeira, the barque's first landmark, lay six hundred miles ahead.

This was not safe sailing. In the quartering seas, it was easy to broach to—the ship driven beam-on to wind and waves—easier than running dead downwind. But after sixteen days of hand-to-hand combat, Captain McMillan was ready to break off the engagement and run south-southwest, no matter what. He would drive this iron tub like a clipper. He stood beside the two helmsmen, ready to lend his weight to theirs to keep the barque's course true. He was smiling for the first time on this passage that any seaman had seen.

"G'won your best!" he bellowed at his ship.

His pale wife came up the companionway steps from the aft cabin and stood in the doorway, smiling too, watching her husband watch his ship. The mate, standing wide-legged with a hand on the lee jigger rigging, stared aloft and forward at the sails, as tense as the canvas itself, never taking his eyes off them.

CHAPTER SIX

The Ocean Paths My Home

How slightly soever many esteem sailors, all the work
to save ship, goods and lives must lie upon them.
CAPT. JOHN SMITH

I can sing a true song about myself,
tell of my travels, how in days of tribulation
I have often endured a time of hardship,
how I have harboured bitter sorrow in my heart
and often learned that ships are homes of sadness.
"The Seafarer," TRANS. BY KEVIN CROSSLEY-HOLLAND

I joined the barque *Europa* in Horta, on the Azorean island of Fayal,
as one of a dozen "trainees" who would sail aboard the last leg of
the vessel's homeward voyage from South America and the Antarctic to Holland. I had last seen the square-rigger at dock in Ushuaia, Argentina, after my own short, cold foray into the waters round the Horn.
It had been useful research to look at the details of gear and sails in this
scaled-down replica of a late-nineteenth-century barque, but that
hadn't been enough. Back at my desk, writing about a voyage under
square sail, I realized I needed at least a short passage on the real thing.
It would help avoid errors, and provide more colour and context, if I
could watch real sails being set, brought in and furled, see how they
were trimmed, and observe (and participate in) the procedure during a
tack or when wearing ship. Perhaps even go aloft myself seventy or

eighty feet and do some work. Such an excursion would, of course, be nothing like a voyage under square sail 115 years ago on an under-manned cargo carrier, eating salt horse and hard tack for four or five months at a stretch down into the Southern Ocean and round the Horn. But this two-thousand-mile passage felt necessary, a way of understanding the roots not just of my story but of my ancestry as well. And it was, after all, the North Atlantic in early spring; the benign summer climate patterns were still weeks away, and heavy weather was almost certain. In fact, the barque's route would take it north, hunting for strong wind in which to run its easting down.

As always, it was a mug's game to predict what will happen at sea. We did get some wind, although most of it was from dead ahead. We sailed and motor-sailed (using the ship's twin Caterpillar diesels) north for two days, looking for the strong gales that would speed us east. They were out there, a string of them, marching across the belt of the westerlies one behind the other, but we would have had to sail as far north as the latitudes of Iceland or Scotland to get close to them. A big high-pressure system, a premature summer model, had settled over the eastern North Atlantic off the Iberian coast, Biscay and the Channel. It was strong enough, and sufficiently entrenched, to deflect the lows as they rolled eastwards. They bounced off the high like billiard balls and caromed north, far away from the *Europa*'s track.

We turned northeast and found some wind: for two days, we beat under full sail, at nine and ten knots into force seven, with gusts into gale force. Lifelines were rigged, and some green water broke over the rail and across the decks. We even took in some sail for a while: the flying jib, the spanker topsail and a big mizzen staysail. Together with some of the other trainees (the *Europa*'s motto for work aboard was "Anything you may, nothing you must"), all middle-aged Dutch and Germans, I stood four-hour watches, did turns at the wheel and hauled on lines. No shanties, but when hauling with the vessel's professional seamen, we did emit the loud, grunting groans from the belly—the sailor's "unearthly sounds"—that seem to come naturally while jerking upper topgallant halyards.

I went out to the end of the thirty-foot-long bowsprit, which was doing a quick, wrenching oscillation between sea and sky, to handle the flying jib. Old wind-ship sailors often had their wildest times on the exposed bowsprit, but the *Europa*'s had been tamed by netting that was slung underneath to catch a clumsy or unlucky sailor before he hit the water. Nevertheless, I had to hang on hard and move with care as the spar whipped up and down. I developed blisters that burst and burned like fire when I hauled on the hard manilla lines. I got a saltwater-infected cut and hauled on lines some more.

As we made farther into the stationary high-pressure zone, the wind gradually dropped away. We spent the last three days under iron topsails (the diesels), in a flat, blue, hazy calm that looked like the Mediterranean, or Lake Ontario, in summer, rather than the wild Atlantic in rowdy spring. I had gone to sea to sail a square-rigger in a gale of wind, but I ended up on a steamer voyage. I climbed to the mainmast cross-trees, about eighty feet above deck; the barque's hull looked narrow and distant enough even from that modest height. The climb out and around the maintop platform was the traditional rock climber's spider-like crawl out and up at a forty-five-degree angle. At the cross-trees, there was nothing to do but hang on in the gentle swell and sightsee, and reflect on what it might be like up there on some dirty night—or indeed, up 165 feet on the main royal yard of a big four-poster like the *Beara Head* in a storm.

We put into Cherbourg, where the captain had decided to dock to spruce up the barque before heading the last 150 miles home. I jumped ship there, literally—swinging my seabag across the five-foot gap between the barque's side and the pier, then performing a standing jump, to applause, ripping the knee out of my pants and cutting my hands on the gravelly dock. I'd had enough motoring, and had seen and done all the square-rigger pulley-hauley I was likely to. For the first time in my nautical life, I had sought out heavy weather, hoped for it—more or less—rather than doing everything possible to avoid it. A storm or two would have provided useful writing material, and I was confident that the big, iron *Europa* could handle anything. The sea, perverse as usual, had refused to co-operate.

However, I had seen the *Europa*'s crew at work. There were eight seamen, besides the captain, the mate, the engineer, the cook and the steward. They were as heterodox as the old square-rigger men: one from each of Canada, the United States, New Zealand, Argentina, Scotland, the Falkland Islands and Germany, together with six from

Holland (the latter including the officers). These were professional seamen, aged from twenty or so to their early thirties, many of whom had shipped aboard some of the few other working square-riggers still sailing for purposes of training or tourism. They were as agile and confident aloft as I was clumsy, clinging and careful. They took pride in their skill at ropework and sail-handling. And they saw themselves as the inheritors of an old and glorious tradition, a handful of young men carrying on as much of the old skills and knowledge as working a comfortable, well-provisioned, diesel-auxiliaried vessel with GPS and radar would allow. They went to sea under sail for some of the same reasons as the old Cape Horn seamen: adventure, the opportunity for hard and dangerous work, the glamour of it all. None sailed out of economic hardship or necessity or as part of family tradition, the most frequent motivations for going to sea in the days of sail.

Perhaps the biggest difference was that the loss of a seaman from the *Europa*—a slip off the footrope or a fall over the side—would be considered a tragedy, triggering an investigation, an inquiry and, perhaps, the refusal of the safety rating necessary to keep the vessel sailing. The destruction of the barque itself would be an utter catastrophe. In contrast, the loss of men from a nineteenth-century square-rigger was so common that it was expected and usually occasioned nothing more than a note in the log. Each time a ship and crew went mysteriously missing, it was the equivalent of three or four times the number of the *Europa*'s complement wiped out, month in and month out, year after year.

Life aboard the *Europa* could be tough and perilous—working the deck or going aloft while sailing across the Drake Passage to Antarctica was no easy go in any century—but it was, nevertheless, comfortable and safe to an extent that would have been incomprehensible to Benjamin aboard such a vessel as the *Beara Head*. To him, the *Europa* would have seemed a sea-going Fiddlers' Green, the gentle paradise where the souls of dead seamen rest. The *Europa*'s crew had experienced just enough of the Southern Ocean and the seas around Cape Stiff to develop the keenest appreciation for what the seaman's travails and suffering must have been like aboard the *Flying Cloud*, the *British Isles*, the *Slieve Roe*—or for Conrad aboard the *Torrens* or any of the hundreds of thousands of forgotten men aboard the bygone wind ships.

I talked to the *Europa*'s crewmen. I told them I was writing a book about the old Cape Horn seamen, and these modern sailors each reacted in the same way: they ignored my gratuitous information about the book and began to talk about the men. In one way or another, they all asked the same question, and it was a striking echo of my own query months earlier as I looked out *Baltazar*'s Plexiglas bubble at the hurricane ripping the water's surface to shreds in our little bay of refuge a few miles from the Horn: How did they do it?

On the evening before the Battle of Waterloo, the Duke of Wellington, the Dublin-born Arthur Wellesley, rode amongst his resting regiments. At one point in his inspection, he pointed to a soldier and remarked to Creevey, one of his commanders, "There, it all depends on that article whether we do the business or not. Give me enough of it and I am sure."

Even allowing for the Iron Duke's acerbic irony, there's no doubt that Wellington regarded his men (a large number of whom were also Irish) as the proverbial scum of the earth. The soldiers were an asset, like horses or guns, which he would apportion and use as necessary during the following day. Yet everything depended on them. They had a simple task: to stand firm in their squares in the face of the French cavalry and guns until the attackers grew weary of their assault or the Prussian allies arrived.

It was an odd mix of a view: common soldiers were contemptible but essential. (It was an ancient perspective, and it would persist through Balaclava, Gettysburg, the Somme and Verdun, Dieppe and the Yalu.) Without these lower life forms, it would be impossible to do the great and necessary work of society and the empire. The same could be said for the workers in the new industrial factories and mills, and for coal miners. It was no different for seamen.

There seems to be a foolish and self-defeating contradiction in despising and ill-treating men upon whom everything depends: the ratings and gunners in the Royal Navy, behind whose warships the British nation, and later its empire, consolidated itself, flourished and dominated; seamen in the merchant marine, who were equally necessary for the survival of a maritime nation and were treated almost as badly as their naval counterparts. Perhaps it had to do with invisibility. Miners were out of sight below ground, languishing in the same category of filthy and unseen necessity as stun gun men ankle-deep in slaughterhouse blood were to the sensitive and educated meat eater. Seamen were invisible too, of course, because they were at sea. They "kept" the sea in their coasters or deep-sea merchantmen, and that kept them out of view. Landsmen thought that when sailors did come ashore, it was only to use the scandalous facilities of the shoreside ghettos of Sailortowns or docklands, where they did their whoring and drinking. In any case, seamen were anonymous, alien, a notorious race apart.

Historical records say a great deal about soldiers, but much less, often nothing, about sailors. When Henry V transported his army to France before Agincourt, he needed fourteen hundred ships of various types to carry his ample "band of brothers"—six thousand men-at-arms and twenty-four hundred archers—across the Channel. We know the precise numbers and who many of them were. Yet there's no record of how many sailors were involved, even though at least forty thousand were needed to handle the king's ships for crossing the Channel—"the perilous narrow ocean," as Shakespeare described it.

Seamen were in that category of humans that traditional "history" either ignored or caricatured. Historians have a difficult time saying very much about people who were mobile (often nomadic), poor, illiterate and prone to dying in large numbers relatively young. Perhaps only itinerant agricultural labourers were in the same obscure order of men.

When sailors did get a mention, the reviews were invariably scathing and remarkably consistent over several hundred years. Elizabethan seamen were "insolent, superstitious, feckless and blasphemous"—a rabble in need of discipline. They were "unacquainted with civility. . . . Although his hand is strong, his headpiece is stupid. He is used therefore as a necessary instrument of action." But some admiration for these strange savages also creeps into the accounts and estimations of landsmen. Even in days of early, sudden or violent death, the sailor stands out as someone who is more familiar with it than anyone else. The soldier faces death occasionally in war, but otherwise, he carries on his regulated but peaceful life, subject only and always to disease, which was far more deadly than battle anyway. The sailor, however, faces death daily, simply by virtue of doing his job on the dangerous sea (although disease was his most frequent killer as well). This has armed him with "a kind of dissolute security against any encounter." He is tough, hardy and strong in ways, and to a degree, that landsmen can hardly imagine. He is part of a brotherhood that stands together no matter what. "His visage is an unchangeable varnish; neither can wind pierce it, nor sun parch it. . . ." He has an "invincible stomach" that could digest iron. Mostly, though, the seaman must be controlled, driven, kept down. He's like a stiff-necked horse, said Richard Hawkins, which the horseman must ride hard, break and bend to his will. Francis Drake lamented that his sailors were "so unruly without government." And the privateer-admiral's "government" was severe.

Jail was preferable to a ship, said Samuel Johnson, even, apparently, to the charnel houses of his time. Johnson's mind was concentrated on the conditions of seamen because his black servant had been the victim of one of the notorious press gangs that, from the early eighteenth century onwards, stalked the country's ports, and even farther inland, its rivers and canals, lifting any man between the ages of eighteen and fifty-five who "used the sea" (a broad interpretation was given to the phrase) for indefinite service in His Majesty's Navy. It took all the doctor's influence to get his man back. Boswell didn't record Johnson's only visit aboard a ship—the man-of-war *Ramillies*—but it was words that made the greatest impression on the lexicographer. He was disturbed by the fact that everyone on board, even the first lieutenant, swore like . . . well, sailors. When Johnson was told that some men went to sea voluntarily, he replied, "I cannot account for that, any more than I can account for other strange perversions of the imagination." He was being funny, but he meant it.

"Poor sailors," said Capt. Thomas Pasley in 1780, echoing John Smith two centuries earlier, "you are the only class of beings in our famed Country of Liberty really *Slaves*, devoted and hardly used, tho' the very being of the Country depends upon you." Pasley was thinking of naval press gangs too, but his words were not much of an exaggeration in the case of merchant seamen of the time, nor those of the nineteenth century either.

History has its views from above and those from below. It depends where you are on the scale of things: English governor or Irish subject;

conquistador or Aztec; winner of the war or loser. Seamen certainly thought about how they were treated and viewed by their dependent country, and sometimes their voices do ring out—with surprising eloquence, and even with nobility. During Drake's Cadiz expedition of 1587, there was a mutiny against the admiral's "government" aboard the *Golden Lion*, one of Drake's chief warships (it carried three hundred sailors, thirty gunners and one hundred soldiers). The mutineers stated their grievances in a letter to their captain. The complaints would be familiar and true in any century, the seaman's endless lament: "What is a piece of beef of half a pound among four men to dinner, or half a dry stockfish for four days in the week, and nothing else to help withal: yea, we have help—a little beverage worse than pump water. We were pressed by her Majesty's press to have her allowance, and not to be thus dealt withal—you make no men of us, but beasts."

In the great naval mutiny at Spithead in 1797, a parlous time for England in the middle of its war with France, the mutineers, who refused to work their ships off moorings and put to sea, produced a document of responsible and cogent complaint. Their requests were limited, the seamen's delegates wrote, "in order to convince the nation at large that we know when to cease to ask, as well as to begin, and that we ask nothing but what is moderate, and may be granted without detriment to the nation." What did they want? The list seems pathetic in its small petitions; it is the universal list from the "down under" side of history. Their wages were so low, the sailors said, that they were unable to support their families; they objected to the "pursar's commission" of two ounces out of every pound of food, and to the lack of fresh vegetables and meat when in harbour. They wanted their sick to be better attended to, though they would settle for having them attended to at all. And last, "that we may be looked upon as a number of men standing in the defence of our country; and that we may in somewise have grant and opportunity to taste the sweets of liberty on shore, when in any harbour, and when we have completed the duty of our ship, after our return from sea."

The men upon whom England's existence depended did not request an end to flogging, or even an amelioration of its worst abuses (presumably they considered that demand immoderate and unrealistic), and punishments remained severe—at least until Wellington got the job done at Waterloo, and the navy could relax its brutal discipline a little (but only a little). Most of the requests of the Spithead mutineers were granted by Parliament or Admiralty orders, including the pay increase. The mutineers also won a royal pardon. Their last request—that when they were in harbour, they might be allowed to go ashore once in a while—was refused. Their commanders feared, with justification, mass desertions. Having made up their ships' complements at great effort through press-gang kidnapping, they were reluctant to see crews evaporate while on shore leave. What pressed man in his right mind would return? The only solution was to hold men aboard for the duration of their service. Samuel Johnson was not speaking metaphorically when he compared a ship to a jail.

The government's reasonable response to its seamen's moderate petition had dire consequences. A month later, men in the North Sea fleet, encouraged by the outcome of Spithead, mutinied at the Nore, and their petition was not moderate. They too wanted improved living conditions, but they also put forward a string of ill-defined and confused, yet clearly radical, demands for political changes. These constituted mutiny veering towards revolution, a terrifying prospect for an English government just shorn of its American colonies by force and contemplating the earthquaking effects of the French Revolution a few miles across the Channel. Some of the force of the Nore mutiny also came from disaffected Irish sailors—the United Irishmen rebellion of 1798 was brewing—who, as always, made up a good proportion of the English armed services. The government's response this time was not gentle. It exploited the differences between the men seeking a better life aboard ship and the "politicals." There were no negotiations or concessions; the mutiny was suppressed, its leader hanged and many men brutally punished. Four months later, the same sailors under the same officers, working and fighting in conditions that were even more cruel as a result of the mutiny, won the crucial and decisive battle against the Dutch fleet at Camperdown.

In 1800, one seaman wrote to his wife: "Where I am it is a prison. . . . We are looked upon as a dog and not so good. . . . There is so much Arbatory [*sic*] Power that a man must not say his Soul is his Own. . . . They flog them every day. . . . Only one shilling on Shore is better than a dozen on board of Ship."

The successful end to the Napoleonic Wars, with their string of decisive naval victories, did bring one benefit to seamen, especially those in the navy: they became popular and admired as the country's defenders through the long period when the army, by contrast, had been slogging away in Spain (under Wellington) and not doing much else. In fact, apart from Waterloo, the army had a bit part in the dramatic struggle between England and Bonaparte; it was the navy that really did the job. This was the period of full flower of the legend of Jack Tar, the indomitable, swashbuckling English sailor, heroic saviour of the realm. It was an image helped immeasurably by the achievements and personality, and early death, of the great Admiral Nelson, whose sailors, pressed or not, followed him with the primitive blind devotion of men for a beloved leader. His aura descended onto his seamen as well.

The public's newfound warm feelings towards sailors did not result in much improvement in their conditions, however. Jack Tar might have been a hero, but he remained feckless, and needed what might be called a firm guiding hand from the captain and his officers. Captains were still God aboard ship, and a vengeful one. They had "hearts as hard as stone / Who flog men and keel-haul them, and starve them to the bone," wrote Charles Kingsley, a champion of the working classes. John Masefield, the popular elegist of the last days of sail, had no illusions about life aboard ship, especially on a man-of-war: "perhaps no place has contained more vice, wickedness, and misery within a narrow compass." Winston Churchill put the matter more pithily in his famous dictum on the real traditions of the Royal Navy: "rum, sodomy, and the lash."

Even the revered Nelson carried out flogging punishments like any other captain of his time, although there is some contemporary evidence that he disliked it more than most. One account quotes him as saying how difficult it was "to endure the torture of seeing men flogged." Flogging was not abolished until 1886, the year after the *Beara Head*'s voyage. While Benjamin was suffering enough on his hard-labour limejuicer, it was still theoretically possible for naval seamen to be legally flogged with the legendary, lacerating cat-o'-nine-tails—perhaps my great-grandfather aboard the steam corvette HMS *Thunderer* for some violation of ship's regulations? In fact, however, the use of the "cat" became a "suspended" punishment—even though it was still authorized by the Articles of War—soon after 1860, a year in which one thousand men, out of a navy of fifty-four thousand, were flogged. For the next twenty-five years, the cat, although unused, was hung in a prominent place aboard warships, like a crucifix on a Roman road to deter the uppity slave.

The propensity to whip men like vicious dogs or recalcitrant horses wasn't restricted to the navy, although a rope-end replaced the cat aboard merchant ships. Dana's lifelong mission to improve the lives of American seamen began aboard the *Pilgrim*. He watched the captain (a man admittedly at the psychopathic end of the scale of ship's officers, but not uniquely so) tie up and flog two seamen, one for supposed "impudence," the other for questioning his shipmate's punishment. One man shouted out, in his agony, "Oh, Jesus Christ!"

The captain responded: "Don't call on Jesus Christ! He can't help you. Call on Frank Thompson! He's the man!"

Dana's "blood was cold"; he turned away in sick disgust. "A sailor's life," he wrote, "is at best but a mixture of a little good with much evil, and a little pleasure with much pain."

✻

The master-at-arms was the enforcer aboard naval vessels. The archetypical example is Claggart in Melville's *Billy Budd*. No seaman of the story's time—the very year of the Spithead and Nore uprisings—and for many years to come, would have had much trouble with the idea of this officer as the essence of evil. When Billy in his innocent outrage strikes Claggart dead, the blow reverberates with meanings: good against evil; the common seaman against the tyranny and ruthless discipline of the quarterdeck. It was perhaps even an instinctive reaction to the homoerotic claustrophobia of the crowded man-of-war, as embodied in Claggart's tortured and contradictory obsession with the fair foretopman.

Billy Budd goes back to the "time before steamships." Among other things, the story represents Melville's nostalgia for that period, before industry and war changed the world. Melville saw what a modern industrial war was like during the four years of the American Civil War, which both recapitulated the methods of past conflict and foreshadowed the "rational," efficient wars to come: trenches, mass mobilization, machine guns, submarines, armoured warships, guerrillas, prisoner atrocities, scorched earth, bloody attrition, the indiscriminate wasting of men. Melville thought that the warship incarnated the intrusion of war and industry into the world of the pure and infinite sea. In *White-Jacket*, he wrote: "life in a man-of-war . . . with its martial formalities and thousand vices, stabs to the heart, the soul of all free-and-easy honourable rovers." In *Israel Potter*, he describes how "the martial bustle" of a warship disrupts "the solemn natural solitudes of the sea."

If Claggart is the instrument of Billy's death, Captain Vere is his real executioner. Vere has no regard for the sea or seamanship, or for "the ocean, which is inviolate Nature primeval." With the whiff of mutiny still in the air, his allegiance is to discipline, war and the necessary destruction of the Handsome Sailor, pressed into service from his merchantman the *Rights of Man*. Vere dies ashore, where he really belongs. Billy Budd, the universal, unknown seaman, for whom the sea itself is his grave and only monument, goes home:

> *Fathoms down, fathoms down, how I'll dream fast asleep,*
> *I feel it stealing now. Sentry are you there?*
> *Just ease this darbies at the wrist, and roll me over fair,*
> *I am sleepy, and the oozy weeds about me twist.*

Billy Budd was pressed off his merchant ship by a man-of-war on the high seas. This was not an uncommon practice, and the Royal Navy, in its hubristic power and reach, extended it to ships of other countries as well. So many seamen were unceremoniously scooped off American ships by British naval boarding parties that it became one of the causes of the War of 1812. However, it was the shoreside press gangs that did most of the work. From 1710 until the general sigh of relief after Waterloo, the Admiralty authorized the forcible recruitment of seamen. (This was an ancient practice, in any event; in the fleet of Richard the Lionheart, pressed men who deserted the king's service were liable to a year's imprisonment.) A voluntary register set up fourteen years earlier hadn't worked; nor had volunteer bounties. As war, desertion and disease whittled down the crews, the government first refused to release men already serving and then resurrected impressment. By mid-century, historians estimate that one-third, and perhaps one-half, of naval recruits were in fact pressed men. The navy's demands fluctuated with war and peace, but at the height of the

Napoleonic Wars, in 1805, as much as one-third of the navy had been mustered by impressment. The actual numbers of nominally free Englishmen doing forced labour afloat were very high. During the Seven Years War, in the mid-eighteenth century, the navy's need for men rose by 70 per cent, and eventually 180,000 men, one-tenth of the adult male population, were enlisted.

Anyone who stood a reasonable chance of deserting and getting away with it jumped ship. The supply of men had, therefore, to be constantly replenished. In the late 1770s, with the prospect looming of a combined French-Spanish assault on England, disease and desertion reduced the numbers of crews so dramatically that the press gangs had to find as many as two men for every one in active service just to maintain manpower. The government suspended *habeas corpus* in cases of impressment (not that the ancient prerogative remedy had made much, or any, difference for sailors anyway). The press gangs' motto: "All as says they be landlubbers when I says they baint, be liars—and all liars be seamen."

The life of the merchant seaman during this period was inescapably bound up with the navy and its practice of impressment. Merchant sailors were the main fodder for the press gangs, ashore or at sea. The merchant service was supposed to be a reservoir of trained seamen upon which the navy could draw. The idea was that merchant sailors would, in time of war, join up voluntarily, induced by signing bounties and patriotism. Many men did so, or made a career of the navy from the start. That this happened illustrates how bad conditions could be ashore. Dispossessed Scots or English farmers or hungry Irish might enlist simply to make sure that they got something to eat every day, or that they made enough meagre pay to keep a wife and children alive. The need for subsistence food and pay, as well as the signing bounty, often overcame the deterrents of loss of conventional liberty and the various brutal punishments so dear to the navy's heart.

Even though the merchant marine was the navy's nursery, both merchant seamen and shipowners caused problems for the scheme. The seamen were understandably reluctant to move from the harsh-enough world of the commercial vessel to a brutal man-of-war. And in fact, men often avoided the press gangs by heading to sea on foreign-going merchant ships: out of England, safe enough. Shipowners took a contradictory position: they asked for government favours because they were the source of naval manpower, but then they complained when press gangs absconded with their men; in wartime, they wanted naval convoys but refused to release the men necessary to man them. This reluctance to provide the navy with the men it needed was characteristic of the English moneyed classes generally, and it made the manpower problem an intractable one.

While merchant seamen were always in danger of being kidnapped and slung aboard a man-of-war, they also had to contend with the civilian version of impressment: getting shanghaied. The term "shanghai" originated in the United States, where American press gangs got sailors drunk, or drugged or blackjacked them, and forced them aboard ships. The next port of call for some men was Shanghai, then the busiest Chinese port. Shanghaiing didn't really become common before the effective end of impressment, but it persisted long afterwards.

The two methods of kidnapping, either sponsored or tolerated by the government, were inextricably connected. As a practical matter, the press gangs manned both warships and merchant vessels, the latter inadvertently. The threat of impressment provided manpower for civilian vessels in the form of deserters or press evaders.

There was a philosophical nexus too. Every time a press gang randomly assaulted and restrained a man on the street, carried him off by force, held him against his will and put him aboard a warship for a long period of service—sometimes with the assistance of soldiers, who were used to cut off the retreat of eligible men heading for the hills—it eroded the notion that this was something you simply could not do to a citizen of Great Britain. Because it was obviously something you *could* do. Only in tough, grimy ports like Liverpool and Newcastle did the gangs meet their match in the congregations of merchant seamen, themselves skilled enough with the knife and the knuckleduster, who resisted and chased the gangs out of the docklands. Still, the lesson was clear: force prevailed, not any notion of rights; war and national necessity always trump rights. After the experience of more than a century of brutal impressment, the public, and sailors themselves, were inured to the idea of forcible seizure and service, and shanghaiing aroused little indignation and no surprise.

In fact, the so-called crimps, who continued to shanghai seamen until nearly the end of the nineteenth century, had their origins in the period of impressment during the desperate days of the Napoleonic Wars. The press gangs became so rapacious that merchant shipowners had to hire agents—the crimps—to sequester potential seamen and keep them out of the hands of the gangs. Crimps used money and booze to encourage men to ship aboard merchantmen. Sometimes, however, they worked the other side of the street, gathering men together for merchant service and then betraying them to the press gangs for the signing bounties.

Crimping became easier, and more widespread too, because of changing attitudes toward seamen, and because of where they lived and worked in the years of peace during the nineteenth century. The heroic aura of the war faded quickly. No man is forgotten faster than the soldier or sailor when the war's over. In fact, the impulse is to put them away, out of sight, killers and violent men, necessary enough at the time, but scary embarrassments in peace, reminders of the hatred and fear. Jack Tar and Merchant Jack were no exceptions. In 1849, Melville wrote in *Redburn* that sailors were "shunned by the better classes of the people, and cut off from all access to respectable and improving society . . . generally friendless and alone in the world."

They were cut off from society and physically cut off too. The extensive docks built in London and Liverpool, which Melville admired—each one "a small archipelago, an epitome of the world"— were so enclosed and self-sufficient that they became "docklands" populated by seamen, longshoremen and coal-whippers, as well as whores and tavern owners. These neighbourhoods became, in fact, separate working-class areas detached from the greater society, and avoided by it. The seaman's invisibility while at sea was duplicated in the docklands ashore. In these noxious, violent ghettos, the crimps— Shanghai Brown, Paddy Fearnaught and all the rest—flourished. They set up and operated from the boarding houses that sailors lived in while ashore.

By 1869, according to the British consul in Marseilles in a report to the Board of Trade, crimping had become a worldwide system. Its growth and permanence were encouraged by the need for men caused by the expansion of the British and American merchant fleets, both sail and steam, and by the desertion of seamen to Australian or North

American goldfields. Government and shipowners saw crimps as a necessary evil, the peacetime version of the press gangs, both of them justified in subordinating the fundamental rights of many men to the economic requirements of a few.

Reformers made frequent attempts to end crimping, mainly by campaigning against the "advance note." The sailor didn't get paid off in cash until several days after he was discharged at the end of a passage. In the meantime, he availed himself of women, booze and a boarding-house room on the crimp's credit. The seaman was soon forced to get an advance note on his pay to keep the crimp happy, but the amounts spent by the long-deprived man fast exceeded the meagre money coming to him. Deep in the crimp's debt, he had to ship out again, or acquiesce in getting shanghaied, to pay it off. Reform legislation in 1875 proposed the abolition of the advance note and its replacement by allotment notes, payable only to wives or "sweethearts." The crimps lobbied with vigour and defeated the bill, which would have ended their rewarding business.

Nothing could stop crimping because, ultimately, it served the purposes of the government and the shipowners. Crimps didn't get far with steamship sailors, who were an altogether more sober and sensible crowd. It was the square-rigger tars, determinedly un-sober and non-sensible (and often combining the two to become insensible), who provided crimping fodder. Crimps disappeared only when the deep-sea wind ships, and the marginal men who sailed them, faded away.

Using impressment as a solution to the problem of manning ships was extraordinarily inefficient because forced-labour seamen very sensibly absconded at every chance they got. And the navy also faced the ancient difficulty aboard ship: keeping men free of disease and alive long enough that they had a chance, eventually, to fire a few shots at the enemy, or in the case of merchantmen, to get to port. This wasn't easy. Every part of society during the seventeenth to nineteenth centuries was, of course, subject to the sudden horror of many diseases. The causes of illness, and the mechanisms of infection and transference, were discovered only slowly and painstakingly during this period. Sailors were especially vulnerable because they were forced to live in conditions that were ideal for a number of diseases to flourish. Over-crowding, poor hygiene, inadequate ventilation (none at all in bad weather) and bad and insufficient food encouraged, in fact guaranteed, outbreaks of typhus (known variously as ship, jail or camp fever, but they were all the same thing) and scurvy. Yellow fever (black vomit, yellow jack) and malaria (ague) were problems in tropical ports. These diseases cut down far more than battle. Together with desertion, they devastated the navy's numbers year by year. It became statistically un-usual for a man to be killed in battle.

The disparities are astonishing: during the Seven Years War, almost 134,000 seamen were lost to disease or desertion, while 1,512 were killed in action; during the reign of Queen Anne, the navy at any given time averaged 9,000 men sick out of 50,000 in service; from 1774 to 1780, the navy raised almost 176,000 men, the majority by impress-ment, and lost 42,000 of them to desertion; between 1792 and 1815, 81 per cent of the total naval casualties of 104,000 were the result of dis-ease (including a small number in accidents), while only 6 per cent of losses were to enemy action (twice that were lost in wrecks, foundering or fires, as the sea itself remained a hazardous constant).

Scurvy was the seaman's disease, his very own, and personal, afflic-tion. It disabled three men for every man it killed, and that made the actual situation aboard ship far worse than the statistics suggested. The numbers of men the navy lost to scurvy don't include the multitudes who would survive but were incapacitated. They could neither work the vessel nor fight.

Scurvy has haunted ships for centuries, ever since voyages became long enough to trigger the effects of vitamin-C deficiency (the precise cause, discovered only in the twentieth century, even if remedies were stumbled upon much earlier). For the first five or six weeks of a pas-sage, living on a diet of salt meat and biscuit and without fresh vegeta-bles or fruit, the sailor suffered only the perils of the sea. Then scurvy's insidious onset began: pimples broke out on the gums, and teeth loos-ened and fell out; dark blotches appeared on the skin; old sores reap-peared; fatigue grew into profound lethargy, and a man became so weak and vulnerable that a blow from a rope end, even a sudden move-ment, could kill him. The prostration of large numbers of seamen eventually put the ship itself in jeopardy: soon, there weren't enough men on their feet to handle heavy weather or prevent shipwreck. Half of all wrecks took place because crews crippled by scurvy could do nothing to save the ship.

The history of any deadly disease is sad enough. The story of scurvy might be the saddest of all, however, because its actual cure was suggested almost as soon as it appeared but was forgotten or sidetracked for more than two centuries. It was not the finest hour of medicine or the new, supposedly rational, empirical science.

The first recognizable alarm bells about scurvy appeared almost immediately after the beginning of long voyages of exploration: Jacques Cartier describes what is clearly scurvy during his trip up the St. Lawrence River in 1534. The earliest account in English is by Sir Richard Hawkins of his Pacific passage in 1593. Because the symptoms first appeared at about the time he was crossing the equator, he ascribed them to the change of air, a theory that accorded with the prevailing idea of disease: that it was caused by bad vapours and malign climates. The disease is "the plague of the sea and the spoil of mariners," he wrote. Then he made a remarkable guess that had nothing to do with the nonsense spouted by scientists. Hawkins, the rough seaman, was the one who practised the emerging scientific method of observation, experimentation and deduction. He had seen what happened when his men ate certain kinds of food, and he stated his logical conclusion: the best way to cure and prevent the plague was with sour oranges and lemons. The fruit was "a certain remedy for this infirmity."

Seven years later, Hawkins's inspired conjecture was confirmed by Sir James Lancaster on his voyage to India. Only his flagship, of all the ships in the little fleet, remained free from scurvy. For some reason, Lancaster had taken some bottles of lemon juice aboard with him, and he gave each man a spoonful every morning. The evidence was unequivocal, and Lancaster laid out the prescription with slightly askew precision: "The juice worketh much better if the party keep a short diet and wholly refrain salt meat, which salt meat, and long being at sea, is the only cause of the breeding of this disease." In fact, any amount of salt meat was all right so long as they downed their juice.

No one ashore paid any attention to what the sailors had to say, however, or they did at first and then forgot to, or more important and learned men thought they knew better than a seaman, even a captain and a peer. For whatever reasons, Hawkins's suggestion and Lancaster's confirmation sank from sight in England and didn't surface again for another 150 years. Plenty of time for hundreds of thousands of seamen to suffer and die. (There was an analagous situation with malaria. In 1572, an observant, wide-ranging captain of a merchantman repeatedly drew attention to the association of mosquito bites with outbreaks of malaria. No doctor or government official took any notice of these "outside the paradigm" observations, or the subsequent similar reports by seamen, until towards the end of the nineteenth century.)

Finally, in 1753, the naval surgeon Dr. James Lind published *A Treatise on the Scurvy*. It was dedicated to Lord George Anson for the very good reason that in the course of his circumnavigation from 1740–44, more than half of the nearly two thousand men who set sail with him died, mostly of scurvy (a high casualty rate even by the standards of the time). Lind had read the *Observations* of Sir Richard Hawkins, and he carried out controlled dietary experiments to test the Elizabethan sailor's hypothesis. Lind confirmed beyond doubt that a poor diet caused scurvy, and that one rich in citrus fruit, onions, vegetables or lemon juice cured it and prevented its recurrence. (In his rugged experiment, twelve scurvy cases were kept on a salt-meat diet, with each pair given different daily supplements: elixir of vitriol, vinegar, salt water, garlic and mustard, and two oranges and one lemon. The two orange-and-lemon men recovered; the other ten died—a conclusive demonstration, if not one that a modern medical ethics committee would have approved.) He followed up three years later with *An Essay on the most effectual means of Preserving the Health of Seamen in the Royal Navy*, which included his views on the prevention and treatment of typhus and malaria, as well as scurvy, and his original, and correct, ideas on hygiene and disinfectants.

Sailors were very interested in all this. Unfortunately, the Admiralty and the hidebound College of Physicians were not. Part of the problem was that Lind was a naval doctor—not quite up to snuff, not the right sort of chap. His ideas, therefore, could not be taken seriously. In fact, they were officially ignored for another forty years. Individual ships' captains and surgeons, however, found out about Lind's suggestions through scuttlebutt and were impressed enough to give them a try. In spite of the Admiralty and the medical experts, the incidence of scurvy began to fall.

In fairness, there were two supportable reasons for the Admiralty and the medical establishment to ignore Lind's findings: the Great Navigator himself, James Cook, thought that there were better, and cheaper, ways of treating scurvy; and in the fifteen years between the publication of Lind's treatise and Cook's first voyage in 1768, fourteen additional works proposed a wide range of remedies. Nearly all of them were nonsense, but some—more rum, for example, or the use of malt—were endorsed by eminent people like Sir James Pringle, president of the Royal Society. The malt hypothesis also got a lot of support from Cook, who forced it on his recalcitrant crews. (He made them eat sauerkraut as well, which the conservative salt-meat-and-hard tack aficionados detested, but it did them good, unlike the malt.) They didn't get scurvy, but that was because of Cook's insistence (considered a little cranky, even fanatical, at the time) on fresh meat and vegetables, not to mention washing one's body and clothes once in a while. Lind's work got lost in all the confusion.

Only when two of Lind's young acolytes grew older and achieved positions of some influence did the citrus-juice remedy for scurvy finally break through. One of them, Sir Gilbert Blane, persuaded the Admiralty to supply lemon juice to sailors on a nineteen-week voyage in 1793. Not a single case of scurvy appeared. Within two years, naval seamen were getting a daily issue of lemon juice and sugar to mix in with their grog. By 1805, Nelson's fleet was ordering thirty thousand gallons of lemon juice at a time.

This should have been the end of scurvy, but it wasn't. Merchant seamen continued to die from it because the government did not make it mandatory for shipowners to give citric juice to their crews until 1844, ninety years after Lind's treatise and forty years after the navy had adopted the remedy. Lemon juice conclusively worked for naval seamen, but merchant-owners were quite prepared to watch a percentage of their crews fall ill, and die, rather than take the trouble, or spend the money, to prevent it. Even after 1844, some shipowners watered down the mandatory juice to save money, and men got scurvy as a result. The 1851 Merchant Shipping Act allowed cheap citric-acid crystals to be used in place of lemon juice. These were almost useless, and even real lemon juice lost its efficacy if it was stored a long time. More men got scurvy. Around the middle of the century, owners substituted West Indian lime juice for Mediterranean lemon because it was much cheaper. But it had far less antiscorbutic effect. The number of scurvy cases doubled and doubled again. Finally, the Board of Trade accepted a doctor's suggestion that adding 10 per cent alcohol to lemon or lime juice would preserve it—and adding 15 per cent alcohol would guarantee that sailors would drink it. This practice was formalized in law in 1867.

An 1875 Arctic expedition that used lime juice was crippled by the disease. Eventually, lemon juice was universally reinstated. It cost more, but both shipowners and the government finally, if grudgingly, acknowledged that it was a necessary cost if it kept men from dying. Even seamen might be worth the expenditure. The legacy of the lime-juice era was, of course, the sobriquet "limejuicer" for English ships and "limey" for an Englishman, even though American vessels adopted antiscorbutic measures soon after the English. (European ships depended more on decent and varied food.) The names stuck even after the return to lemons. Limes made their last appearance in the Scott Antarctic expedition. Scott took along lime juice, instead of lemon. He and his companions died of a combination of exhaustion and scurvy, the former exacerbated by the latter.

It's close to unbelievable, but the disease persisted in sailing ships into the twentieth century. It showed up here and there, not infrequently, among the crews of long-voyage vessels whose owners or captains hadn't provided sufficient antiscorbutics, and British ships—the notorious "hungry-gutted limejuicers"—were the worst. The seamen had little in reserve to resist the malady in any event. They were fed rotten, inadequate food while aboard ship, and many debauched themselves in alcohol and whoring ashore. Venereal disease debilitated them further, as did all the other diseases to which the wide-ranging seaman was susceptible: malaria, typhus, yellow fever, beri beri, tuberculosis, smallpox. Their bodies had no chance to recuperate after the effects of four or five months at sea, especially if they signed aboard another ship, or were shanghaied, too soon after a previous passage. Even their tough bodies, a prereq-

uisite for a man to last long in sail, broke down under these combined insults. A wind-ship sailor was an old sailor by the time he was
forty.

One example of many late scurvy incidents: In 1897, the *T. F. Oakes*
took eight months to get to New York from Hong Kong, and it arrived
with the entire crew dead or incapacitated by scurvy; only the captain's
wife and the second mate were able to manage the ship.

The diet aboard British ships in the late nineteenth century ad-
hered to the Board of Trade's so-called Liverpool Scale, which set out
the seaman's daily whack: the amounts of salt meat, hard tack, pea
soup and a few other items each man was entitled to. The grossly defi-
cient scale wasn't modified until 1906, when sailors were promised
twenty items of solid food in place of the previous four. These included fish, potatoes, dried or compressed vegetables, oatmeal, con-
densed milk, suet, butter and even marmalade or jam—almost a
livable diet. The salt-horse ration was cut in half, and fresh potatoes
had to be provided for the first eight weeks of any voyage leaving a
home port between September and May. Fresh meat and vegetables
were mandatory in port "when procurable at reasonable cost."

Even then, however, there was a way out for miserly, or just plain
evil, shipowners: they could substitute for "reasonable cause." On a
long voyage, therefore, the sailor's diet could, and often did, legally re-
vert to the Liverpool Scale, as long as the seaman was paid an extra
fourpence to a shilling a day. Reasonable cause usually required only a
log entry as justification, and if seamen wanted their meagre compensation, they almost always had to go to court to get it.

Conrad's description of the seamen of his time, of Benjamin Lundy's
time, and the work they did, is often shot through with distorting nostalgia and romanticism. It's part of his "quarterdeck view," the officer's
perspective (unlike the fo'c's'le outlook of Dana and Melville), but he
also had writerly reasons for choosing to remember seamen in the way
he did: the elevation of the wind ship to an icon of an earlier, purer,
more virtuous age before the time of polluting steam. Seamen were its
loyal attendants. Ships are entrusted to the care of their crews, he
writes, "to watch and labour with devotion for the safety of the prop-

erty and the lives committed to their skill and fortitude through the hazards of innumerable voyages." When the crewmen went to that "long watch below" in large numbers in the process, death became the fervent guarantor of authentic danger.

The ouster of sail was a technological revolution, the sea-going chapter in the book of the Industrial Revolution. Like all revolutions, the disappearance of the ancient technology of sail produced nostalgia for the old regime—in Conrad and amongst people who remain connected with the sea or ships. The square-riggers have become utterly romantic symbols of simpler, better, more virtuous times, and of machines and technology, inherently beautiful, that had reached a state of perfection. "They mark our passage as a race of men / Earth will not see such ships as these again."

There's something to this. The ships were undeniably beautiful in their elegant and well-ordered complexity; they did indeed represent a technology, a kind of complicated machine, that had almost reached its most perfect, exquisite form.

"Men and their mournful romanticisms," laments Margaret Atwood, as poet. There's a corrective view as well: the beauty of the last wind ships belied their brutish nature. They were sailing ships given over to giantism. They had to be big because the only way they could pay for themselves, and pay marginally at that, was to carry and warehouse large quantities of the bulk cargoes that made up their last commercial hope. The other necessity for a profit was a reduced crew. Not that these hard men made much anyway: just a few pounds a month. But the economics of the last wind ships were nickel-and-dime (or penny-and-farthing); a few extra quid here or there and the owners' thin margins evaporated. The vessels became as big as could be handled by the muscle power of short-handed crews of ill-fed, often desperate men. It was the Armstrong Patent method: strong arms did the work.

The simple fact is that sailors withstood hazards, hardships and casualty rates that are almost unimaginable to us now. They were men who, in their capacity for living rough and accepting short life expectancies, seem as strange and distant as the soldiers in the mud and killing of the First World War's trenches—we just don't know how they could have done it.

Many things were also difficult and dangerous to do on board because of the neurotic conservatism of everyone concerned: designers, owners, captains. This was especially true of the British. They resisted labour- or life-saving innovations: a donkey boiler to provide steam power for strenuous deck work; a brace winch to reduce the labour involved in hauling yards around one way or the other; bridge decks and connecting catwalks to get men off the low-lying main deck in heavy weather, keeping them a little less wet and much less likely to get swept overboard; nets along the main deck above the bulwarks to strain the sailors out of the sea before they were lost forever. The latter were mandatory on German ships, but the British long eschewed them. Just do as we've always done; none of this newfangled stuff.

These innovations weren't false economies, something to coddle the crew with no payback to the owner. The winches and donkey boilers would have helped make up for crew shortages, but because the new equipment was very slow in coming, things were harder for the men and the ships: less efficient handling resulted in longer passages and more shipwrecks. Masters often carried insufficient canvas because a tired, short-handed crew might not be able to reduce sail quickly enough when bad weather hit. A fast, clean passage round the Horn, which depended on a dogged determination to take advantage of every shift of wind direction or change in velocity, was far more difficult. Captains often couldn't manage this because they didn't have enough men.

All the labour-saving improvements were eventually incorporated into square-riggers, but (in British ships especially) years, and many drowned and maimed sailors, after they could have been. No wonder that for some seamen, it was good riddance to life before the mast: work hazardous in the extreme; atrocious food and living conditions; intellectual and social isolation; subjection to arbitrary and often violent discipline. At the very least, sailors were bound by the Philadelphia Catechism: "Six days shalt thou labour and do all thou art able, / And on the seventh—holystone the decks and scrape the cable." Or the naval variant: "On the seventh—thou shalt labour a bloody sight harder." After an exhausting struggle aloft or a month-long war with the Horn, even the most dedicated wind-ship sailor might think it a good move to "leave the sea and go into steam." It was a phrase both derisive and wistful, and many men were glad to do it.

Men went to sea under sail for many reasons and in spite of the hardships. Usually, it was to make a living or because they were shanghaied into it; sometimes it was because of family tradition and the pull of ancestral occupations. But there were those other motives as well, ones that were essentially romantic and aesthetic. "There is a witchery in the sea," writes Dana in *Two Years before the Mast,* "its songs and stories, and in the mere sight of a ship, and the sailor's dress, especially to a young mind, which has done more to man navies, and fill merchantmen, than all the pressgangs of Europe." He describes the "irresistible attraction" of ships, "the passion for the sea" called up by the mere creaking of a block.

If there's an ideology of manhood, Jack Tar was one of its intensely appealing symbols. His outlaw attraction had to have become even more compelling as the nineteenth century ground along and men were drawn more and more into the humdrum, indoor routines of the factory and the office (good family men with wives and children to support, a pint in the pub after work, church on Sunday). The image of the sailor—half ruffian, half outlaw; a tough, free-wheeling world wanderer—was not necessarily something most men wanted to emulate. But they knew, or believed, that he was out there, roistering with his mates in port and contending with the sea with his crewmen-brothers, away from women and their demands, yet with access to women for his own needs, living an adventure and free from domesticity.

Of course, once the seaman was at sea and subject to the routine of the ship, which was far more irksome than almost any job ashore, he began to understand the nature of the sea, its hard, testing antagonism. The romantic impulse falls away, as Dana puts it, "like a fine drapery." The young sailor sees that his new life "is but work and hardship after all."

However, there is still the possibility of acceptance. Conrad describes the moment when he is purged of his illusions; he sees the "cynical indifference of the sea to the merits of human suffering and courage." He was a seaman aboard the iron ship *Loch Etive* when it came across a derelict Danish brig in mid-ocean. The brig had been dismasted and raked by the big waves of an Atlantic storm. For two weeks, night and day, its nine crewmen had pumped to stay afloat. When they tumbled aboard the rescue boat in which Conrad was an oarsman and curled up like near-dead dogs on the bottom boards, he understood how the sea, the true sea, had shattered them. At that moment, he was able to look cooly at his chosen life: his fantasies were gone, but the sea was fascinating still. "I had become a seaman at last."

The seaman also kept going back to sea because he valued the sailor's traditional attributes. It was a hard life, sometimes grim in the extreme, but you had to be a man, the genuine article, to do it. The true sailor belonged on the decks and in the rigging of wind ships. It was where he could acquire and use the ancient craft of the mariner. The term "able seaman" is deceptive. It sounds like mere competence but in fact denoted a man who could be counted on absolutely to do what had to be done on board a ship at any time, under any conditions, and who had mastered that extraordinary set of skills with which to do it. Someone who could withstand the intimidation of storms and great waves, and keep heart and a tranquil gut when all hell was breaking loose. The able seaman knew what devotions the ship required and how to perform them: "Must hand and reef and watch and steer / And bear great wrath of sea and sky."

Of course, other seamen persisted in sail because it was merely the force of habit of hidebound men unable to change. Or of men so feckless, and alcoholic, that they were unable to resist the cycle: shanghaied onto a ship; at the other end, desertion into the arms of the unscrupulous crimps who got them drunk, provided women, took whatever money the men had left and then shanghaied them onto another ship; and so on, time after time.

> *Old Swansea Town once more, fine girl,*
> *You're the girl that I adore,*
> *And when my money's all spent and gone*
> *I'll go round the Horn for more.*

Historians have lately become more interested in Jack Tar, some as part of the rise of social history during the last half of the twentieth century. The traditional approach of "great man" history or political-military study—Napoleon or Hitler, the Battle of Trafalgar or the kings and queens of England, and so on—has slipped into near disrepute. Now we're interested, with justification, in how people, all classes of people, lived: the details of their private lives, what they thought, what their relationships were with each other and with the state. In the past, sailors appeared only in literature, in the works of the great sea writers—Dana, Melville, Conrad—and in that of many other writers too, from James Fenimore Cooper to Jack London, Ernest Hemingway and Peter Matthiessen. Now seamen have become a legitimate subject for historians and social scientists as well.

Sailors certainly made up a clearly defined and distinct group of workers. There was no confusion about what they did and where, and that they suffered in the process. The suffering might often have been at the whim of the sea in storm or calm, but other humans—brutal officers, parsimonious owners, an uninterested government—made it much worse than it had to be. The relationship of seamen to those other individuals and groups has become, therefore, grist for the scholarly, and ideological, mill. A variety of gazes have been cast back onto the iron men.

Marxist or leftist analyses, for example. In an otherwise useful and persuasive argument that Jack Tar, or "Bachelor Jack," had become a caricature towards the end of the nineteenth century, the writer asserts that "caricature had a purpose in extending capitalist control of labour and legitimizing the hegemony of a ruling class." Another academic sees the seaman aboard ship as embodying "capitalist social relations." He writes: "Removed from home and the once-binding ties of kin and family, a collective body of seamen were now assembled and confined aboard the merchant ship. Their experience of work at sea—broad cooperation within a complex division of labour, coordinated and synchronized effort within a routinized and regimented labour process, continuous shiftwork as organized through a watch system, close supervision and harsh discipline—insured the primacy of work over life at sea, and hence a kind of alienation that was expressed in their characterization as 'hands.'" This may be an interesting summary of one view of seamen, even if it's designed, with Procrustean motivations, to fit a theory. (The idea of the sailor as an exploited worker has been a literary theme as well—for example, in the work of Jack London and Eugene O'Neill.) But it's certainly a long way from Conrad's "great passion for the sea"—or from Billy Budd, who, in irons and soon to die in the foretop, will, like Milton's Lycidas, be "mounted high" in "kingdoms meek of joy and love."

Feminist and gender scholars have become interested in square-rigger seamen as one of the last repositories of traditional male virtues (and vices), living in the segregated, intensely masculine society of the ship, cut off from women and family, acting out the great (and suspect) male fantasy of escape, sanctuary and redemption in adventure and hard, dangerous physical struggle in the great expanse and wilderness of the sea. Odysseus was merely the first in a long line. Even after Homer finally brings the hero home, other poets—Tennyson and Kazantzakis—set him off, and set him free, once more.

There were certainly women on board many sailing ships: the wives of captains, who sailed with their husbands and gave birth, raised children, occasionally acted as co-navigators (logarithmic and sight-reduction pillow talk?), enduring bad weather and long isolation. The alternative was to spend a few months with their men every couple of years. The wives of whalers always had children spaced three or four years apart, the length of a typical whaling voyage. There were also a few female pirates—Anne Bonny and Mary Read were two of the most famous—courtesy of the ferocious democracy of the nautical *demi-monde*. Some women went to sea on merchantmen, and warships as well, by passing themselves off as men. Women were also present in a different sense: in the affections of sailors for mothers and wives, even whores (although the shanties seem to contradict that), or in the homespun, feminine details of life aboard—sewing of clothes or sails; body ornamentation with jewellery and tattoos; meticulous, orderly stowage (even if it's for a survival purpose, rather than its own sake); the protocol of officers' meals in the captain's cabin.

There's no doubt that some men chose the sea, in part, as a means of getting away from women and, perhaps, into the intimate company of men. There was a strong strain of misogyny in these seamen, which shows up sometimes in their scanty diaries. "Capt. woman sick and I wished she would dye," wrote a whaler in 1871. "She walks the deck with majestic grace. With her cherry picker nose all over her face," rhymed a sailor aboard the *Ivanhoe* in 1867.

Men before the mast might have resented the captain's wife because she was a part of the relative comfort and privilege of the aft cabin and didn't let the seamen forget it, or because she provided a sexual outlet for the captain that was denied to them. More often, however, the problem with a hen frigate was more mundane: there were restrictions on dress, or the lack of it (no naked rumbustiousness in the doldrums rain), and on the cussing and blaspheming that was every sailor's birthright. Or perhaps the woman's presence upset the balance, often precarious, among the fo'c's'le personalities, threatening the crew's solidarity. She might have favourites among them—very young seamen or apprentices, for example, to whom she extended favours, or induced her husband to extend them. Captains' wives often tried to soften the regime of violent punishment, an effort not unwelcome to seamen, but an intrusion on tradition nevertheless, and harmful if it was applied selectively. The seamen's resentment of female presence and interference was similar to the resistance of soldiers, especially those in combat units, to the enlistment of women: females subverted, complicated and threatened to destroy the ancient brotherhood of males doing what they regarded as theirs, and theirs alone, by nature.

Apart from these occasional feminine presences, however, the maritime workplace was the exclusive realm of men. And of men who lived in each other's pockets, cheek by jowl in the fo'c's'le, isolated in their small groups for months at a time, mutually dependent for companionship and support in their perilous world. It could be described as a hyper-masculine society, saturated with exaggerated male behaviour: stoicism taken to extremes, harsh physical discipline, drinking to excess at every opportunity, fighting amongst themselves, sleeping with strange women. There's the whiff of overcompensation in all this. In similar male "total institutions" (those in which contact with the outside world is limited and the inmates are subject to all-pervasive control), such as jails or barracks, a certain amount of homosexual activity takes place. The navy is known for this, at least anecdotally— Churchill's "tradition" of sodomy. The homoerotic aspects of life at sea are clear: young, strong bodies; close physical quarters; the bonding ef-fects of co-operative work and shared danger; the naked skylarking; the leisure-time dancing or wrestling in which some crews were al-lowed to engage; the transvestism of the crossing-the-line ceremony, when sailors dressed up as Neptune's queen and her attendants.

Yet in all the voyage literature of life and passages under sail, there is not a single overt reference to homosexual behaviour. Academic studies speculate that it must have taken place; it's just that there's no direct evidence for it. There are intimations, however. One writer (more writer than seaman) who made an early-twentieth-century voy-age describes the crew's eruption into joyful horseplay during a tropical squall, the men, "with one accord, like the male chorus of some gay, naughty musical, tossing off their clothes until they were mother naked." Later in the soapy free-for-all, "amidst the welter of thrashing bodies," the men's clothes float in the water of the stoppered decks, "gently bobbing and bumping up against you with a sort of secretive pleasure that was positively dangerous with its suggestion of furtive, cloying touch." There's also an indirect observation about the ample size of one small man's genitals. Discipline is momentarily suspended as even the mate joins in, naked and frolicking. The rain ends, and the captain turns away aft; "the mate gets it, goes as hard as iron all of a sudden, and all the high jinks and horseplay on the foredeck suddenly freezes into coldness."

Another early-twentieth-century sea writer, Bill Adams, celebrates the homoerotic brotherhood of the sea. All hands are called to handle sails on a hot, squally night, and they tumble up with little or no clothes on (no hen frigate). Two of them are aloft two hundred feet on the skysail yard as dawn breaks. One of them "rose upright on the foot-rope and rested a huge hand upon the shoulder of the naked lad beside him." They look out at the day breaking.

"Suddenly, waving an arm, the sailor encircled an arc of the watery waste.

"'*The sea—the sea!*' he shouted.

"The apprentice's lips moved. Though with the close of that long night his young body was now grown chill, his wide eyes shone.

"'Our's, Manus!' he cried."

Later, when more warm rain fell, "We shouted, we laughed, we jested. We played leapfrog, and slapped one another's glistening bodies, flat-handed."

The sailor-narrators have cleaned up the act. They do the same thing with language, never including blasphemy or cursing more extreme than "damn" or "bloody" (though they often leave blanks for our imagination). Part of this is the reticence of the time. Most writers wrote during the Victorian and Edwardian periods, or came of age during them. They weren't about to reproduce the language for which sailors have been notorious since time immemorial, and which so offended even Samuel Johnson, let alone describe homosexual encounters in the fo'c's'le.

In a more diffident age, it might also have been that acting out one's impulses happened far less often than now. Feelings then were more latent, less patent. There was a lot of the "goosing in the locker-room shower" horseplay involved as well. Yet intuitively, it seems reasonable to assume that some square-rigger sailors were "fore-and-afters," the sailor's term for those who preferred homosexual relations. If they had a term for it, presumably it took place. Why should ships have been exempt from the situational homosexuality that was, and is, a feature of life in other similar circumstances? The usual behaviour of sailors storming ashore after a long period at sea—the wild drinking, fighting and whoring—was partly a flamboyant assertion and reaffirmation of heterosexuality: I'm a sailor and I'm okay.

Melville proposes that everyone shares Ishmael's feelings about the sea. He delivers a cascade of sea image and metaphor. There is magic in water. "Were Niagara but a cataract of sand, would you travel your thousand miles to see it?" The image we see in all rivers and oceans, says the sailor, cetologist and hunter, is "the image of the ungraspable phantom of life; and this is the key to it all." The unsounded ocean itself is life; "all deep, earnest thinking is but the intrepid effort of the soul to keep the open independence of her sea." Anything but "the slavish shore." The true sailor expatriates himself to nationalize with the universe. The voyage is everything. "Yes, the world's a ship on its passage out, and not a voyage complete. . . ." Water and meditation are wedded forever; "in landlessness alone resides the highest truth, shoreless, indefinite as God." In the Southern Ocean, Ishmael watched the albatross as it "arched forth its vast archangel wings . . . and in those forever exiled waters, I had lost the miserable warping memories of traditions and of towns."

In all the casting back to him, the sailor has been many things: ruffian, Jack Tar, expendable article, able seaman, exploited worker, epitome of masculinity. In Melville's gaze, the sailor, detached from the ignoble land, voyaging on the infinite sea, is the most worthy of men: a seeker of truth.

W. H. Auden's *The Enchafèd Flood* sets out the elements of the Romantic attitude to the sea that Melville typified. Every man of sensibility and honour desires to leave the trivial land to go voyaging, which is the true condition of man, and the sea is the place where "the decisive events, the moments of eternal choice, of temptation, fall, and redemption occur."

We still know that the ocean is important: cargoes must be shipped over it, fish caught in it, its overarching role in creating and changing climate understood. But the Romantic view of the sea and sailors has dissipated. It's not a vital part of our consciousness any more. With a few exceptions, the sea has lost its power to create images, metaphors, stories. People no longer think of the sea as a place to go to find the truth and save their souls.

That became impossible with the advent of the Industrial Revolution at sea. There's an essential contradiction between Auden's summary—the field of action for the "typical Romantic Marine Hero"—and the steamship. The oily ingenuity of powered, brute-force machines—constructs of isolated mechanical movements endlessly repeated—took the place of human muscle, cunning and resourcefulness. There is no opportunity on the bridge or in the engine room of a steamer to hear the voice of the sea that seems "to speak of some hidden soul beneath."

We're detached from the sea. We don't need to travel on it any more; we can, but it's not necessary. Cities on the sea have lost their maritime connections. Who can write about any city, now, what Melville once wrote about New York? "Your insular city of the Manhattoes, belted round by wharves as Indian isles by coral reefs—commerce surrounds it with her surf," he said. "Right and left, the streets take you waterward." There are no more docklands or Sailortowns. There are no more seamen under sail.

In order to get his ship from port to port, the sailor-truthseeker, or even the mere, unadorned square-rigger seaman, needed the established masculine qualities: fatalism; stoicism; physical courage; acceptance of strict, cruel hierarchy; anger as an enabling response to fear; a self-regulated domesticity. The corollary is that when square-riggers, and the seamen who sailed them, disappeared, so did a reservoir of traditional masculinity. These values began to seem more and more antique, unnecessary, even unhealthy and harmful. Because the sea is pre-eminently the realm of men, it came to seem less necessary and more remote as well. When we lost the wind ships and their seamen, we began to forfeit the sea itself.

CHAPTER SEVEN

The Circle of Her Traverse
of the Sea

O voyagers, O seamen,
You who come to port, and you whose bodies
Will suffer the trial and judgement of the sea
Or whatever event, this is your real destination.
T. S. ELIOT, "The Dry Salvages"

Two days after Benjamin's first gale blew itself out, the *Beara Head* picked up the Portuguese trades.

The wind had blown sour on them for weeks. When the Biscay gale finally moved on, the ship wallowed for a day in the big, leftover swells, the light and variable wind not enough to give it more than steerage way in the slop and spill of seas. Then a light northerly got them farther south by a hundred miles in the fast-calming water; the warming sun dried out men and gear. Now the trade winds appeared just where they should have at the time of year, but their reliability was, nevertheless, a pleasant surprise. Wind rarely enough showed up where and when it was supposed to that it was a cause for celebration when the pilot charts were right. The barque bowled along in the heel of the trades, full canvas again, sailing by and large and making ten knots on the port tack.

The Greek found the body when he went to relieve himself in the fo'c's'le toilet halfway through the forenoon watch. He sat on the seat over the opening straight down to the sea and, as he shit, contemplated the Clerk's swinging corpse. The man's face was contorted in the typical manner of the asphyxiated; his ragged suit was more white with salt stains than black, and he reeked. But this wasn't the smell of death—the job had been done not long before—nor merely fo'c's'le odour, "the comforting stench of comrades." The suicide hadn't washed in three weeks, and he stank of feces and piss where he had soiled himself in terror of the ship and its mate. This was the mate's doing, Kapellas thought. That bastard had just killed his first man of the voyage, but maybe not his last. The Greek finished up and went aft to report the event to the murderer himself.

The two mates and the captain came to look at the dead man. It was unusual for the captain to forsake his territory aft for any reason, and it took a calamity to draw him to the ship's bow. Both watches were on deck in the fine trade-wind weather, chipping, painting, overhauling lines, repairing gear damaged or chafed in the Biscay gale. In a temporary suspension of discipline, the entire crew abandoned work and crowded forward behind the officers.

Benjamin peered with the others into the small, dim space of the heads. He saw that the Clerk had hanged himself from an iron beam with a piece of gasket line, out of which even he had been able to fashion a reasonable knot, enough to take his meagre weight. Benjamin felt sick, appalled, guilty. He'd seen death before—two infant sisters, and his grandfather laid out like a serene manikin in the front parlour while the family drank whiskey, told stories and sang the old songs around him—but nothing like the Clerk's grotesque, twisted grimace. Guilt roiled Benjamin's guts. He had done nothing to try to help the man, to break through his isolated despair, to protest his treatment. None of them had. The Clerk was like a plague victim or a leper. The seamen manoeuvred around him as if he would contaminate them somehow and bring down the mate's insane anger on them all. The Clerk had borne the assaults alone and in silence, until he could no longer tolerate them. Benjamin felt the chill of remorse and his own vulnerability on this hard-case barque.

"He's put a hand on himself, right enough," said the captain. "He was a dead horse the minute Fearnaught brung him aboard. The man wasn't worth a traneen. Cut him down and parcel him, Mr. MacNeill."

"Aye aye, sir."

The elderly Yankee sailmaker, John Page, sewed a canvas bag for the dead man; he'd done it twenty times before, he said. By six bells of the forenoon watch, an hour after Kapellas found the body, the crew had gathered by the lee rail below the poop, the captain had read the burial service, the Clerk had been tipped into the sea and all hands were back at work. The dead man had no belongings to auction off, but his bunk became another drier space to stow sea chests clear of the slosh of heavy-weather water across the fo'c's'le deck.

"The goddamn mate scuppered him," said Kapellas.

No one disagreed.

No one was surprised either. They had a hot ship and a bucko mate, and that was that. He was a Bluenose and they were worse than Yankees. MacNeill had been a seaman and then a mate on Canadian schooners running down to Brazil in the fish trade, which was notorious for its tough drivers. They carried sail to the last and used up men like horses. Even yellow-jack epidemics in the Brazilian ports didn't keep them away. Limejuicers had to report the deaths of seamen resulting from assaults by mates and captains—unlike Yankee vessels, and maybe Bluenoses too. Every sailor had heard scores of stories about Yankee mates killing men with impunity, using belaying-pins, knuckledusters, revolvers. The Clerk had killed himself, so the mate was safe from a formal inquiry; the captain noted the suicide in the official log (this was distinct from the deck log, which recorded weather, courses made good and sail handling). These were laconic enough chronicles of voyages; if nothing went seriously wrong, an official log was often empty. However, a suicide qualified as an event worth noting.

"That mate ain't done with killing yet, neither," said Kapellas.

"Maybe," said Russell. "The Clerk hadn't got any sand. If he'd have stood up to the mate, he would have backed down. The man wasn't worth a bosun's damn. He couldn't go aloft. It would've broke a snake's back to follow his wake when he was on the helm. He had Cape Horn fever."

"All the same, no need for the mate to ride him down all the time," said Maguire, the labourer.

"Maybe not. But he still scuppered himself."

"The man was pressed," said Benjamin, the first time he had ever spoken up during fo'c's'le wrangling. "All he ever was was the last hough in the pot. Does a pressed man deserve to be hounded to despair?"

"It's just the way she goes aboard ship," said Russell. "The captain's king and the mate's his executioner, if he needs one."

There were such things as happy ships, although fewer of them as the century wound down and sail fell under close siege. Bigger, heavier vessels, pared-down crews, skimpy gear and provisions, dirty bulk cargoes given to shifting or burning, the disappearance of passenger liners under sail and the civilizing effects of carrying what Conrad called that "honoured bale of highly sensitive goods"—all of these coarsened life aboard the wind ship. Nevertheless, at the time of the *Beara Head*'s voyage, there were still happy ships. One imagines that Conrad's *Otago* must have been one of them when he was captain; the *Torrens* too, since it was a passenger carrier and Conrad was the mate. There were many others. The ships *Thomasina MacLellan* (*Tommy Mac*, for short) and the *British Ambassador* kept their crews year after year. The *Tommy Mac*'s sailmaker stayed aboard for ten years. Capt. A. Smith of the ship *MacCallum More* kept it up like a yacht, and he looked after his crews with the same care. So did Capt. W. D. Cassady of the beautiful little ship *Greta*. He fitted out his cabins in ash and black walnut, then filled them with pictures, flowers and his collection of curios from around the world. He kept orange trees, geraniums and other flowers in his mess room and trained Scotch ivy round the pillars. The captain's care with his quarters extended to the fo'c's'le; "full and plenty" was the standard (although only in comparison with the standard stingy whack). The *Ivanhoe* spent years carrying passengers on the London-Melbourne run in the 1870s and 1880s. The owners' motto was "no stint" for passengers and crew (although the phrase had a somewhat more limited meaning for the seamen). The lives of its sailors were so stable that many of them married, settled their families in Melbourne and stayed aboard for years. When a vacancy came up, fifty men applied. The *Ivanhoe*'s charmed, and charming, life held for forty-seven years, until it was wrecked at Honolulu in 1915. Capt. William Porter of the *Ennerdale* died on board his ship during a passage from San Francisco to Queenstown in 1881. He had been like the proverbial father to his men, and as the mate read the funeral service before the ensign-draped body slid into the sea, almost every man aboard wept.

Just after dawn on May 27, after twenty-two days at sea, the *Beara Head* passed to the west of the Portuguese island of Madeira, having made close to two hundred miles during the previous twenty-four hours. The ship stayed well clear of the land, as *Ocean Passages for the World* advised, avoiding the squalls and calms of the water close in. The captain got a shore bearing to confirm the accuracy of his sun sights and ensure the chronometer's dependability. The trade winds were holding at twenty to twenty-five knots, strong for the time of year, and the barque was making the most of it, surging along with a bone in its teeth.

On deck after breakfast, Benjamin watched as the ship passed by the distant high island. Reared in the cold rain and mists of the north, where summer was often nothing but warmer rain, and having spent all his sea time in the narrow, grey-green waters of the British islands, he was enthralled by the trade winds, this intimation of the tropics: warm wind; a sea of blues—navy, ink, indigo, cobalt, prussian, turquoise—mutating from one to the next with the curl of waves; breaking crests brilliant in the sun; the high, puffy trade-wind cumulus, like scattered cauliflowers.

Clouds are aesthetic correspondents of the weather. Ugly skies mean ugly times for the sailor. The pretty cumulus are guarantors of fine times in the only way that the seaman really cares about: nothing hazardous is in the offing; for now, he can relax.

In the trade winds, ships relax as well. Yards squared, the barque sailed almost dead downwind, crojack and mainsail clewed up to let clean air get through to the foresail, but otherwise every sail pulling, the wind direction constant enough that little or no brace or sheet pulley-hauley was required.

Benjamin loved his stints on the helm. With its long, slender shape and loaded hull deep in the water, the ship tracked along like a loco-motive, almost sailing itself. The helmsman merely superintended the vessel's self-rapt progress. Looking out at the blue sunlit sea and the ship's grey-white canvas stretching before and above him, feeling the gentle pitch and roll, Benjamin thought that nothing in his life had ever been, or maybe ever would be, as fine as this hypnotic, ardent, serene, endless motion through the sea, its ecstatic elements brought together and guided by his own hands on the wheel. With the strange relativity of time and pleasure, his hour-long tricks went by in minutes. Then it was back to chipping iron, and those hours were hours long.

When Madeira dimmed and disappeared astern, the barque held its course a little south of southwest, towards the Cape Verde Islands.

The *Beara Head* was happy in its headway through the kindly sea, but its crew was not. There seemed to be a malign correspondence at work: the less enmity of wind and sea, the less amity aboard. The various personal wars, gone cold during the struggle to make southing, hotted up in the gentle weather. There were the usual irritations of men living and working in close quarters day after day, but everyone expected these; it was the universal run-of-the-mill stuff of families and work-places. It was different, however, in the case of two pairs: Grey, the cockney, and Russell, the Yankee; MacNeill, the first mate, and Jagger, the second.

On the *Beara Head*'s twenty-third day at sea, Grey tried to kill the Yankee. The ill will between the two had never let up. Grey was irritable in all things while he was at sea; it was his habit, and nothing could break it. His fear and loathing of the water never lessened. Even the genial warmth and seas of the trade winds didn't mollify him; he could still barely look past the rail. Russell's self-assurance and humour—did the Yankee mimic Grey, as well, when he was working aft or aloft?—his assumption of the fo'c's'le leadership as if by right, the attraction of the other seamen and their glad deference—all this rankled Grey, and more and more, enraged him. During the gale, he had put it aside. He had worked with Russell, and the others, in necessary unity. The ship in its need had the power of reconciliation, but it was temporary.

Russell did indeed imitate the cockney whenever he had the chance, the man's obsessive, harsh anger made comic in its pop-eyed exaggeration. Apart from that, he tried to tolerate Grey and made sure he never turned his back on the man. On other ships, the Yankee had been forced to do this—keep someone, or a few men, always in his mind, so that everything he did, everywhere he jogged or climbed aboard ship, always included an awareness of the other man, a constant calculation of the danger he might pose at any moment. He avoided getting next to such a seaman tailing on to a line, or on a yard; at night, he kept track of his movement on deck and, especially, aloft. During some of his earlier voyages, Russell had been able to maintain this uneasy watch and distance until the end of a passage, when the crew dissolved in drunkenness, shanghaiing and desertion. Other times, he had had to fight, although never to kill. He didn't resent this; it was merely the way she goes aboard ship—like the mate riding a man down to his death.

In the middle of the gravy-eye, the port watch was trimming sails to a wind shift. Russell called to Grey to slack off the tack as the seamen were squaring the foreyard. It sounded to Grey like an order, and it pulled the trigger. Usually the loudest of men, Grey stared at Russell for a moment without a word. Then he turned and walked to the deck house, took his knife (the one with the point) from its hiding place in his donkey's breakfast and walked back towards Russell, holding the knife out of sight behind his hand and arm. The Yankee waited, appearing almost indifferent except for a slight crouch and a tensing of muscles. Benjamin and the other seamen of the watch stood motionless too. This confrontation seemed preordained, irresistible; to interfere with it would be a violation of the inevitable order of things. The cockney knew how to use a knife, and the Yankee was not an agile man. Grey lunged, stabbing Russell once on his left side and slashing his right arm twice as he tried to protect himself. All this in near silence and moonlight. The quick attack released the other seamen from their fatalistic stasis. Kapellas and Anderson jumped on the cockney and dragged him down, avoiding the blade. Benjamin remained paralyzed, his old fear of confrontation and his first view of a blade sliding into human flesh filling him with fear, rooting him to the barque's deck. Russell's blood flowed out in a steady thin stream, not in spurts or gouts; he had defended himself efficiently, and the wounds were superficial.

Things were still quiet enough—this was fo'c's'le business and nothing to do with the officers—but the mate's ear on the poop caught the familiar grunts and thumps of combat above the static of wind and sea. He stamped along the deck without a word, his silence one of the most astonishing things the men had yet experienced aboard the *Beara Head*. "Let 'im up," he said to the seamen restraining the struggling Grey, breaking, finally, the wordlessness of the previous minutes.

As Grey began to rise up snarling bare-teethed like a dog, the mate kicked him several times, once hard in the head, flattening the man onto the deck again. He ordered Russell aft to the poop, where the captain would patch him up, and accompanied Grey, in the grip of his mates once more, to the fo'c's'le deck, where he handcuffed him to a pipe in the bosun's locker.

Russell spent a week in his bunk and had light duties for two more weeks after that; three more scars were added to his marked, tattooed body. Grey remained in irons on half-rations, sleeping on a blanket on the dirty iron deck of the storeroom. The isolation seemed to calm him; he grew gaunt but quiet. The sea was invisible from his impromptu cell.

The violence was contagious. Less than a day after Grey's attack, the mates clashed.

They had never reached an understanding or accommodation; the first mate's personality foreclosed that. He was blank as a rock, adamant and obdurate, as if there were two worlds: his own, expansive and singular, and the lesser one in which all other men lived, inconsequentially. Jagger, for his part, had never found his place or voice aboard the *Beara Head*. MacNeill condescended to him at best, often insulting him with his offhand instructions; even the captain's solicitude—the watchful commander trying to ensure his young officer's morale and authority—rankled. The captain thought he still needed special treatment. Jagger was a competent junior officer—he had his Board of Trade ticket—but he lacked the furious energy, the fierce will of a true wind-ship ramrod. He knew it himself: when he led his watch aloft in the gale in the dark, he could feel the noise and wind and vicious motion sapping the fight out of him. He quailed under the sea's assault. He had always wanted to command a sailing ship; yet he realized, more and more, that his destiny was in steam, enclosed in the cosy bridge, his only responsibility making sure the vessel got to port on schedule, leaving the stokers, firemen and engineers to sweat around the brutish engines.

It happened easily. On a change of watch, the mate pushed past Jagger at the top of the poop steps, shouldered him aside as if he wasn't there. The second mate screamed: "MacNeill!"

The mate stopped and turned. Jagger stepped into him and shoved hard. It was like shoving a rock: MacNeill hardly budged. He stared at Jagger for a moment without much expression and then, with astonishing speed and ferocity, hit him hard on the side of the head. The second mate collapsed to the deck, almost unconscious. The mate went below into the aft cabin and reported to the captain what had happened.

The two men came back on deck, where Jagger was now standing, groggy, trying to focus his eyes. Captain MacMillan ordered his mates to shake hands. They did so, after a long pause and without looking at each other.

Later, when Jagger was on deck with his watch, the captain, in the privacy of the aft cabin, told his mate to show his junior the respect he deserved.

"With respect, sir. He don't deserve much respect."

"Mr. MacNeill, he's not just a gossoon; he's second mate of this ship and runs the starboard watch. Maybe he deserved to be hit a dunt, but that's it. I want peace between you. No more of this!"

The mate stared at his paddy captain for a few long seconds before replying, but the habit of compliance remained intact.

"Aye aye, sir," he said.

Mealtimes in the aft cabin became more strained than ever. The mates ate fast and in silence. If the captain asked a question, they answered, but no more. The captain's wife, still pale in spite of the trade-wind sun, replied to her husband, but with the self-consciousness of someone overheard; she no longer tried to draw out the second mate. She had been horrified by the Clerk's death and the fighting. She saw the voyage stretching months ahead of her, an ordeal she could not escape. Living in the mate's proximity—his small cabin was steps from the captain's—was like living in the same cage with a predator.

The captain had his reservations about MacNeill. He had decided to look the other way when his mate laid about him without his captain's say-so, strictly speaking a breach of law. What could you expect from a Bluenose? Frontiersmen with the half-savage brutality of the wilderness. He had seen captains and mates who were plain evil, but

MacNeill was simply acting out his evolutionary destiny. He was still a lower form of life. That was what Darwin and the other monkey-men had gone on about, wasn't it? Captain McMillan was a hard case himself. Nevertheless, he was aware, deep in his innards, of a slight swirl, a *frisson*, of fear of his first officer's irresistible force. He had to trust the restraints on MacNeill's behaviour of two-thirds of a lifetime of ship's discipline.

In the meantime, Valparaiso was a very long way ahead, and at meals, the captain did the talking, as if it was all up to him to keep going some kind of civilized discourse among the afterguard. But his efforts became monologues: on the genealogy of Islandmagee; on the finer points of old Irish dancing; on Home Rule and the evils of Parnellism; on his life aboard previous ships, past gales, an Indian Ocean cyclone he had lived through. The others received this stream of opinion and information stoically.

The helmsman and other seamen working on the poop had seen the mates' set-to. Fights aboard ship were commonplace, but they were rare between officers. The hands seemed to have a common response: glee that Jagger had dared confront the mate (it raised him in their estimation as nothing else he had done); and resignation in MacNeill's obvious impregnability.

Benjamin was happy that Grey was out of the fo'c's'le. The cockney's permanent anger had seemed to corrode the air of the deck house, feeding Benjamin's old anxiety about violence. Now he slept better. The fo'c's'le card games and conversation went on with an easier amiability, albeit a little forced at first, as most of the men (the Finn and the Dutchman were impervious) recognized it was because of Grey's absence. Their good humour had the feel of schoolboys released from a master's grip, but as the days went on, the relaxation became unremarkable, habitual. They still had to give the flat-earth Finn a wide berth and jolly along the Dutchman, who, without these efforts, became morose and was hell to get on deck for his watch; but that, too, became a matter of course.

Benjamin realized with surprise, and some pride, that he wasn't really afraid of the mate. Except for his perverse vendetta against the Clerk (Benjamin still felt the sting of the man's suicide, the way they had all abandoned him), MacNeill acted impersonally, handing out insults and blows with the rope end as if he was merely whipping some necessary inanimate objects into line. Benjamin knew that he was on the way to becoming a good square-rigger sailor; he thought that the mate saw it too. The man could go berserk, but there was no doubt he knew his business. Benjamin was able to absorb the voice and the whip with serenity, more or less.

"Who is not a slave? Tell me that," said Ishmael. Even if he gets thumped and punched about aboard ship, it's all right. Everyone gets served in the same way, either physically or metaphysically, "and so the universal thump is passed round, and all hands should rub each other's shoulder-blades, and be content."

What a peculiar first mate Conrad must have been. The man who would delay going on deck to order the topgallants taken in because he was waiting for a literate passenger's verdict on the draft of *Almayer's Folly*, his first novel, and who became John Galsworthy's intimate. Surely no captain or seaman had come across another officer like this gracile Slav with his syrupy accent and unpronounceable name. "I had never seen before a man so masculinely keen yet so femininely sensitive," wrote his first editor. He stood out from other captains, a Mauritius sugar-exporter wrote after doing business with Conrad. He dressed like a dandy, "everything well cut and stylish" compared with the standard white ducks and straw hats. "He wore a black or gray derby slightly tilted to one side, always wore gloves and carried a cane with a gold knob." His colleagues called him, with irony, the Russian Count.

Another writer and former seaman, who knew Conrad when he was working on *Almayer* and *An Outcast of the Islands*, thought that the Pole "was too highly strung, too sensitive and nervous, if I may put it so in all friendliness, for a sailor to put any great confidence in him as a commander."

The assessment is not surprising, but it was wrong. Not that Conrad wasn't sensitive and highly strung. But so was Horatio Nelson, the seasick admiral, whose men nevertheless followed him into a dozen sea battles with the blind, tribal love of warriors for their leader. A tender, empathetic disposition was no impediment to leading men—so long as it was accompanied by practical skill and high competence.

Conrad had those too. He qualified as a second mate in 1880, when he was twenty-two, only two years after joining the British merchant marine and learning English (although he had spent several years as an apprentice seaman on French ships). He earned his first mate's ticket two and a half years later and his captain's less than two years after that, about as fast as anyone could pry the tough qualifications out of the Board of Trade. Galsworthy had a long opportunity to watch Conrad at work aboard the *Torrens:* "He was a good seaman, watchful of the weather, quick in handling the ship; considerate with the apprentices."

Conrad had only one command—the pretty little iron barque *Otago*—but he became a captain when sailing-ship berths were growing scarce compared with the numbers of qualified officers. More important, he was beginning his drift away from the poops and bridges of ships to the writing desk, his experience at sea providing a fecund context for his stories of the mysteries of the human heart.

Nevertheless, when Conrad did go to sea, he went as a proper ship's officer. He was captain of the *Otago* for only fifteen months, but that was under difficult circumstances and in tricky waters. In 1888, three years after Benjamin's voyage in the *Beara Head*, Conrad took the ship out of Bangkok with half the crew, including the mate, sick with yellow fever, dysentery and cholera. The barque sailed to Singapore, Australia, Mauritius, Durban and ports of the Malay Archipelago: Surabaya, Macassar, Timor, Benjarmassim—the country of Almayer, Lingard and Tuan Jim. The barque's owners had complete faith in Conrad's abilities to pilot their vessel (as if it was uninsured and his own) through the reefs, balky winds and maddening calms of the South China, Celebes, Java, Andaman and Banda seas and their narrow connecting straits.

When the *Otago* was due to sail from Sydney to Mauritius, Conrad wrote to a friend that "all of a sudden, all the deep-lying sense of the exploring adventures in the Pacific surged up to the surface of my being." On an impulse, he wrote to ask the ship's owners if he could take the barque to its destination by way of the constricted and difficult Torres Strait, between Australia and New Guinea, and the Arafura and Timor seas into the Indian Ocean, instead of the usual route around the south of Australia, a hard, but much safer, windward slog. To his surprise, the owners agreed, providing that, in his judgment, the seasonal calms of the Arafura had not yet set in. To make sure he had time, Conrad left Sydney in a southeast gale so strong that neither tug nor pilot would venture out beyond Sydney Heads. The *Otago* had to sail itself out the narrow harbour entrance. Nine days later, Conrad "put her head at daybreak for Bligh's Entrance, and packed on her every bit of canvas she could carry. Windswept, sunlit, empty waters were all around me, half veiled by a brilliant haze. And thus I passed out of Torres Strait before dusk settled on the waters." The voyage and its description encapsulate the unique abilities of Captain Korzeniowski: the bold, skilful, decisive work of a good seaman told in the words of a true and natural writer.

The *Otago* was "a high-class vessel," Conrad wrote, "a harmonious creature in the lines of her fine body, in the proportioned tallness of her spars." Nevertheless, he resigned from the ship, to the profound regret of its owners. He told them that he wanted to visit Europe again. In reality, he wanted to avoid a return passage to Mauritius. He had fallen in love there, but his marriage proposal to the beautiful Eugénie Renouf had come too late; she was already engaged.

His resignation from his first command became a crucial juncture in his life, a literal watershed: he would begin to shed the sea. *Almayer's Folly* was in his head, and it was time to write it. He sailed to England as a passenger on a steamer. However, he needed to make a living, and he found out that he couldn't do that at home. After six months working in the Belgian Congo (which provided its own inspiration), he signed on as first mate of the *Torrens*, the composite iron-and-wood passenger clipper, on which he made two voyages from England to Australia. It was its topgallants he thought about as he waited for the passenger's verdict on the unfinished draft of his first novel. After that, he left the sea for good, although with ambivalence and reluctance.

With his psychological insight and his high moral sensitivity as a writer, Conrad could not have been anything other than a fair and decent manager of men as a mate. He must have been tough when necessary—that was a condition of running crews on square-riggers— but he could never have been a bucko mate. It's obvious that he found ship life congenial; in fact, it suited him to the core of his soul. He loved, and was deeply comforted by, the order and hierarchy of the wind ship. A place for everything and everything in its place—and everyone too: the captain on the weather side of the poop; the mate to leeward or in his "satrapy," forward on the fo'c's'le deck; each line on its pin; the routine of the watches; the same orders for each action each time; duty, faithfulness, fealty; "fellowship in the craft and mystery of the sea." Like the soldier, the seaman values the strict rules of brotherhood because of their profound utility; they will keep him alive when his job is at its most daunting and chaotic.

Conrad was an officer above all else, having worked before the mast for only a few of his twenty years at sea. His intelligence and ambition marked him as officer material, although he got to the poop deck through the hawse-hole, as a seaman, and not the cabin windows, as an apprentice. His experience and temperament, and his identification with Britain's class demarcations, gave him his characteristic perspective. He always looked forward from the poop, imposing his discipline, allocating the necessary tasks aboard ship. His sympathies were with the afterguard and its responsibilities, rather than the seaman and his fo'c's'le burdens.

As with the ship, so it should be ashore. In *Joseph Conrad and Charles Darwin*, Redmond O'Hanlon writes: "Conrad's quiet heroes . . . command the symbolic ship of society, guide their vessels safely through the worst that nature blindly can do either from within or without, the brute or the storm, and they carry out their life's work upon the surface of the sea, and sail home, as Marlow says, to 'touch their reward with clean hands.'"

Conrad was a seaman and a writer, like Melville and Dana, but he was a "sea brother" with different ideas of brotherhood. Melville read Dana and celebrated their affinity. When they came ashore to write, the two men looked at the ship and the sea—both as facts and as metaphors—from the point of view of the seaman, not the captain or mates. "When I go to sea," said Ishmael, "I go as a simple sailor, right before the mast, plumb down into the forecastle, aloft there to the royal mast-head." The two Americans wrote from the democracy of the fo'c's'le: "Let us all squeeze ourselves into each other; let us squeeze ourselves universally into the very milk and sperm of kindness," said Ishmael. Conrad couldn't bear this sort of propinquity; he couldn't even read Melville at all. He had no sympathy with ideas of universal brotherhood beyond the necessary and temporary bond of men at sea in hard circumstances. He couldn't understand a man who would "marry" Queequeg the cannibal. Perhaps it was a matter of writing after, rather than before, Darwin published his theory. Conrad abhorred the savage and the primitive and the possibility of sliding down—back down—of degenerating, into it. Melville's natural and cheerful acceptance of Queequeg as an equal was opposed by Conrad's horror: Kurtz, lost in the heart of darkness.

Dana made one long voyage and then stayed ashore to become a lawyer and bureaucrat. He spent years advocating the improvement of conditions for American seamen aboard ship. His interests and outlook were practical, and *Two Years before the Mast* is a straightforward chronicle of the life of a seaman (although it is notable for its author's literary skill and because it was the first voice from the fo'c's'le to find a wide audience). Because for Dana the sea was merely the sea, he also wrote about the monotony of long passages and the lack of intellectual stimulation among his shipmates. His voyage left its mark, though, as all voyages do; afterwards, he said: "I was a sailor ashore as well as on board."

Conrad and Melville might have differed in their point of view about shipboard society, but their own brotherhood was clear in how they felt about their time at sea. Melville, as Ishmael, wrote that "a whale ship was my Yale College and my Harvard." Conrad could say the same about his two decades as a sailor: "The sea was to be all my world, and the merchant service my only home for a long succession of years." Elsewhere he wrote: "I was, in heart, in mind, and, as it were, physically—a man exclusively of sea and ships; the sea the only world that counted, and the ships, the test of manliness, of temperament, of courage and fidelity—and of love." The sea was his first passion, and he gave himself up to it until, at thirty-six, he entered into his second, when he began to write his first words for publication. He owed everything, he said, to the "murmurs and scents of the infinite sea."

The day after the *Beara Head* left Madeira astern, near thirty degrees north latitude, twenty degrees west longitude, the trade winds fell light. The captain took advantage of the near calm to change to the fair-weather canvas, the old, supple and patched sails, good enough for the trades and the fickle winds of the doldrums and horse latitudes.

The *Beara Head* had carried its newer, heavy number-one sails longer than usual. The Biscay gale had given way abruptly to the Portuguese trade winds. In the day's calm afterwards, the big leftover seas made the operation too difficult. Now this uncharacteristic lull in the trades gave them their chance to take off the heavy canvas and store it below, where it would be protected from chafe (and, although they didn't know about it, the wearing effects of the sun's ultraviolet rays) until it was needed again in the southerly belt of westerlies. Then the crew would go through the entire procedure again. Every square-rigger on a north-south route did this. It was essential preventive maintenance for its engine.

Both watches were called up at four bells of the morning watch, or 6 A.M. They didn't wash down the decks—a daily task in fine weather—to keep them, and the canvas, dry. The old sails were hauled up from their locker and laid out, hard work because the bigger sails—the courses and topsails—each weighed the good part of a ton. The number-one canvas was even heavier; in taking it off, the crew began at the top of each mast and worked down. A gantline (a line rove, or threaded, through a single block) was taken aloft to the masthead. Cavers the Elf, nimblest of all, cast adrift the royal, the topmost sail, from its yard and bent on (attached) the gantline; seamen on the deck end lowered the sail down. And so on with the topgallants, topsails and courses, except that more and more men were needed for each progressively bigger and heavier sail. On deck, they rolled or folded the canvas and lugged it below to storage wherever the sailmaker wanted it—this was his bailiwick.

The light-weather canvas was sent up using the same gantlines. The sails were stretched along the yards: a few men for the royals and topgallants, an entire watch for the topsails and courses. The Finn, with fifty sail changes behind him, took the head earings (the attachment points for the top corners of each square sail) to the ends of the yard. He straddled its narrow end like a casual high-wire man, one leg hooked around the brace, the other swinging in space, a hundred or 150 feet up—it was all the same to him. On the lower yards, he sat well out over the sea, away from the ship's side. He stretched the sail out along the yard while all the other hands "lighted out" the head of the sail to him.

"Tie 'er up!"

The sail was tied to the iron-rod jackstay, which ran along the top of the yard, with robands (light pieces of line) passed through eyes, while the man at the end did the same with the earing, getting purchase on each turn with a marlin spike. The job had to be done well, although it wasn't as crucial for the light sails as it was for the heavy-weather canvas, since the latter had to bear a great deal of strain.

For Benjamin and the other green men, it was their education in bending on sails and in reeving and leading the running rigging, and they worked under the direction of the able seamen. Most of the time, however, Benjamin found himself on the business end of the gantlines, hauling until his arm muscles almost quit, and sweating the cumbersome sails here and there.

Next, the seamen passed the buntlines around the sail, except in the case of the courses whose buntlines were passed before they were hoisted. Then all sheets, braces, tacks and clewlines had to be attached to the clew (the lower corner) of each sail. The fore-and-aft sails—the jibs, the staysails and the spanker and topsail—were easier: heavy ones lowered down, light ones hoisted up and their lines reattached. Benjamin participated in all these tasks, his admiration for the wind ship's intricate ingenuity growing by the hour.

All this took one long, strenuous day.

The light trade winds lasted for the next several days. The *Beara Head* made no more than a hundred miles in each twenty-four hours, in spite of a helpful push from the southwest-setting Canary current, and on its twenty-sixth day at sea was only halfway between Madeira and the Cape Verdes. Liverpool lay just under twenty-three hundred miles astern in a straight line, but the barque had sailed at least fifteen hundred miles farther in its tacks, drifts and curves. This was a considerably slower passage than the average, the result of the ship's difficulties getting clear of the Irish Sea and its being embayed in Biscay. It could make up time: by a fast passage through the doldrums, for example, a quick rounding of Cape Stiff or an easy run up the Chilean coast (avoiding the calms near Valparaiso, which could stop a ship cold for a week, almost in sight of the desert shore). If two of these things happened, their passage time would be respectable; all three and the captain would have something to crow about as he made the rounds in Vallipo. As it turned out, Captain McMillan would have a different story to tell there.

The trades were behaving more like horse latitudes. The wind stayed northeasterly but light, except in the squalls that swept over the ship a dozen times a day. These kept the watches busy getting in royals and topgallants, and during a few long-lived blasts of forty or fifty knots, they hauled down the upper topsails as well, as the barque ran off to the southwest in ten- or twelve-knot bursts of speed. Heavy rain fell in most of the squalls, and men were told off to collect water in canvas awnings or running off the deck houses to fill the water casks, topping up the dubious Liverpool stuff with pure, sweet rainwater. In the more relaxed routines of the dogwatches, the seamen stripped down near naked—they could go no farther with the woman on board—and scrubbed the sweat and salt out of clothes, blankets and themselves in the cold rain. They latched the washports, and in some heavy squalls, the water mounted ankle-deep over the main deck. They splashed like children and soaped each other's backs. More than anything else, Benjamin wanted to clean his bedding to get rid of the lice and bedbugs that called it home. It was impossible to replace the straw, however, and unless he wanted to sleep on the bunk's bare board, he had no choice but to continue the unsymbiotic relationship for the rest of the passage.

Waterspouts came with the squalls. One day, they were so common that Benjamin, at the wheel, could see three of them at the same time. They formed and collapsed all day, two dozen of them, although none came closer than five miles or so. Most were fair-weather spouts, short-lived and more spectacular than dangerous.

Ships had to fear the bigger, slower-moving storm spouts—as powerful as land tornadoes—which form when columns of wind and water drop down from thunderstorm squalls or the leading edges of cold fronts. One of the biggest reliably seen, the Great Waterspout of

1896, formed over Vineyard Sound, six miles off Martha's Vineyard, Massachusetts. A few early photographers recorded it: a rotating funnel 145 feet thick and stretching almost 4,000 feet up, and a seething, boiling parent cloud hovering 16,000 feet above the sea. In 1898, a spout that formed off Eden, Australia, was described as being thirty times as high as a clipper ship. It was a good guess. Theodolite measurements put its actual height at almost a mile above sea level, although its diameter was only ten feet or so, making it unusually skinny; some funnels have measured as much as seven hundred feet across.

According to *The Ocean Almanac*, Lucretius described waterspouts with accuracy in the first century B.C.: "It happens at times that a kind of column lets down from the sky into the sea, around which the waters boil, stirred up by the heavy blast of the winds, and if any ships are caught in that tumult, they are tossed about and come into great peril." In the later, degraded thought of less rational centuries, writers believed the spouts to be sea dragons or great serpents, says the *Almanac*. Some Arabs regarded them as manifestations of *jinni*, powerful spirits that assumed a variety of forms. The sailor in his slow-moving ship could neither run nor hide. To scare the dragons away, according to Samuel Purchas, they would "take new swords and beat one against the other in a cross upon the prow." Or one crew member would kneel by the mainmast, holding a knife with a black handle and reading a passage from the gospel of St. John. At the right moment in the reading, he would turn towards the spout and slash the air with the knife to cut the spirit and let the water fall back into the sea. Other sailors tried to scare away the spout-spirits by sprinkling vinegar into the sea, beating drums and gongs, stamping on the deck. William Dampier, the explorer and privateer, signalled the transition to modern times when he tried the more blunt and practical method of peppering spouts with gunfire to try to break them up.

Lucretius was right: a spout that hit a ship tossed it about somewhat. *The Ocean Almanac* records that, in 1761, a tornadic spout off Charleston harbour plowed into a squadron of naval vessels, sinking five of them and dismasting the rest. The crack iron barque *Lillian Morris* was hit by a waterspout five hundred feet in diameter; it destroyed everything higher than the deck houses and blew a sailor overboard to drown. The nocturnal spouts were the worst; the seamen could hear them (a roar like a dozen line squalls getting closer), but they couldn't see them—denying the sailor's desperate need, as Conrad said, to see! Even steamers couldn't avoid these marauders. In 1923, a big, savage spout hit the White Star liner *Pittsburgh*, destroying a good part of the ship's superstructure. Waterspouts were like lightning strikes in the wind-ship seaman's spectrum of hazards: he could do nothing about either of them, trusting only to luck and the sea's amplitude to avoid their cataclysmic ambush.

One day, a tornadic spout formed off the *Beara Head*'s starboard side, about three miles off. Two dark-clouded rain squalls merged, and the sea beneath them boiled. The wind sucked sea water up into the air—the watching sailors could see it clearly. The green-black cloud spiralled down, tapering into a funnel until it met the sea's surface. The funnel drew up more and more water and began to writhe from side to side as its circumference grew. The cloud overhead became blacker and spread out like an angel-of-death mushroom. The watchers began to hear the roar of the thing, as loud as a big waterfall. The entire apocalyptic mass was bearing down on the near-becalmed barque, its slow, stately movement beginning to look like a drawn-out execution.

"Clew up royals and to'g'ants, Mr. MacNeill. And stand by your tops'l halyards," said the captain.

"Aye aye, sir!" replied the mate, his voice even louder and more energetic than usual.

"What in blazes is it, boys? What in hell's goin' on?" Grey shouted blindly from the bosun's locker, the only man not on deck. The captain's wife, under a sun parasol, stood by the cabin skylight, watching their nemesis approach.

The spout kept coming.

"Let go tops'ls! Clew up lower tops'ls and courses!"

The seamen, guts churning like the spout itself, had never worked faster. The sails were clewed up snug to the yards and furled. Benjamin thought the situation was absurd; he had feared he might die at sea, but in a fall from aloft in a gale or a foundering off the Horn, struggling hard for his life. Not like this, at the whim of this dark monster, sitting waiting for it like a slave the swinging axe.

I'm bound away to leave yer,
Goodbye, my love, goodbye.

Paddy sang a slow hauling song.

I never will deceive yer,
Goodbye, my love, goodbye.

The man could sing anytime at all, Benjamin thought. This sounded like a death song.

The air went completely still and became more humid. The spout trundled closer in a dead-straight line towards the still ship, the noise of wind and falling water growing louder. All hands kept by their stations, spread out along the decks by the clewline, buntline and halyard-pins, watching.

When it was within a third of a mile of the barque, the spout abruptly faltered, the funnel twisting off to one side and breaking apart. This changed the course of the overhanging parent cloud, and it began to swing off to one side of the ship. The funnel split in two, its spinning, airborne water falling back into the sea, now a real waterfall, with a sound like hurricane-force wind through the rigging. Then the cloud enveloped the *Beara Head*, and rain and returning sea water fell so heavily that the men had to shield their faces to breathe. The residual wind struck like an intense line squall over the quarter, and the ship heeled like a boat.

"Bear off!" the captain yelled to the helmsman, and he jumped to the wheel to help wrestle it over.

The barque ran off to the east, its jibs, staysails, spanker and the wind resistance of the tophamper driving it along as if it was running its easting down in the forties. In ten minutes, the wind slacked off, fell light and veered back to the northeast; the rain stopped. The helmsman brought the ship back onto its course, and the crew made all sail again, sweating but exuberant under a hot sun. In half an hour, there was no sign that the spout had ever existed. The men dispersed to their jobs. The captain put his arm around his wife's waist and guided her down the companionway steps. The mate took up his position on the leeward-side poop rail, where he could watch the softly pulling sails and the labouring men. Benjamin returned to serving a mainmast shroud. He had a headache.

The second mate saw the shark first off the stern. It was about ten feet long and had an unblemished, glossy dorsal fin that sliced neatly through the water, making its own small bow wave. No permission or orders were needed; they hated sharks beyond anything else in nature. Jagger bent a piece of salt horse to a big, iron hook attached to a link of chain and a heavy line. He trailed the bait over the stern and, together with three men from his watch, tailed on to the line and waited. The shark struck the lure and was hooked within a minute. The second mate and the three seamen hauled the fish close along the leeward side, everyone else crowding round. The mate stood by with his revolver in hand. For once, he and Jagger were brothers-in-arms. Jagger ran the line through a block and slipped a running bowline down around the shark's head and along its body so that it tightened round its tail. An entire watch hauled away and brought it over the rail and down onto the deck. They dragged it tight against the bulwark with the line to try to control its desperate thrashing and lunging. The captain's dog barked hysterically, snapped and drooled. The mate stepped forward and fired two bullets into the shark's brain. The animal's body continued to whip and flog for another minute, the men hanging on hard to its restraining line, then it twitched and stilled. The steward cut its throat, disembowelled it and cut off its tail. Later, he carved out the jawbone to be cleaned and mounted. The tail was hung up in the rigging as a warning for other sharks (until it began to stink too much to be endured even for that protective purpose). Then, following custom, half a dozen men hauled the remains up and over the rail and into the sea. They flushed its guts out through a washport with buckets of sea water and scrubbed the deck where the fish had been killed. The idea of shark soup occurred to nobody. The captain ordered a lookout to see if the carcass would attract more of its kind. None appeared, however, and in an hour, the seamen returned to work.

"The accursed shark," Melville called it. For the fish, not a trace of the sympathy he had for the leviathan mammal he admired even as he hunted it. The shark embodied the sea as an "everlasting *terra incognita*." It swam mostly hidden, a menacing killer beneath the blue beauty of the sea's surface—the subtlety of that deception! You could see it, too, in the "devilish brilliance and beauty" of sharks, those "most remorseless tribes." The shark and the sea both: "No mercy, no power but its own controls it." The shark aids man in his recollection of "that sense of the full awfulness of the sea which aboriginally belongs to it." When the land swallows people up forever, it is a natural, or preternatural, disaster, "yet not a modern sun ever sets, but in precisely the same manner the live sea swallows up ships and crews."

A brief variation on Melville's "Cetology," a *selachology:* Sharks are one of the oldest living forms of vertebrate life. They have no ribs, a skeleton of cartilage, separate gill slits and strong hinged jaws containing multiple rows of teeth that replace themselves as they are lost or worn down. A shark's skin is so abrasive that it scrapes off any human skin with which it comes into contact. There are at least 250 species of shark, maybe as many as 350. It is hydrodynamically efficient: its fins act as stabilizers, rudders and brakes; the powerful sweep from side to side of the caudal, or tail, fin and the laterally flexible body propel it. It can only move forwards. Sharks can smell one part of blood diluted in one hundred million parts of water, and they can detect tiny amounts from more than a quarter of a mile away. Two-thirds of their brain weight is used for smelling. They hunt down the trail of the blood like a land predator does its prey. The sharks that attack humans are the hammerhead, tiger, bull, blue, mako, oceanic whitetip and great white. A shark's eyes glow in the dark, like a cat's. The jaws of an eight-foot shark can exert a force of twenty tons per square inch. Even when they've been disembowelled, sharks keep swimming and attacking. The French name for shark is *requin*, says *The Ocean Almanac;* it's derived from the Latin *requiem,* because if a sailor falls into the sea among sharks, his comrades can only repeat for him the prayers for the dead.

When a whaler brought its kill alongside late at night, wrote Melville, the tired men waited until the next day before cutting in the whale. Sometimes, however, sharks gathered in their multitudes. The men tried to fend them off with whaling spades. They killed them by the scores and hundreds in bloody massacres. On such a night, said Melville, a man unaccustomed to the sight "would have almost thought the whole round sea was one huge cheese, and those sharks the maggots in it." The bloody, dying sharks snapped at each other's entrails, but they gobbled their own disembowelled guts as well, until they "seemed swallowed over and over again by the same mouth, to be oppositely voided by the gaping wound."

With the lighter trades and the proximity of the doldrums, more ships appeared around the *Beara Head*. Crewmen could see three within the arc of horizon visible from deck, and a man at the royal masthead could have seen a few more. Some were tracking south like the *Beara Head*, bound round the Horn or the Cape of Good Hope. For now, vessels for either destination followed the same path; they would diverge only after making farther down the South American shore. Others were on reciprocal courses, homeward bound to ports like London or Liverpool with bespoke general cargo, or, if they were carrying bulk, to Queenstown or Falmouth for orders—telling them where they should go to discharge their warehoused and oft-traded grain, nitrates or copper ore.

The watches had seen distant ships almost every day of the voyage, except during the gale, when their field of concentration contracted to the shape and direction of the big seas around them. Seven or eight hundred sailing vessels a year might round the Horn from east to west, and because of the common knowledge of wind patterns, they followed the same narrow routes. The modern sailor's experience of empty ocean for days or weeks on end was unknown to Benjamin. The shipping lanes were still sailing lanes, and the traffic was constant. Occasionally, vessels would pass close enough to signal their identity to each other. A homebound ship, when it arrived, would report the outbound vessel's position and continuing existence; it was the only news of the latter's progress until the owners received the captain's telegram from the ship's destination. Unless he sent word from somewhere else to report damage and inform the wincing owners of their evaporating profits—or unless no word came, the ship's lengthening silence a kind of news, final and sorrowful, but hardly unusual.

Two days south of the Cape Verdes, the *Beara Head* ran into a flotilla of northbound vessels: barques, ships, a brig or two and a long-haul three-masted schooner (a Yankee, by its profile). Everything but steamers; they had no business in these waters, among crews intent on the intricate workings of the wind. The homebound ships had piled up in the doldrums and had been released en masse by a breeze. They would soon scatter. "It is the calm that brings ships mysteriously to-gether," Conrad wrote; "it is your wind that is the great separator." He had once seen a hundred vessels becalmed within sight of his own near the Azores, each one pointing in a different direction, according to the vagaries of cat's-paws and stray currents. It was as if "each had medi-tated breaking out of the enchanted circle at a different point of the compass." A wind came, and the next day, Conrad could see only seven ships and a few more distant specks. The day after that, one or two.

The homebound ships, which were beating into the northeast trades on a reciprocal course to the *Beara Head*'s, came over the horizon top-down: the mastheads first, then the upper yards and sails appear-ing one by one, growing bigger, until the hull was visible, a speck un-der the tall spread of canvas.

"You see the way they come up on us?" said Benjamin to the Finn. "First the royals and to'g'ants, then on down to the hull. That's be-cause the horizon's a curve, and it's a curve because the bloody earth's a ball. What more do you need to see that?"

The Finn was not convinced. Gravity was the missing item in his stock of ideas.

"The ship don't sail uphill," he said stubbornly. "It slide back down. Farther back, it fall off the earth. And the seamen fall off the ship. Don't be ploddy stupid, paddy."

A few miles off to starboard, too far to signal, a barque missing its fore topgallant and royal masts passed by with a jury-rigged jigger on which was set a makeshift spanker and topsail, scars of Cape Stiff and the Southern Ocean. Later, a lean ship with the profile of a clipper passed close enough to hail. A seaman ran up the ensign, and the other vessel displayed one too. Captain McMillan used his speaking trumpet from the poop.

"What ship?"

"Ship *Slieve Roe* out of Frisco."

"Where are you bound?"

"Dundee."

"How many days out?"

"Seventy-two. What ship's that?"

"*Beara Head* from Liverpool."

"Where are you bound?"

"Valparaiso."

"How many days out?"

"Thirty-one."

The second mate held up a large blackboard on which he had chalked the *Beara Head*'s latitude and longitude. The *Slieve Roe* did the same. The latitudes corresponded to within a few minutes (one minute of latitude equalling one nautical mile), but the *Slieve Roe*'s longitude differed by thirty miles. Captain McMillan and the mate agreed that the other vessel's chronometer was wrong. It might not have had a chance to corroborate its longitude since rounding the Horn, or perhaps passing Staten Island, a good month ago, maybe not even then if the weather had been thick. It was a good-luck speaking for the homebound ship; its need to know its exact longitude would grow more urgent day by day as it passed by the landmark archipelagos to the north and on into the narrow waters of the Channel approaches. As the vessels sailed away from each other, boys on both ships dipped ensigns in farewell.

Thirty-one days out and an anniversary came due. "Dead-horse day" marked the end of the first month's advance, payable when the ship's articles were signed. Seamen hardly ever saw that cash, or else it passed through their hands and into those of the crimps with lightning speed (instead of to wives or sweethearts for whom it was intended). Sailors had the distinct impression that they were working their first month at sea for nothing. Dead-horse day celebrated the beginning of their real earnings on a passage. It's not clear how a dead horse became associated with unpaid labour—some obscure agricultural origin, perhaps; a case of shore usage transferring to the sea rather than the more usual reverse route. Or maybe it was because, in older days, ships took close to a month to get to the horse latitudes.

In the first dogwatch, the *Beara Head*'s crew formed a rowdy procession beginning under the fo'c's'le deck. The starboard watch had constructed the "horse," a canvas effigy stuffed with wood shavings, straw and a few holystones; it looked vaguely like a large dog. Paddy and two of the second mate's seamen dragged it along the deck, with Cavers the Elf perched on top, dressed in a bright red jacket and a sou'wester sewn into the shape of a jockey cap. The rest of the seamen and boys stamped along behind. They dragged the horse back to the break of the poop, from which the captain, his wife and the mates

watched, smiling. Even the mate, many of the men were surprised to
see, although it was a smile in character: sardonic, contemptuous. As
they marched, they sang the traditional dead-horse song:

> They say, "Ol' man, yer 'orse will die,"
> An' we say so an' we 'ope so!
> They say, "Ol' man, yer 'orse will die,"
> Oh poor ol' man!

There were seventeen verses.

The procession wound back to the foremast. The doctor conducted
a post-mortem.

"Are ya dead, ya dirty jughead?" the cook shouted at the lumpy
canvas.

He probed it several times with a long galley knife, then kicked it
viciously.

"You've swindled your last mattalow, ya bastard! Take that! And
more o' the same!"

He administered more probes and kicks, then straightened up.

"I pronounce you well and truly scuppered," he said solemnly.

Some of the men tied a sling around the officially dead horse and
hoisted it, with the Elf hanging on, up to the foremast main yard.
The Elf climbed up onto the yard and, as the seamen sang and
cheered, cut the line. The effigy dropped into the sea. The crew ran
aft to keep it in sight in the ship's wake, and they watched, still
singing and shouting, until it became waterlogged and the holystones
dragged it under.

They assembled at the break of the poop, a noisy, exuberant, shoving crowd. The steward had brought out a flask of rum, and he doled
out a good-sized tot to each man, all swigging it from the same pewter
cup. With its deck full of men giddy with the unaccustomed jolt of alcohol, the *Beara Head* became a happy ship itself for the duration of
the first dogwatch, and to their surprise, through the dusk and quick,
tropical darkness of the second as well. The captain and the mates held
to the poop, downing their own grog, and left the seamen alone. The
captain's wife stayed on deck too, sitting in a chair by the companionway, sipping a glass of port in the mild trade wind, gazing out at the
light of stars and phosphorescence.

On this moonless night, the ship's wake flashed and shimmered like an undulating wave of jewels. For a few hours, no work was done by anyone aboard the barque, except for the series of tipsy helmsmen. On the main deck, the men yarned, sang, played music—there was a tin whistle, a few harmonicas and the Greek's banjo mandolin—and sang shanties and forebitters. The men even danced a few "ram-reels" together, moving with precise courtliness. Still, it all had a desperate, rushed feel to it, like a party held by hard-labour prisoners on furlough who knew that their freedom was temporary, and that the worst of their sentence lay ahead.

On its thirty-third day at sea, the *Beara Head* lost the northeast trade winds; within an hour, the wind faltered, fell light and withered away to a breath that boxed the compass. Even the swell subsided, and in a few more hours, the stilled barque's only movement was a slight roll on the sea's gentle heave.

They had found the doldrums.

The Meteorologist:

> *A belt of low pressure occupying a position about midway between the high-pressure belts—the horse latitudes—found near 30 degrees to 35 degrees in both hemispheres. The tradewinds are the flow of air "downhill" from these high-pressure areas to the low-pressure doldrums. Except for slight diurnal changes (occurring four times every day), the atmospheric pressure along the equitorial trough is almost uniform. The wind is usually very light and is variable in direction. Hot and sultry days are common. The sky is often overcast, and heavy showers and thundershowers are frequent.*

The Poet:

> *Day after day, day after day,*
> *We stuck, nor breath, nor motion;*
> *As idle as a painted ship*
> *Upon a painted ocean.*

Another Poet:

> *How still,*
> *How strangely still,*
> *The water is today.*
> *It is not good*
> *For water*
> *To be so still that way.*

The Landsman:

> *No weather is ill*
> *If the wind be still.*

After a month of perpetual motion, the calm was eerie, unnatural. Benjamin was surprised at how quickly it irritated and spooked him. Only Urbanski, with his strange tolerance for this weather, or the lack of weather, was content. At the captain's word, the mates lost no time in running the watches up and down the deck, bracing yards and hauling on sheets and tacks, trying to coax a few cables of distance out of the wayward drafts. Here was where the ship had to break through fast to the southeast trades—a tantalizing hundred or so miles farther south—if it was to have any chance of a decent passage time. Sailors had to be patient always, but these calms taxed their perseverance. Even a contrary wind made the ship move; they could trim the sails and then stand down for a while as they waited for the next tack. In the doldrums, they worked constantly, sweated, drank up their water whack by mid-afternoon and went thirsty the rest of the endless, sticky day. And the ship made only a few miles, fifteen or twenty if they were lucky.

The problem for the square-rigger wasn't only no wind—it was also too little wind. To move its great dead weight, and the drag of the weed and barnacles beginning to foul the hull, it required wind velocity that was above the average in these waters. The breeze might waft at three or four knots, not enough for a big wind ship to ghost along, the way a smaller schooner could, for example.

If there wasn't enough wind to move a ship, the sailor wasn't completely powerless. He had superstition to fall back on—an honourable and sensible category of conviction for a man at sea—and action too. Superstitions, unlike beliefs or faith, require the adherent to do something: whistle, stick a knife in the mast, perform some sort of ceremony. "Superstition is a quality that seems indigenous to the ocean," said James Fenimore Cooper (famous for Hawkeye and *The Last of the Mohicans*, but also a notable sea writer in his time). The act of carrying out the ceremonies of superstition can have the most profound purpose: to soothe the fear of death and oblivion. Or it might have the more mundane goal of changing weather in the seaman's favour: to induce wind to blow or a waterspout to dissipate. In either case, the sailor, his life always in jeopardy and helplessly dependent on weather, has every reason to be superstitious. Or at least, he has every reason not to ignore superstition, to give it a chance. "When you come to the end of the line, bend on another one," say the old sailors. It may help; it can't hurt.

Whistling on board ship was a delicate matter. It teased wind out of a calm, especially if you stuck a knife in the mast and whistled softly to avoid offending the wind spirit. But heaven help the man who whistled when the wind was blowing already. That would bring a storm, even a hurricane, down on them all for their trouble. Spitting into the wind had the same result. The *Beara Head*'s iron masts were knife-proof, and the seamen were supposed to have blunt knives in any event. Forget the knife, then; they could still whistle—if anyone had the breath to do it with all the damned pulley-hauley. It was best if all the boys on board whistled together. No landlubber should be allowed to whistle on deck, however, which caused problems on clippers with jovial passengers or emigrants aboard. Likewise, no black cat could be allowed to romp on deck; they carried gales in their tails, and a blow was sure to follow. However, sailors could still throw the head of an old broom, or a coin, overboard; that might bring wind.

Some superstitions have fallen out of favour: for example, the French method of flogging the cabin boy with his lacerated back pointing in the direction they wanted the wind to come from. Finns were sometimes crucial to getting wind. They had weird powers and could tie the wind in a bag with three knots; the wind grew stronger as each knot was untied. Dana tells a story about a Finn. A ship had laboured into headwinds for two weeks after the captain had ridden down a Finnish seaman. The captain told the Finn that if he didn't lay off the headwind, he'd put him in irons in the forepeak. The seaman refused and was put in irons with no food. After a day and a half, he had had enough. He did something—no one knew what it was—that changed the wind, and the captain let him up.

The doldrums was also a place where men supposedly went mad. In the heat and ominous repose of its calms, Melville's sure wedding of meditation and water might become too much of a burden. The true nature of a man's isolation, and the sudden awareness of how much of his confinement in the small space of the ship still had to run, could turn his mind to obsession and delirium. Senses that had too little of their normal fill of things to look at and hear might compensate by creating the consoling opposite of what the sailor could see around him: soft, cool, green fields, for example, in place of the alien, burnished and opaque surface of the sea—Coleridge called it "that round objectless desert of waters." This was so frequent and specific a delusion that there is a name for it: calenture.

One old sailor described seeing one of its victims. In 1898, he was a port official who boarded a ship in the Thames that had just arrived from Australia with wool. The vessel had had a very slow passage, with days of flat calms. On deck, he saw a large, wooden cage that had been erected over the main hatch. There was a man inside who beckoned him over. One of the crew said to the official: "He is mad, but he won't hurt you." He explained that during one of the calms, the man had suffered the delusion that the sea was a lush green field, and he had jumped overboard for a run. The crew had saved him with great difficulty. The seaman continued to resist strenuously, and he tried to jump overboard again and again. The only way to restrain him, and to save his life, was to build a cage and imprison him within it. The official talked to the mad seaman, who seemed completely normal but worried about his future. What captain would sign him on after this?

No one has ever reported a case of calenture aboard a steamship. It must be a temporary madness that occurs only aboard sailing ships, with their days or weeks of stillness, sensory deprivation and forced contemplation.

The *Beara Head* drifted in the calm for four days, making no more than ten or fifteen miles a day, most of that in the brief wind that came with the squalls, each one following a pattern: a dark cloud bore down on the ship, and the wind hit first ("When the wind comes before the rain, you may make ready to sail again"). The watch ran the braces or sheets down the deck, hauling the sails round to catch the twenty or thirty knots that lasted for a few minutes, barely time enough for the helmsman to bring the ship around on course. The rain hit, a hard, cold shock in the heavy heat. Some men were told off to top up the water casks. The seamen stripped near naked to shiver and get the sweat washed off them. Within ten or fifteen minutes, the rain stopped, the wind died away and the barque lost steerageway once again.

On the fifth day, the ship went nowhere at all. The hands chipped and painted, slushed down the masts with the leftover galley grease, tended to rigging and lines, but all in slow, sweaty motion, a desultory going through the motions. The captain paced the poop; he could walk where he wanted because there was no weather side. His wife sat in her chair by the companionway under a small awning and read a book. Even the mate seemed to have lost the energy to hound the men.

On the second day of dead calm, the captain ordered the boats over the side. The *Beara Head* carried four: two on davits on either side of the steps at the break of the poop and two on skids on either side of the galley deck-house. They were the way to get ashore in an anchorage and the dubious means of saving the souls on board if the ship foundered, was run down or got driven ashore. On this day, they were put to yet another use as they took tow lines from the barque's bow and ranged ahead of the mother ship in a ragged line. Benjamin found himself in one of them with the mate at the tiller and five other oarsmen from his watch: the two Yankees, the Elf, the Finn and the Dutchman. The captain ordered even the idlers—the doctor, the carpenter, the sailmaker and the steward—to the oars. Only Grey, in irons, was excused the labour.

"Give way, now, you scum of the earth," the mate ordered.

But he spoke mildly and ironically, almost, Benjamin thought, with solicitude for them and their ordeal ahead. Benjamin had never before had to pull hard on an oar for hours at a time, but he'd done enough of it to know what to expect.

It was seven bells in the forenoon watch (11:30 A.M.), and the sun was high. A glare of moist heat poured down on them, deepening the tan of Caucasian skin, reddening even more the pale Irish. The men in their thirties and forties all showed the blotches of benign skin cancer, spreading on hands, arms, faces and torsos. They started to row. It took three or four minutes of steady pulling before the *Beara Head* began to move. With all boat crews working steadily, the ship got up to a knot or so; that appeared to be its top speed under this muscle power, enough to throw up only a suggestion of a "white ash breeze" (the oars were made of that wood). Within the circle of the visible horizon lay a dozen other vessels, not many compared with some doldrums gatherings. They were too far away for the men to see whether any of them had boats in the water too.

They rowed without a break for an hour, then each boat crew in turn took five minutes to rest and drink a little water while the other three kept pulling. It was easier to keep the barque going than to let everyone rest together and have to get it moving all over again. In this way, the oarsmen worked through the high, hot sun of the afternoon watch and the cooling air of the first dogwatch. The mate ordered all boats alongside for food. They ate cold pea soup, bitter coffee sweetened with molasses, salt horse—trying not to inhale at the same time they bit into the rank meat—and hard tack. They broke up the biscuit pantiles and knocked each piece sharply on a handy hard object to shake out the squirming weevils. The captain doubled the water whack for as long as they worked like galley slaves, and they were able to drink what they needed.

After half an hour's rest, the mate ordered them into the boats again. They rowed on, through the second hour of the second dog-watch, all four hours of the evening watch and into the cooler, silky darkness of the gravy-eye, the men at the boat's tillers steering by stars low down on the hazy horizon. The still darkness was disconcerting; it muffled everything like a vast, murky bowl that had been clamped down over the ship and its close surround of sea. It deprived the senses of their accustomed stimuli—there were no points of reference for the eye—and the men felt disoriented, off balance, as if they had drifted into some strange otherworld whose odd laws of matter and motion only daylight would interpret for them. The barque was a huge shadow behind them. They felt exposed in the little boats, half an inch of pine between them and the bottom, two thousand fathoms down. It was almost like swimming above an abyss. Anything could happen: a hole might open up in the sea; perhaps they would row into an unknown veiled island just ahead; one of the great sea dragons that old sailors feared, a giant squid, a toothed sperm whale with a long memory and a grudge against small boats, or God knew what unknown beast, might rear up out of the dark, and smash them and devour the men. Overhead, stars glimmered. The oarsmen could see Polaris low down to the north, and to the south, behind them, the Southern Cross. Occasionally, lightning forked and flickered around half their horizon, but none of the squalls it presaged came close.

Even for tough, lean, horny-handed seamen, it was hard labour. Their daily food consumption barely sustained them for normal duties, let alone this steady, aerobic work. Their energy ebbed, the force of the oars gradually declined, the boats and the mother ship slowed down. They had rowed for thirteen hours. At four bells of the gravy-eye (2 A.M.): "Way enough, Mr. MacNeill!" the captain bellowed from the poop. "Hoist the boats up! That'll do all hands!"

He knew he had got all he could out of these men for now. He'd rest them, then feed them and turn them to in the boats again sometime in the forenoon watch. He would wait only as long as it took to refuel their muscles; then he would start them rowing again. And he would keep doing that until they rowed the ship right out of this goddamn black, windless, son of a bitch of a hole.

"I'll be buggered if I can row another fuckin' cable," said Urbanski as he flopped onto his bunk. "This is more roughing it than I signed on for."

"I thought you like this up-and-down wind, Yankee," said Kapellas.

"Go to hell, ya goddamn dago!"

"Belay it, boys. Sleep in now," said Russell.

As Benjamin crawled into his bunk, consumed with muscle pain and cramps, he realized that he couldn't straighten his fingers or thumbs. His hands had frozen into a compact curve that fit perfectly the circumference of the oar. During the hours he had spent aloft in the Biscay gale, fisting the topsails with bloody hands, he had thought nothing on the voyage would be more difficult. He had been wrong; rowing for many hours at a time with two dozen other men to tow a

five-thousand-ton ship through the sea at one knot was more difficult.
This was only the first of the times he would be wrong.

As it turned out, the captain didn't have to wring more production
out of his weary human machinery. All hands were called out at eight
bells (8 A.M.) for a breakfast of coffee and hard tack (broken, tapped
and shaken). While they were eating, spread around the main deck on
hatch covers or leaning on the rail, they felt the first breaths of moving
air. It came and went, but it was always from the same direction: south
or a little east of south. The southerly winds had found them at last.

No one had a chance to finish eating. The captain got them hop-
ping at the first cat's-paws. The boats, which had been slung out over
the water, ready for lowering, were brought back in and secured. The
apprentices went aloft to overhaul the buntlines. The seamen braced
the yards round and trimmed all sails until the barque was ghosting
along on the starboard tack in the fast-rising southerly breeze. They
headed east-southeast on a course that, if they stuck to it, would land
them near the mouth of the Congo River in a few weeks. This was the
necessary diversion prescribed by *Ocean Passages*. The wind would
soon track round to the southeast as the trade winds consolidated. The
ship would turn onto the port tack and get a good slant on the trades,
allowing it to keep to the east of the islands of Fernando de Noronha
(190 miles off the Brazilian coast) and the South American shore itself
(at Cabo Branco and farther south). At the southern limit of the trade
winds, said the sailing directions, the wind would draw almost due
east. The barque would have a beam-on soldiers' wind down the coast.

The liberated, reanimated barque sailed full and by on the starboard tack at five knots. Even that modest pace seemed swift after
the sluggish inchings through the doldrums. The *Beara Head*'s crew
savoured their ship's renewed headway before the kind trade winds.

The first few weeks aboard a square-rigger for green hands, shanghaied landsmen and first-voyage apprentices were a confusing nightmare. Men and boys were hazed and ridiculed; the mates shouted at
them in novel and unsettling barrages of insults and scorn. They were
made to do very dangerous things, and as Ishmael noted with genial
resignation, they got thumped about a bit as well. Their old-salt shipmates sent them to find the galley downhaul or the sky hook, to the

bilge to polish the main rivet and to the sextant to ask for a prayer, and they had felt the mate's anger and his rope-end in the process. It was a merchant-marine boot camp on the fly, like sending new recruits into battle and teaching them under enemy fire how to be soldiers. It had felt like that to Benjamin from his forced climb to the mainmast truck within an hour of stepping aboard the *Beara Head*, to the frantic running up and down the deck during the long beat down the Irish Sea (hauling here and pulling there with only the vaguest idea of what he was doing), to the icy, bowel-voiding fear of furling topgallants and topsails in the Biscay gale. This had been his rough initiation, but it was merely part of the utilitarian need of the ship to have certain things done quickly. There had been no rite or ceremony. That came later in the voyage, when the ship passed south across the equator.

Crossing the line was an occasion of separation and induction. It marked the passage into a new hemisphere—the estrangement from home and familiar waters was complete. Zero degrees latitude was a kind of point of no return. The physical version of that point came much earlier, probably when the ship cleared the Channel, certainly after it had weathered Biscay and Finisterre (the so-called end of land). But the equator was the psychological point that emphasized the ship's isolation and jeopardy. It had made it this far, but the worst lay ahead. Give thanks for the sea's indulgence and plead for more in the months of sea-time remaining. Crossing the line meant all these things, and so it also signified that the green hands and the pressed men and the boys in the half-deck had achieved a nautical milestone of sorts: if they were here, so far out and away from the world, thirty or forty or fifty days at sea, they had become sailors. It was a good time to recognize that. For the seamen, however, there were two reasons for the ritual: it was a good excuse for a bit of a hooley, just to break the monotony of the passage and shipmates already too well known, and it was another occasion on which the captain was honour-bound to administer more booze to the fo'c's'le.

This literal rite of passage, and initiation, is an old one. It used to be held much closer to home. Early Mediterranean mariners went through a ceremony when they passed one of the landmarks that meant they had truly left home and were out on the perilous sea—most notably, when their fragile little ships passed through the Pillars of Hercules from the enclosed middle sea into the endless northern ocean. It remained a religious ceremony: the seamen prayed, ate fish, tossed coins overboard as a maritime gift for the gods. The main character in the ritual was King Neptune, and it's tempting, and probably accurate, to find the origins of the rite in classical Greece and Rome in sacrifices, or other obeisance, to the sea-god Poseidon, or Neptune. Later, farther-ranging European sailors celebrated when they crossed the Tropic of Cancer and, as voyages became longer, the equator. The crossing-the-line ceremony, as we know it now, began with French ships in the seventeenth century, when they passed from the northern to southern hemispheres or the reverse, and there are English accounts of it by the beginning of the eighteenth century.

It might have been a popular event for sailors before the mast, and for passengers on liners or emigrant ships, but captains and mates viewed the ceremony a little differently. It could get rough: initiates' heads and faces were shaved, not gently, with dull, rusty iron hoops; brushes soaked in Stockholm tar and white lead were shoved into their mouths; they were dunked, manhandled and insulted. Anyone who resisted got even worse treatment. That was all right for new men, but officers were subject to it as well. A new second or third mate, for example, on his first trans-hemispheric passage, was subject to King Neptune's uncouth attentions. An officer could certainly stand on his rank and refuse (although he would suffer for that in the long run), or bribe his way out of it with a bottle of alcohol, as could passengers, if they were aboard. On some ships, the captain himself paid in strong drink to have the ceremony cancelled, or at least toned down to a more genteel celebration without the physical hazing. That was part of the tradition as well, another aspect of sea-cunning in the sailor's constant search for ways of cadging booze from the afterguard or passengers. In the old days, before merchant ships became effectively teetotal (except for special occasions), seamen might extort gallons of liquor from squeamish passengers and first-time men. On Cook's 1768 voyage, the crew listed everyone who was crossing the line for the first time, including animals, and demanded brandy and wine from the gentlemen (or else they would be shaved and ducked). On troop transports to India, British soldiers had to pay a gallon of rum a man to avoid the ordeal. Almost all paid, and the seamen gargled with booze for weeks. (Those were days of more relaxed discipline on merchant ships.) Darwin was initiated aboard the *Beagle*. The doctor shaved and ducked him, but he didn't stuff the shaving brush in the young scientist's mouth because he was a gentleman. Darwin recalled the ceremony as being disagreeable.

For a captain, the crossing-the-line ceremony was particularly threatening: while King Neptune (his part played by any of the seamen) was on board, the sea-god obviously outranked the captain. It was a temporary reversal of power in the ship's hierarchy. For a few hours, the fo'c's'le could lord it—with just the right amount of careful humour—over the poop. Most captains tolerated this, but it was too much for some, and by the middle of the nineteenth century, the practice had reached its lowest ebb. Heaving to while the festivities went on wasted time, and waiting for drunken sailors to sober up so they could work the ship again wasted even more. The habit of discipline was ingrained enough that the captain's simple refusal to have King Neptune on board settled the matter, although there was often resentment amongst old hands who had gone through the ritual and didn't like the idea of new hands avoiding it. Dana was pleased that he missed initiation when he crossed the line in 1834, and he wrote that "this ancient custom is now seldom allowed, unless there are passengers on board." However, the ceremony revived towards the end of the century, perhaps as some compensation for the greater hardships for small crews handling big ships, and to create a safety valve to release some of the resentment at tighter discipline. The initiation fostered cooperation and harmony amongst the seamen, and that was also in the interests of captains. When they had fewer men than they needed for virtually every aspect of the ship's operations, the more teamwork the better.

The new men aboard the *Beara Head* got the full treatment. The Elf played a diminutive King Neptune, with a wooden crown made by the carpenter and something resembling a cloak. Paddy was his queen, Amphitrite, with a hideous yellow-painted, mophead wig and two iron cups strapped on as breasts. The burly Irishman loomed over the Elf, looking like his crudely transvestite bodyguard rather than a consort. Russell and the Dutchman, bare-chested but wearing skirts of old canvas, acted as their attendants. The neat doctor, in his usual clean apron, was an unlikely villainous surgeon and barber.

Neptune emerged from the fo'c's'le and walked aft with his entourage, one of his attendants blowing hard on an old metal horn that was supposed to be a conch, and that produced two alternating, wavering, off-key notes. The king demanded to know what ship this was. The captain, bare-headed and respectful, in a teeth-gritting way, came down the poop steps to the main deck and gave the ship's name, hailing port and destination. The captain's wife stood on the poop, watching with scandalized fascination; the mate stood beside her with his sardonic twist of a smile. The second mate was below the poop with the other initiates; he had decided not to evade the humiliation, but he looked apprehensive and unhappy. All would normally have been nude, but they were wearing trousers, long, rolled up or cut-off—more deference to the woman aboard. It had caused a lot of grumbling about hen frigates. First they hadn't been able to go arse-naked in the sweet, cleansing doldrums rain, and now they had to run athwart the old tradition.

The veteran hands clustered round with buckets of water and other ominous, unidentifiable substances. One of the king's attendants—Russell—read from a scroll. He exhorted Neptune's subjects to admire the beauty of the king's hairy-chested queen. He read out the three rules that would see a man handsomely through life: never heave anything to windward except hot water, ashes, and piss once you'd rounded the Horn; never drink small beer when you can get strong; never roger the maid when you can roger the mistress. The captain looked grim, and his wife blushed deeply. Russell commanded all those who hadn't yet crossed the line to appear before him; if they passed their initiation, the king would give them certificates so that they wouldn't have to go through it again.

"Let the ceremony begin," hollered the Elf-king.

When it was Benjamin's turn, two seamen grabbed him and pushed him forward. They dumped buckets of water over him. They held him while the doctor shaved his head and straggly beard, the rough barrel hoop cutting him with every stroke. One seaman daubed a mixture of tar, paint and linseed oil over his torso and finished by shoving the brush into his mouth. The stuff was awful; it made Benjamin gag, and he almost vomited. They finished him off with more buckets of water, whacked him with an oar and drove him to join the other miserable initiates by the mizzen mast.

Jagger, predictably, got the worst of it all: more cuts—his head streamed blood—more of the tar mixture and a double brush in the mouth. The buckets of water somehow thumped him as they were emptied, and the oar struck hard; he was bruised and bloody by the time he reached the sanctuary of the mizzen. He looked close to bursting into tears. The four apprentices were treated more gently; the doctor took some care shaving them, and they were spared the tarbrush in their mouths. The sprightly boys were the only initiates who appeared to find any fun in the business.

"Youse are now clean of the dirt of the north," yelled the Elf-king. "I accept you into my kingdom!"

His attendant, the Dutchman, handed sheets of paper to the tarry, near-retching graduates. Each certificate repeated exactly, in childlike letters, the king's words. Benjamin and the others took some time to try to clean themselves up, although it would take days to scrape off all the oil, tar and paint.

The steward distributed cups of grog, several to each man this time. Everyone was soon tipsy or outright drunk. The captain and his wife went discreetly below, as did the mate, while the carousing, singing men shouted and danced and told loud stories about past ceremonies that were real initiations, by heaven, not like this modern, off-the-wind shite. A few seamen lamented, with drunken solemnity, that the supposedly all-powerful king could not order the captain's dog flung over the side, gutted, garrotted or otherwise disposed of.

Benjamin had been genuinely frightened by it all—there had been bullying and sadism in the air—and his cuts and bruises wouldn't disappear for weeks. He knew that he had gone through a kind of certification that separated him even more from landsmen, and from narrow-waters or North Atlantic sailors too, but it all felt premature; he had been designated a real sailor too soon. It seemed presumptuous with Cape Stiff still to come, as if all this might jinx him and the other green hands when the Horn got its hooks into them. That was where he would become a true sailor. The only question in his mind was, A live or a dead one?

❋

"Eight bells, you sleepers! Show a leg, there! Tumble out! D'ye hear the news?"

"Go to hell!"

"I swear those bastards are sounding the bells early."

"Goddamn!"

"Who'd sell a farm?"

"When I was on the *Golden Fleece*, the bloody mate'd do it himself. Never a change of watch without the bastard kicking every man he could reach. The best one was *Clan Ferguson*—d'ye remember her? A pretty little barque. Happy ship. The doctor had hot coffee standing by for each watch. Didn't matter what she was doing on deck, he always had the coffee ready and we got time to drink it. Good stuff, too."

"Aw, shut the hell up, will ya? D'ya see any bloody coffee here?"

"Turn out, ya goddamn Dutchman! I never saw a man harder to unmoor."

"Oh, I met with Napper Tandy / And he took me by the hand, / And he said, 'Boy, how's old Ireland, / And how does she stand?' / 'Well, she's the most distressful country / That ever man has seen . . .'"

"Pipe down, Paddy. How in hell can you sing before sun-up?"

"Let's go, boys. Someone drag that fuckin' Dutchman out, for Jesus' sake."

The port watch made its way aft. The *Beara Head* was forty-eight days out, nearing twelve degrees south latitude, three hundred miles off the port of Salvador on the Brazilian coast, making nine knots on the port tack, the southeast trades blowing steady and strong. The tired starboard watch waited grumpily for their tired replacements to muster with them at the break of the poop. Every seaman and boy aboard, except the idlers, still asleep at 4 A.M., mingled momentarily.

"Relieve the helm and lookout!" ordered the mate, his voice a rasp.

Benjamin walked forward to the fo'c's'le deck; Urbanski went up the poop steps.

"About bloody time! One-nine-five. Sail to starboard about twenty minutes ago. Maybe two, three miles off."

"One-nine-five. All mine."

The relieved helmsman came down to the main deck.

"That'll do the watch," said the second mate wearily.

"Don't sink the ship, you lubbers!"

"Good night, ladies!"

"Now you can walk off your pride-o'-the-morning-light there, boys."

The starboard watch went below. The port watch distributed themselves along the weather side of the main deck, leaning, yawning, stretching, strolling up and down. The two apprentices, with their special dispensations, curled up under the poop steps and fell asleep.

The mate stood by the wheel for a few minutes, staring at the compass, before taking up position by the jigger weather shrouds. He peered up at the sails. Never could trust that fool of a second mate to set things up all Bristol. He could see aloft clearly on this night of near-full moon. Check the set of sheets and tacks. Did some of those buntlines need overhauling? He yawned and looked away to starboard; ship out there somewhere, heading south too.

On the fo'c's'le, Benjamin took a quick scan around the moonlit sea. Then he leaned against the foremast fife rail, took hold of a belaying-pin with each hand, closed his eyes and immediately dozed off; he'd mastered one old sailor's trick at least—how to sleep standing up, like a horse.

Seven bells, and after three and a half hours of sleep, less the time taken to piss, smoke or talk, the starboard watch was called up on deck. In the fine trade-wind weather, the crew ate there and watched the sun climbing higher, the sea blueing.

"Burgoo! What's the occasion, Doctor?"

"No occasion, boys. Just the results of a little game of chance with that goddamn crook of a steward. I won some oatmeal. Weevily as hell, but they're cooked now; you'll never notice them. Gives it more body. There's some molasses too."

The heat of hard tack, porridge and coffee mingling with the quickening warmth of the sun revived the seamen. It made the coming hours of work a bearable prospect.

"All hands aft now, d'ye hear! Look alive!" The mate wasting no time.

The captain was on deck too, checking the course, staring up at the sails and round at the golden blue horizon. He was sleeping well these days, storing it up against the wakefulness of the westerlies and the Horn.

The men gathered under the poop in an unaccustomed mood of friendship and well-being.

"Now that was decent grub. If we had that every sun-up, we'd be better men for it."

"That doctor's one of the best I've seen, and I've seen fifty of the bastards. He's got a heart in him. This don't feel at all like a hungry-gutted limejuicer right at the moment."

"Mr. Jagger, take your gangs for'ad and finish off those rigging screws and the capstan, if you please."

"Aye aye, Mr. MacNeill."

The mates treated each other with courteous correctness these days. Jagger knew that he couldn't afford another run-in with the mate; the captain wouldn't stand for it. And the bloody Bluenose might kill him next time; he didn't know his strength. As for MacNeill, the captain was right about him. The Nova Scotian's habit of obedience would keep him civil and in line. Nevertheless, the mate had promised himself that if the sonnywhacks put a hand on him again, he'd lay out the pipsqueak cold as a kipper, no matter what the captain said, and take the consequences. No man—he didn't give a damn who he was, or where—laid a hand on the MacNeill.

"My watch, carry on chipping. Urbanski, take the goddamn green micks aloft and get started on scraping the mizzen royal and to'ga'nt mast and yards. See the carpenter when you need paint. Check those fairleads while you're about it too."

"Aye aye, sir."

"You boys, lay aloft and overhaul buntlines, all of 'em. And be sure you leave enough slack."

The apprentices ran up the ratlines, eager for the freedom of the yards for half an hour. The Yankee and the two green micks, Anderson and Benjamin, climbed up more slowly with various scrapers. It was pleasant enough work, no fear aloft on the swinging yards in this gentle weather. And like the boys, they were happy to be clear of the deck, and the mate, for a while. The Yankee worked on the royal mast and Benjamin and Anderson straddled the wooden yard, scratching at the old paint, which came off in flakes that fell and then fluttered off to leeward in the warm wind.

Now at ease aloft, at least on days like this, Benjamin looked around at the trade-wind scene and the horizon, which was forty miles away from this high vantage point. He could see two ships that were invisible from deck. It was easy to imagine the great jungled continent itself just beyond. He found himself longing for the sight of land, and not just any land: America, his new world. He breathed deeply, the air fragrant with salt and the ship's tar and paint. He could never imagine going back home after the experience of this space and light. The narrow, cold, rainy streets of Carrick; the damp little houses crowded together in back-to-back rows; the cold, cramped minds of the people who lived in them; peelers round every corner; British soldiers squatting in their barracks, ready to sally out at the next eruption of trouble. He thought of the sailor's perennial exclamation when things were rough: Who'd sell a farm and go to sea? Well, he had his own version: Who'd go back to Ireland once he'd gone to sea?

Perched on this high yard, the tropic sun like a fire on his shoulders, the line and its ceremony behind him (he still looked like a Prussian recruit, and his bruises hurt like hell), the real adventure (the Horn and the West Coast) ahead, these deep-sea sailors already showing him respect and friendship, Benjamin had the sudden thought, like a bubble bursting, that this was a one-way passage. He might sail back to home waters eventually, but only because the ship was bound there; from now on, Queenstown and Liverpool and London were just ports he might call at. And maybe he'd stay out in this wider world anyway, aboard ship or ashore, the East, America. He was becoming a seaman; he had just become an emigrant as well.

"Friend Ben, I'm thinkin' we should jump this old iron wagon in Frisco and runthegither. It'd be good crack. We've had enough of that ramguntchagh mate already. We could go north, mebbe Canada. There's sealing up there—a pile of money."

"Ye know, I was just thinkin' on the sod myself. I'm thinkin' I'm damned if I'll go back to Carrick, or even to the bloody country itself. Sure why would you ever want to when you've seen all this?" Benjamin gestured out towards the surrounding world of blue sea and sky, white cloud and wave crests, tropical sun climbing fast, rush of trade wind through the rigging, crowding white sails bellied out into firm, smooth curves, the narrow hull below pushing aside a halo of white-water as it moved through the sea.

"But jumping ship? I don't know. We signed the articles, so we did."

"Bollocks to the articles! Sure there's a crowd of dozened eejits in every port ready to sign onto a homebound ship. We owe them nothin'."

"Mebbe. We'll see."

They scraped down the mizzen royal and topgallant mast and yards until eight bells of the forenoon watch. There was no need for sail trimming, not even a tweak, in the steady trades. The seamen were free for maintenance. And the mate was beginning his preparations for the stormy westerlies and the Horn itself, still thirty-five hundred miles away, but making its imminence felt.

"Aloft there! Lay below for some grub!"

Benjamin, Anderson and the Yankee climbed down to a midday meal.

"By heaven, this old horse stinks more every goddamn day," said Russell as he tried to chew the salt meat without actually smelling it. "Doctor! Time to broach a new cask."

"Not till this one's all down your holds. The old man's a reg'lar tar-tar about the grub," said the cook.

"On Greek ships, we get olives and wine every skoff'm," said Kapellas. "First chop. Not this dead horse every time."

"Ah, belay your jaw!" said Urbanski. "I'm sick of all your palaver about goddamn dago ships. How do you separate the men from the boys in the Greek navy? With fuckin' crowbars."

"There's fore-and-afters everywhere," said Kapellas reasonably. "Maybe even aboard here."

"Maybe," said Urbanski, looking hard at the Greek.

Kapellas stared back.

"I can't wait to drop anchor off the West Coast," said Paddy. "Them Vallipo gals know how to get you up all night. When I was on the *Cape Comerin*—Do you remember her, Jimmy? She went missing in '83, Glasgow to Buenos Aires—*pampero*, maybe. Anyway, we waited three months for cargo in Vallipo. The old man let us ashore every second day. Most of the boys jumped, but I stayed on. Sure why not? She was a pretty ship. I got fed, and I had a high old time with the *pisco* and the gals. It was both sheets aft and stuns'ls set every bloody day. Had to go into drydock a few times—them goddamn dagos are quick with the knife—but it was a good run, I'll tell you."

"It's a good Sailortown," said Russell. "Better than Cally-o. There's nothing there except fuckin' dust and Indian gals. You've got to chip the goddamn dust off the royal yards when you get out to sea. Soogie-moogie the decks for a week."

"How do you think it'll be this year at the Horn?" asked Maguire. "It's winter down there, ain't it?"

"Ja, dead winter," said the Finn. "Bad place for little green roast-beefs."

"Horn very bad place anytime," said the Dutchman in one of his rare bursts of loquacity.

"Aye, it'll be winter right amidships," said Russell. "The worst fuckin' time. I reckon we'll see both sides of the Rammerees this trip. We'll rough it for a month, I'd bet."

"A month to get around?" asked Benjamin. His anxiety showed.

Russell laughed.

"A fuckin' month if you're lucky there, young paddy."

Benjamin heard the yelling and the wind at almost the same moment. The voices were familiar: one of the men on deck calling up the sleeping port watch to shorten sail. Benjamin woke up at the first few words; their tone and timbre sounded more urgent than usual. As he struggled out of deep sleep, he recognized notes of desperation that knotted up his gut in a second. Something was going to clobber the ship. The last words of the summons disappeared in a sudden blast of wind. There was no preliminary growl, growing and building the way a normal squall sounded when it bore down on them. This wind hit like a roaring wall, without a peep of warning; Benjamin felt as if he had been stone deaf and suddenly regained hearing in the middle of a hurricane. It was as loud . . . no, louder than the steam locomotives from Belfast to Carrick that used to overtake him as he walked the nearby path, overwhelming him with their unimaginable speed and noise—the sense of inexorable power and the vibrations they produced in his chest almost made him cry—and leaving sparks and stinking smoke spurling in their wake. The sound, and power, of this wind was like that. Then he was flung out of his bunk and brought up hard against the iron of the deck-house bulkhead, his right arm and head burning with pain. He was aware of liquid running down his forehead and into his eyes; he wondered where the water was coming from. Someone's body was tangled up in his. He could hear shouting around him, but only faintly in the tumult of the wind. He hauled himself onto his feet. Nothing made sense. He couldn't see anything in the dark fo'c's'le, but that was normal when the watch was called up. This time, he couldn't seem to figure out where the door was or which way he should go to get out on deck. Water was coming from somewhere, flooding into the deck house, and in five seconds, it rose over his ankles and kept rising. Then, suddenly, he understood everything: the stuff running into his eyes was blood; he was standing on the bulkhead, and the deck was almost vertical; water was pouring in through the door, flooding the deck house; the ship was on its side, and it wasn't coming up again; they were off the River. *Pampero!*

The River was the Rio de la Plata, the River Plate, between Uruguay and Argentina. It is one of those places that deep-sea sailors were obliged to pass by, and whose prickliness earned it the distinction of a single, simple name, like the Cape (of Good Hope), the Horn and the Bay (of Biscay). They are places where seamen suffered and died in their greatest numbers. The Plata, with its wide estuary and shallow water stretching a hundred miles offshore, isn't in the same category as the others—it doesn't have their reliable and prolific deadliness. But the River can hit hard sometimes.

The *pampero* is one of those homely local winds that blow in their season—products of chance combinations of mountains, desert, sea—big and singular events with obscure, multifold origins. The *harmattan*—dry, hot, coming out of Africa into the Atlantic from December to February—peels off human skin; the *sirocco*—from northern Africa over Italy—creates languor and fatigue; the *kamsin*—a hot, dry Egyptian desert wind—blows from March to May. All the exotic names: *chinook, Etesian, mistral, soland, monsoon.*

The *pampero* is a dry wind that blows from June to October, either from the northwest or southwest from the Andes, across the pampas (hence its name) and out to sea. It brings rain, hail and cold, and lasts two or three days, sometimes longer. The shorter it lasts, the more violent its wind. Over water, it can retain its strength and even augment its fierce force. Ships 360 miles at sea have been blasted by these winds of the mountains and plains; the *Beara Head* was 120 miles offshore when the *pampero* ran it down.

Benjamin found the deck-house door by following the stream of water rushing through it. When he got to the opening, however, there was nowhere to go. If he stepped out, he would slide down the steep slope of the deck and into the sea. Instead, using rivet heads and angle irons as footholds, he scrambled blindly up the front of the house and onto its side. Perched there, he looked around him and, not for the first time in his short bluewater career, believed that he was about to die. Not that he *might* die, or that it was possible or likely, but that death was imminent and certain, merely a matter of a very short time. He could see that the barque was lying over on its port side at a seventy-five- or eighty-degree angle. Seas were breaking over the windward side; water covered almost half the deck below his refuge. The yard ends appeared to be in the water, and there were no sails left set that he could see, just strips of shredded canvas snapping in the wind that screamed over the prone ship at hurricane force or more, pinning it down. Or maybe, he thought, panicky now, the coal had shifted, and the barque would never come back up again. Beside him in the dark, two or three other shapes were clinging on with the same silent immobility.

Elsewhere on the ship, men were making a lot of noise. Suddenly, Benjamin became aware of the mate's voice. Now it came into its own; even this mayhem of wind and breaking seas couldn't muffle the man's bellow. It was the most comforting sound Benjamin had ever heard. The mate was shouting for hands to lay forward and get some headsails on her, and for Russell to go aft, "if you ain't drowned," and lend a hand to the helmsman, if he was still aboard. This had two immediate effects on Benjamin: as abruptly as he had thought that he was about to die, he became convinced that he wouldn't, or at least not this night; and right away and without hesitation, he shinned up a line and onto the ship's rail, now a makeshift deck, and, hanging on to shrouds and running rigging, made his way forward.

There were other sources of comfort, like aural beacons of hope flaring up in the dark surrounding tumult: Benjamin heard Russell's acknowledgment. Thank God, he was alive; this was when they needed a seaman like him. From behind him, he heard the captain shouting for axes, for hands to stand by to cut away royal and topgallant shrouds; seamen cursing, shouting, not in fear but in rage and indignation; the mate again ordering the jibs hoisted, "fast, pronto, chop-chop, my goddamn lovely lads, like the devil's up y'r arse!" In the midst of this unimaginable disaster, Benjamin thought, men were working to put things right; the mate's voice was urgent yet calm, not at all as if the ship was near foundered, all of them on a fast, slippery path to Fiddler's Green.

There were voices all over the ship:

"Dear God preserve us; saints preserve us. God save us!" the captain's wife prayed.

"Turn me loose! Goddamn it, cut me loose, boys! Don't let me go down in here!" The cockney in irons.

"Willie, when they hoist the headsails, up the helm hard as we can! Get the bow to pay off." Russell to the helmsman, mouth up against his ear.

"Johnston, lay for'ard with these men to the main and mizzen. If she don't pay off, on my order: chop! Chop for your bloody lives! Just royals and to'g'ants, mind!" The captain to the carpenter.

"Mr. Jagger! Lay for'ard and look at the hatches. If they're breached, call the carpenter and do what you can for 'em." The captain to the second mate.

"Haul away, my boys! Heave the sonofabitch up there! Two men aft to the sheets! Now the inner jib—heave and wake the dead! Go, boys! Don't let this bastard win! This is a livin' gale we've got here!" The mate, mighty-voiced on the fo'c's'le deck.

"Heave, boys, that's it. Up she goes now. That'll do her! Hang on now, or you're gone altogether!" Paddy on the bowsprit. "Ben, you're here, lad. Now, tail on and haul away, that's my beauty!"

Somehow—in the turmoil of lines and breaking seas, the wind not easing a knot, hanging on to the topsy-turvy ship, the land more than a hundred miles away to windward, the boats useless, if the ship goes, they all go—the seamen hoisted three jibs and sheeted them home. Not a man unbloodied—Benjamin got off lightly compared with the blood he sees flowing and coagulating on the men around him.

The mate begins shouting aft three hundred feet over the wind: "Ahoy the poop! Up your helm there! Bring her round! Now! Now!"

Faintly from astern: "Helm up, aye!"

Nothing happens. The barque lies on its side like a derelict, dead or dying. For three, four, five minutes, the twenty-nine men, four boys and one woman on the *Beara Head* wait for the half-exposed rudder to bite as each wave submerges it. Maybe it can overcome the tenacious inertia of the ship's long, half-capsized hull; maybe not. If the ship can't be turned, then the only recourse is to cut away the upper masts on main, mizzen and probably foremast too. Relieving the vessel of their windage should bring it upright again, the prescribed remedy for this otherwise terminal affliction. Then their passage would be over temporarily; they'd head for Buenos Aires and refit. It would be weeks, maybe a month, before they were able to sail again—cargo delayed; other freights forfeited; the captain a man who damn near lost his first command because he didn't see, or hear, a *pampero* coming, even though he was dead off the pitch of the bloody River.

Dana had been less than two months before the mast on the brig *Pilgrim* when his captain demonstrated the way to handle one of these catastrophic winds. Look out for lightning to the southwest, he told his mates and lookouts. If you see it, get sail in, pronto. Dana himself was the one who saw the telltale flashes on the lee bow. It was dead calm, but the crew furled most of their sails "and awaited the attack." Even in the dark, they could see a vast black cloud moving towards them. "It came upon us at once with a blast, and a shower of hail and rain, which almost took our breath from us." The crew was able to let the halyards and sheets run, and the brig turned and ran off before the gale. Soon, they were able to resume their course under reduced storm canvas.

Eight minutes, nine, ten—the *Beara Head* stayed right where it was.

"Hoist fore topm'st stays'l! Get it up there, boys; that's our chance!"

The seamen hoist a fourth jib. It goes up quickly; there are a dozen men to do the job, instead of the usual two, because raising this foresail is the only thing left to do before they chop. Again, they wait: fourteen, fifteen minutes. This is too long; the captain is waiting too long to cut away the masts. He's endangering the ship to save his name, for all the good that'll do him when she turns turtle or the hatches give way. (What force of shifted coal lies against their covers?) The mate's obsessive driving may have saved them. Just a few days ago, two weeks before many ramrods would have got around to it, he ordered the hatches secured and reinforced for the passage around the Horn. Without that extra timber and canvas, they'd be long gone by now. But the old man has to cut away the masts. Now.

The captain's fear for his reputation, his patience and courage, paralysis—everything that combined to keep him still and silent on the poop—paid off. The wind lulled a little for the first time since it struck, just enough for the straining jibs and sometime biting rudder to pivot the bow. As if it had been popped out of a furrow, the barque began to bear off to leeward. As it did so, it came gradually more upright until, when it was almost running off before the wind and seas, it had righted to near twenty-five degrees or so. But that was all; it went no farther. The coal had shifted.

Nevertheless, the crew cheered like men paroled; after what they had just endured, shifted cargo was a mere detail. The ship was in hand again. The crew mustered at the break of the poop, and to everyone's astonishment, they were all there, bloody and bruised; somehow, everyone had managed to stay attached to the ship. The captain, smiling in triumph on the poop, was surprised enough at this universal survival to order the mates to check again. Three men had to go aft, where the captain gave them a double tot of booze before he patched them up. He sewed a flap of skin the size of a hard tack pantile back onto the Elf's thigh. Two seamen from the starboard watch, on deck when the *pampero* hit, needed a broken arm set and cracked ribs bound; they would be light-duty invalids for weeks, leaving their watch two men short partway round the Horn at least. The captain didn't give a damn; his body was suffused with adrenaline—he had saved his ship. With his iron, goddamn nerve, he had kept her going, by God! He was as cheerful as a Plymouth moll when the fleet came in, and he whistled as he stitched up and parcelled his seamen.

Under the mate's direction, the crew clawed down two of the jibs as the wounded *Beara Head* rushed clumsily downwind at ten knots, listing and yawing as if it had developed a spastic illness. They hauled up a jigger staysail and set a main topsail to balance the sail fore and aft. Then they wore ship round onto the port tack and hove to facing the wind, which had eased off to below sixty knots. The seas were growing meaner, but there was a limit to their size; the fetch was no more than 140 miles, not enough distance to build up big breakers. The pressure of wind on the port side acted like a counterweight to the list, pushing the ship more upright. All this took two hours, but they were secure enough.

This *pampero* was one of the short ripsnorter varieties, and so blew itself out quickly. The wind began to ease off within two watches, and twenty hours after its ambush, the wind had dropped to thirty knots. During the time hove to, the seamen slept and ate, and nursed their gashes, abrasions, bruises and sprains. They needed to rest for what lay ahead.

First, they got new sail on the ship. The wind had blown away all the old, fair-weather canvas, bent on when the barque had entered the tropics. They would soon have changed back to the number-one storm canvas in preparation for the westerlies and the Horn, but the *pampero* had done the job for them very efficiently, in less than a minute. Lucky too: if the best sails had been set, their greater strength would have resisted the wind longer, putting more strain on masts and yards, almost certainly bringing some of them down. The old light sails had sacrificed themselves, giving graceful way to the wind's force, like muslin or silk, and giving the spars a chance. Now the crew hauled out the heavy-weather canvas from its lockers and went through the procedure once again of hoisting up the tons of stiff, bulky sails and bending them on. That took one long day, and it was made more difficult by the sharp wind still blowing.

With the ship jogging along under reduced sail before a lightening northeast breeze, the mate ordered the hatches broached. The reinforcing timber and canvas was stripped away and the hatch covers wrestled off. Then the seamen, idlers and boys of the *Beara Head* went below to shovel coal from one side of the ship's hold to the other, flinging it uphill over the stowing boards, twenty pounds at a time, bracing what they had restowed and shovelling again. They worked for two days—stumbling on the uneven, pebbly surface; sinking into it sometimes, as if it was soft snow; sweating and filthy; torn cloth or kerchiefs over noses and mouths to reduce their intake of choking dust; breaking only for food and four hours' sleep; in the end, working like automatons. At four bells of the first dogwatch on their sixty-second day at sea, they finished the job, and the ship sailed upright for the first time in four days.

Things returned to their even tenor, except for one bad sign. The coal was warm and gave off a paraffin smell. The farther down they dug, the warmer it got; it wasn't hot, but it was too warm, too smelly. And their work had stirred it up, introduced fresh oxygen deep down into the hold.

"Blow me if I want to see the sight of another fuckin' shovel in this life," said Urbanski.

"Where you goin', Yankee, you be shovellin' down there too," said Kapellas.

"Aye, well, we'll all be shovelling the goddamn load again before long," said Russell.

No one disagreed with him.

Three days later, a thin trickle of white smoke began to seep out from under the main hatch cover. It wasn't a surprise. The captain and mates had noted the gradual rise in the coal's temperature during their routine thermometer checks as the ship had made its way through the tropics. They had agreed, however, that the readings weren't high enough to induce them to put into port—Rio or Buenos Aires—to unload, cool down, dry out and restow. That was the last thing the owners wanted; their thin profits disappeared the moment the barque turned for land. On the other hand, it wouldn't do them any good, either, if their money went up in smoke. It was a hard decision: coal behaved in unpredictable ways. It might continue to heat up, but even more slowly in the cold air of the south, and they could easily make Valparaiso to unload a warm but intact cargo. Or they might have a full-blown case of spontaneous combustion within a day or so. They had bet on a slow heating, but two days of retrimming, with its resultant flow of fresh air into the holds, had changed the equation.

The Falkland Islands weren't far away; the *Beara Head* could make it into Port Stanley in two or three days. It was a good port, often a refuge for Cape Horners. Ships rounding the cape from west to east at the end of a long Southern Ocean passage from Australia sometimes needed resuscitation. Vessels fighting to round Cape Stiff from east to west—the *Beara Head*'s route—might have to break off the engagement and run east to the Falklands for repairs before they went back to war again. But the captain was damned if he'd run for land now—not after staring down that killer *pampero* and not losing a mast or a man in the process; not after making a hundred miles' southing even while that fo'c's'le rabble (still, they'd been stout of heart when the ship needed them—they were mostly British seamen, after all) had heaved half a thousand tons of coal up to windward. The barque had made more than five hundred miles since then, and it had picked up the westerlies. The captain was already twenty days behind where he wanted to be at this point in the passage; he'd had a suicide, two assaults and two men from the starboard watch wounded, and the Horn was still ahead. But he had kept the damn ship going, kept the sea, and apart from a nearly full suit of old canvas, the ship had suffered no expensive damage. It was getting hard enough to make a living in sail these days. Only the captains with the best records would keep their commands or get new ones, or get decent ships at all. He couldn't afford to put in now.

No surrender! He would keep going. He would batten everything down, seal the ventilation pipes and starve the little red bastard glowing and growing there below, snuff it out. You couldn't do this with a wooden ship; it would burn like a barn full of hay once the coal reached the critical temperature. That was one of the advantages of an iron vessel: it could be battened down until it shone like a red star at night, but you could keep it sailing. He had heard of ships with cargoes burning for weeks making port with their sides red-hot, deck plates buckled and paint blistered off, even the paint on the lower masts bubbling and peeling.

Everyone knew the story of the *Ada Iredale*. Back in '76, her coal cargo began to burn and her crew abandoned her about two thousand miles east of the Marquesas in the South Pacific. The ship drifted west in the equatorial current for eight months until a French cruiser took her under tow and brought her into Papeete, in Tahiti. The coal was *still* burning, but the damned ship was intact. The crew had jumped too soon. Her captain should have stayed with her, got her to where she was supposed to go, done his duty and made the owners their profit. The fire wasn't extinguished until sometime in '78. The *Ada Iredale* was repaired and rerigged by some Yankees from Frisco, and it sailed for years afterwards. That was a lesson for every captain.

He didn't admit it to himself, but Captain McMillan knew the story of another ship as well, and it taught a different lesson. Everyone had heard of the *Norval* because its crew ended up making an epic small-boat voyage to save their skins. In early 1882, the iron ship neared San Francisco with eighteen hundred tons of coal from Hull, in northern England. The crew noticed a thin stream of smoke oozing out of the fore hatch. The captain and mate took a look below: the coal was cool enough, but there was a smell of gas. The next day, when they opened the main hatch to check on the coal there, they found that the metal stanchions in the hold were hot and more gas was coming up from deep down in the cargo. They began to pour water into the hold, and they kept the pumps going. The fire grew worse, however, and two days later, the coal exploded. The blast blew the heavy hatches off and up as high as the topsail yards, destroying several sails in the process. Two of the *Norval*'s four boats were lowered and laid off the windward side of the ship with three seamen in each. That night, there was a series of explosions that blew off every scuttle and ventilator, bulged the deck like a huge iron bubble, and half demolished the poop and aft cabin. Many seamen were injured by pieces of wood and glass. The next day, the crew lowered the remaining two boats, manned all four and pulled away from the ship, now burning and exploding from stem to stern. They were two thousand miles east-southeast of the Sandwich Islands (Hawaii). That night, the mate's boat, with seven men aboard, disappeared forever. The other three boats, well provisioned and handled with skill, made refuge in the islands twenty days later. At the time, it was one of the longest small-boat voyages in history. (Bligh's famous run across the Pacific after the mutiny on the *Bounty* covered twenty-two hundred miles in sixty-two days.)

Coal cargoes were very dangerous; they had to be watched and monitored with care. Even so, ships often went missing. In 1894–95, fifteen British coal carriers disappeared. That number doesn't include the ships that caught fire and were lost, but whose crews, or part of them, survived. These fifteen just sailed into silence and oblivion. Some probably went down while rounding the Horn—that was part of the normal attrition of ships, no matter what they carried. However, there's no doubt that some of them burned, their damp coal heating up, glowing, breaking into flames, exploding perhaps; the men trying to contain the fire and keep the ship afloat in the storms around Cape Stiff, desperately fighting on two fronts; the outcome a foregone conclusion. Some of the names: the *Colintrave*, Australia to San Francisco; the *Seafield*, from the Clyde to Montevideo; the *Sierra Madrona*, from Liverpool to Rangoon; the *Afon Cefui*, from Swansea to San Francisco; the *Iron Duke*, from Blyth to Iquique; and so on, thirty to forty men per ship.

Many vessels caught fire but made port by luck and the skill and grit of their crews. In 1893, the coal cargo aboard the four-masted barque *Cedarbank*, on passage from New South Wales, Australia, to San Francisco, began to heat up and then burn. For thirty-six days, the crew fought a desperate battle: they jettisoned burning coal; poured water into the hold and pumped it out again; endured poisonous gas, flames and a series of explosions; prepared to abandon ship and then came back aboard to fight on when the weather turned bad. The vessel arrived in San Francisco with the wooden overlay on her steel deck burned away and the hull paint blistered, but it was intact. The crew even managed to conceal the fire to avoid a salvage claim from the tug that towed the ship into San Francisco.

The four-masted steel barque *Pyrenees* was carrying grain, not coal, from Tacoma, Washington, for the British port of Leith in 1900, but the wet cargo was almost as liable to spontaneous combustion. When the vessel crossed the equator, the crew discovered that the grain was on fire. The captain made for Pitcairn Island, hoping to beach the ship there. They arrived, decks steaming like a locomotive, but discovered that there was no suitable place to run ashore. However, the island's mayor, McCoy, one of the descendants of the *Bounty* mutineers, came aboard to pilot the burning ship to a sandy beach in the lagoon at Manga Reva, about 290 miles away in the Tuamotu Archipelago. Even for modern yachts, this is a supremely difficult sea area, with strong currents, unmarked coral heads and reefs, and low atolls that can't be seen until a vessel is almost on top of them. (It became a reasonable proposition to sail amongst these islands and reefs only with the advent of satellite navigation devices.) McCoy was a crack pilot, and the captain of the *Pyrenees* conned the 2,200-ton, 284-foot-long ship as if it was a handy fifty-footer. Together, they got the barque, by then glowing red-hot, through the lagoon pass and onto the beach. The crew later made it to Tahiti on a French schooner. (Jack London based his novel *The Mutiny of the Elsinore* on one of the later experiences of the *Pyrenees*, which was eventually salvaged and sailed on: a mutiny by her crew in 1913 off the Delaware capes on passage from San Francisco to New York.)

Conrad had his coal fire as well. It was just one of the run-of-the-mill experiences of twenty years under sail; he downplayed them all. His life at sea, he said, had not been "adventurous in itself." Yet Conrad went through a collision, a foundering, an injury from a falling spar, a fire and a survival voyage in a small boat. He certainly went through a collection of Southern Ocean gales in a dozen roundings of the Cape and two of the Horn. Any of them could have done him in. All this within the ten years he was actually working aboard ships—as opposed to looking for work. These events certainly add up to a life that any twentieth-century person, even a sailor, would consider adventurous, if not hazardous, in the extreme. But Conrad was right to brush it all off. An itinerant square-rigger seaman or officer could expect to meet all of this, and more, in a working lifetime. They considered themselves lucky if they were alive twenty years on.

Conrad's story *Youth* is a literary transmutation of his own experience aboard the cranky old barque *Palestine*, bound for Bangkok with a load of Tyneside coal. Near the vessel's destination, the cargo began to burn, and it eventually exploded, sinking the ship. The crew took to the boats, including one commanded by the barque's young second mate, Korzeniowski, and made land safely on Muntok Island, in the Bangka Strait, off Sumatra.

In *Youth*, Conrad barely camouflaged the *Palestine* as the *Judea*. The scorched and bloodied crew fights to the end to save the ship: "That crew of Liverpool hard cases had in them the right stuff. . . . It is the sea that gives it—the vastness, the loneliness surrounding their dark stolid souls." The captain refuses rescue by a passing steamer; the *Judea* still floats, still lives. "We must see the last of the ship," he tells the other captain. None of the crew objects; on the contrary, they all consider it the proper thing to do, even though it condemns them to a long and precarious trek to safety in the ship's boats. In that final ordeal, there is more than salvation. It's the young second mate's initiation; parched and burned, pulling at the oars for days with the seamen, he proves himself. "I did not know how good a man I was till then."

In real life, Conrad's passage on the *Palestine* was his second voyage to the Far East. On board the *Judea*, it is his first. His novels and stories about the East and "its secret places" are infused with the murmurings and intimations of what he chose to create as his first contact with it: from the floorboards of a battered open boat as it drifts into a bay at night. A puff of aromatic wind is "the first sigh of the East on my face. That I can never forget. It was impalpable and enslaving, like a charm, like a whispered promise of mysterious delight."

The *Beara Head*'s fire could not be contained. Captain McMillan's little red bastard wasn't deterred by oxygen deprivation. It burned hotter. The smoke sifting out from around the main- and fore-hatch tarpaulins flowed thicker and darker. The iron deck's wooden over-planking became as hot as if it was under a doldrums sun. Now it was clear that the fire could not be smothered. The alternative was to open the hatches, dig down to the fire's heart and pour water into it. The sudden inrush of air might cause a flare-up and make things worse, perhaps unmanageable. But there was nothing else to do—except make for the Falklands, which were only a day away, maybe two in the fading northwesterly wind. The captain had not changed his mind; he would run for land only when all else had failed, and the *Beara Head* was still far from that point.

They opened the hatches. The heat became more intense, but there were no open flames. Thermometer probes showed that there were two hot spots, one near each of the main and fore hatches.

"Dig with both feet, Mr. MacNeill," the captain ordered. "Get it out of there!"

"Rig gantl'n's over each hatch, Mr. Jagger," shouted the mate instantly. He liked the captain's decision. He—MacNeill—wouldn't turn tail for the Falklands either; he would try anything to quell this fire before running off. The stupid bastard should have seen the *pampero* coming. He—MacNeill—certainly would have if he'd been on watch, but that had been the captain's only mistake. Now he was doing the right thing: working these scum to death before giving up. They should all be dead men walking before the ship made into any port except Valparaiso.

"Five men from each watch, lay below and dig! The rest of you worthless sons of bitches, pile it up and pour water on it! And not one, single lump of goddamn coal goes over the side, d'ye hear me? I'll skin alive the man who loses a fuckin' lump!"

The seamen jumped as always, but for the first time in sixty-six days at sea, Benjamin was aware of hesitation, of some men chewing the fat, growling at this dangerous labour. Only the officers knew the ship's position, but the seamen could infer that if the *pampero* had hit near the River, and they had found the westerlies, and it was getting colder day by day, and they had made some good daily runs, then they had to be getting close to the Falklands and a British port. If the wind got up and caught them with open hatches and coal all over the deck, they reckoned they'd be in shite up to their arses.

Growl you may, but go you must! The routine of compliance held them fast, and they began shovelling and hauling. Things were serious, Benjamin realized, when he saw the cockney, going down the fore hatch with a shovel. The captain had released him from irons, proof that the ship needed every hand for its survival.

As so often happened, the shantyman provided the only form of comfort, albeit mild enough, in the face of arbitrary authority. Paddy, tailing onto one of the gantlines, sang:

> *Come all ye bully sailor men, an' listen to me song.*
> *Oh, I hope ye just will listen, till I tell yiz what went wrong,*
> *Take my advice, don't drink strong rum, nor go sleepin'*
>   *wid' a whore,*
> *But just git spliced, that's my advice, and go ter sea no*
>   *more!*

The song rolled on in its interminable variations: verses about Shanghai Brown and an unintended five-year whaling voyage; the sailor's helpless inability to avoid waking up in a squalid bed beside Mary-Ann or Angeline or "a gay young Frisco gal."

As they dug down into the coal, it got hotter. Soon, each digging crew could last only ten minutes in the heat and fumes until they were hauled up on deck, filthy, dizzy and sweating, and replaced by the next gang (barely recovered from its own stint below). The captain gave the men permission to wrap precious scraps of sail canvas around their sea boots to protect against the heat that scorched and softened the rubber. Near the end of each short shift in the hold, the men had to hop from foot to foot to relieve the pain, even though their feet had been hardened by decks that cooked in tropical sun. After ten hours, each gang below could stand it for no more than five minutes at a time. The water that was poured down the hatch to cool things off produced clouds of steam that made the men sweat more and turned the air almost chewable, so thick with humid dust that they could hardly force it through their lungs. One of the apprentices fainted and had to be hauled out and laid on deck for half an hour. He was fifteen years old and malnourished, but he was ordered back down with his gang when he could stand up again. The only men on board not digging and hauling were the captain and the mate, the men injured by the *pampero* and the man whose happy turn it was to take the wheel; even the old sailmaker, Page, laboured stiffly with a shovel, spreading coal down the deck. The captain's dog got in everyone's way, skittering about, snapping at the seamen's ankles and chasing the coal as if it had been flung for his amusement.

Once committed to the work, the seamen went hard at it. With its gaping hatches, the barque was a sitting duck. Their survival now depended on their getting the coal out, the fire snuffed, the coal back in and the hatches covered as fast as possible. And so they dug with furious determination according to the seaman's eternal logic: to save yourself, save the ship. For hours, Paddy, who was given some leeway in his work in compensation, sang the wild stamp-and-go shanties with their runaway choruses. The songs especially helped the men at the gantlines, hauling up the heavy baskets from below, a longer pull by the hour as the pile in the hold shrank down and muscles tired.

> *Wuz ye ever in Timbuctoo,*
> *Where the gals are black an' blue,*

An' they waggle their bustles too,
Way hay an' away we go,
Donkey riding, donkey riding . . .

Wuz ye ever off Cape Horn,
Where the weather's niver warm,
When ye wish to hell ye'd niver been born?
Riding on a donkey . . .

And "Roll the Cotton Down," and "Sally Brown," and "Tommy's on the Tops'l Yard," and "Randy Dandy O!," and "Blow the Man Down," and "Reuben Ranzo." It was a rueful history in song of the deep-sea sailor's feckless cycle: hardship at sea; booze and women ashore; back to sea again, until the ship foundered or stranded or his body gave out. The solution, a purely theoretical one, is always the same: when the ship docks, walk inland right away. Ignore everyone. Keep walking—like Odysseus with his oar—then get spliced to a good woman and stay where you are until the sea is a memory.

Even so, the songs always made the hair stand up on Benjamin's neck and produced a thrilling shiver over his skin, especially the wild sing-outs of the choruses. They served their purpose: they made him feel equal to the sea, for a while.

At eight bells of the evening watch (midnight), the exhausted men were allowed to rest. They had worked for sixteen hours, with a few ten-minute breaks for food. They got water as they needed it—the captain recognized that he had to keep his human machinery hydrated and in working order. The seamen lay down—some below, some on deck—and slept the deep, immobile sleep of babies. The ship looked much as it had two months ago, when they hauled out of the Bramley-Moore dock: a foul scum of coal dust over everything from deck to the tops. The hatches were closed to squeeze down the fire's supply of enabling air.

At four bells in the morning watch (6 A.M.), the mate hollered and kicked the crew awake. They ate salt pork—from a fresh cask the captain ordered broached even though the old one wasn't finished—hard tack, strong coffee and more burgoo, another gift from the doctor from his store of gambling oatmeal. Then they went back to work.

On the second day, the apprentice who had fainted did so again and, pale and vomiting, was excused from more work for the rest of the day. The cockney, weak from the immobility of a month in irons and reduced rations, had a seizure. Without preamble, he collapsed onto the coal, foaming and twitching, and fell unconscious. Two men carried him to the aft cabin, where the captain's wife, anxious but glad for something to do, bathed his head, and that of the sick apprentice, with cold fresh water; later, over the steward's scandalized objections, she fed them soup from her husband's own supply. After six hours, Grey said he felt better and went forward, where the mate put him to work pouring water over the heaped coal. It lined the deck, port and starboard, in a long pile three or four feet high, maybe three hundred tons of it, a tenth of their cargo.

The wind, from the northwest, held steady and light, a "capful." It was luck beyond anything they could have expected. The ship was close to fifty degrees south and only 350 miles from the Horn through the Strait of Le Maire. In these latitudes at this time of year, Maury's pilot charts predicted a mean wind velocity of twenty-five knots, with gales three days out of five.

The *Beara Head*'s crew worked through the second choking, blistering day of fighting the fire. By the end of the second dogwatch, with a cold rain falling but the wind still—blessedly and inexplicably—light, the men digging under the main hatch met coal that burst into open flames when the shovels turned it over. They had found the fire's heart. In spite of the mate's orders, some of the burning cargo had to be thrown overboard. When the men had extinguished the main hot spot, they joined the other gang working through the fore hatch. The source of this separate fire was lower down, and by midnight, they still hadn't uncovered it. They closed the hatches for a second night. The rain had stopped, but everyone slept below. It was freezing on deck in the near winter of these high latitudes, and it felt even colder to enervated men who had spent many hours shovelling coal that was close to 200°F.

At four bells of the morning watch, they began their third day of fire-fighting. They opened the main hatch cautiously, in case the fire had revived; it was still dead, the coal cooler than the day before. Working under the fore hatch was easier; there was enough room for only four men to work efficiently, and that allowed for more frequent replacements and more men to work the deck. The captain decided to start restowing the main-hatch coal. He didn't like the look of the weather.

By eight bells of the forenoon watch (noon), they had found and knocked down the second fire. The captain and MacNeill (not Jagger, who had shovelled with the men and was exhausted) made a series of careful thermometer probes. All clear. After a pause for salt horse, biscuit and coffee, the crew began to shovel the coal on deck—by now more than five hundred tons of it—back into the hold. Paddy's voice had given out the day before; no one had the energy to sing, or to talk.

The captain's intuition was right: the well-behaved wind showed signs of delinquency. A swell made up from the south; the overcast became lower and darker. They would have to shovel continuously until all the coal was restowed or risk losing some of it overboard. The mate drove the dirty, weary crew while the captain watched the seas growing—a little more wind every hour, the men beginning to stagger with the motion as they worked.

Just before dark, the captain ordered the barque hove to; the decks became more stable, and seas no longer broke over the rail. By the light of all the oil lamps aboard, the crew shovelled on until, past midnight, the deck was clear of everything except a salty, sticky black residue. It was a close-run thing. Even hove to, the ship was beginning to take heavy spray and the occasional breaking sea aboard. The mate drove the seamen in his watch—now they *were* dead men walking; he had used them up as they were intended to be used—to cover the two hatches and batten them down with extra canvas and timber: their Cape Stiff rig. Meanwhile, the starboard watch got the barque turned away onto a beam reach in the now westerly wind, clewed up and furled the royals, braced the yards and trimmed the fore-and-aft sails. The *Beara Head* raced south at ten knots. Finally, at six bells (3 A.M.):

"That'll do the watch."

The mate sent his own men below. In an hour, they would turn out to relieve the seamen on deck.

They turned out sooner than that, a scant half-hour later.

Benjamin and his mates tottered out on deck to clew up and furl the upper topgallants and to lower the topgallant staysails. The men were filthy and exhausted. They had been too tired to sluice off the third day's coal dust; it had not been practicable in the dark anyway. The wind was rising steadily, and it had the feel of a wind that would grow stronger—that insistent heft and inexorable augmentation that any sailor recognizes. The crew of the *Beara Head* knew that they had had extraordinary luck; against all odds, the weather had held almost to the hour they got the fire out and the hatch covers back on. But they were too tired to feel they had won any kind of victory. Indeed, all of them were sailors enough now, even the former green hands and the boys, to ever think that. The sea had let them by this time, that was all. It was as if the Southern Ocean had paused to give them a fair start: Deal with the extraneous fire, boys! All clear? Ready? Now here we go!

The fire had also muffled the menacing feel of the Horn drawing closer. The crew had not even noted their crossing of fifty degrees latitude. The number of days a ship took to "round the Horn" was counted from fifty south in the Atlantic to fifty south in the Pacific. Traversing the line signalled that the hazards and contingencies of being at sea were about to intensify. The Horn would distil the normal flow of things aboard ship into its hardest, purest essence.

Both watches stayed on deck past the dawn's reluctant accretion of light through the low overcast. A cold rain began to fall, washing clean the coal-smeared surfaces of the ship. It seemed to purify the vessel for the coming combat. The seamen clewed up and furled the lower topgallants, crojack and main course, and then they took in the flying and outer jibs and spanker topsail. The captain was shortening sail earlier than usual. He had no doubt that this wind would increase; it had all the looks of a snorter. He feared that his men, who had shovelled and hauled coal for almost fifty-five hours over the past three days, might be near the end of their endurance. That wasn't what he wanted with the passage through the Strait of Le Maire coming within the next twenty-four hours—in dirty weather at that—and the Horn in the near offing. But there it was. That was the hand he'd been dealt. He was a driver, but he had to conserve these men. Maybe they'd get round into the Pacific in a couple of weeks; but if it took much longer, he would have to give his crew some leeway. He did so now, ordering sail taken in before the wind piped up strong, reducing the seamen's labour as they hauled on clew- and buntlines and went aloft to furl. After they were finished, he mustered all hands at the break of the poop and, for the third time since clearing Liverpool, issued grog, a double ration, to his strained, gaunt seamen and idlers (and a single small one to his boys).

The *Beara Head* aimed for the Strait of Le Maire like an arrow homing in on its target. The gap between Staten Island and the eastern extremity of Tierra del Fuego was about twelve miles across—not much safe water for a deep-sea wind ship. In a rising gale, and with not much hope of seeing land until it was too late, a prudent captain might have avoided the strait and passed to the east of Staten Island, in clear water. But that added a little more distance. James Cook had written in his journal the first authoritative words on the matter: "With respect to the passing of Strait Le Maire or going round Staten land I look upon of little consequence and either one or the Other to be pursue'd according to circumstances." However, he added, "if you should fall in with the land to the eastward of the Strait or the wind should prove boisterous, or unfavourable, in any of these cases, the going to the Eastward of Staten land is the most adviseable."

The Great Navigator himself spent three and a half days getting through the strait before anchoring to wood and water in Success Bay on January 16, 1769 (he had the good sense, or luck, to be there in summer). Cook proceeded to treat the coastline and the islands southwest towards the Horn as he treated any others: it was new land to be described, measured, sounded, charted. In spite of heavy gales—furious squalls with rain, sleet and fog—he did just that. He spent two or three days coolly jilling around near the Horn itself so that an astronomer on board could establish its exact longitude—crucial information for navigators—using the complicated lunar sights and calculations. Only then did he begin to make his westing. Even so, he doubled the cape from fifty south to fifty south in thirty-three days, a very good time in the bluff-bowed *Endeavour.*

The *Beara Head*'s rising gale qualified as "boisterous," and the big barque was not noticeably more manoeuvrable than Cook's vessel had been. Nevertheless, the captain was going for the strait. The route was a little shorter, and that made a difference for a vessel with cargo. Cook the explorer could afford to be sanguine about an extra half-day; McMillan the freighter captain could not. More important, passing through the strait gave a vessel a better angle on the wind south of the Horn; that might add up to the difference in making westing before some gale, or a string of them, knocked a ship back east of the cape again. In rounding the Horn, making westing was everything. In the

1880s, it was always worth running the strait if you could. It added to the captain's confidence that he had managed to shoot good sun sights two days earlier, even as his crew was snuffing the fire. He thought he was seaman enough to carry two days' worth of dead reckoning through the gap.

It could be a fearsome place. The Admiralty's directions mentioned strong tidal streams; if the wind was blowing against the current, the "very turbulent" seas that resulted could destroy a small vessel and "do much damage to a large one." If a captain was in any doubt about weather or tide conditions in the strait, he was either to heave to and wait for things to improve or to go around the long way to the east of Staten Island. One German captain who had often run the narrow passage described Le Maire: "Just what this current can do to the southbound vessel one first learns by sailing there. This mighty swell of waters which giant forces press through the gate between Tierra del Fuego and the island, this crowding together of millions of tons of turbulent water, creates a sharp piling up of eddies and backwaters in which the largest ship can become unmanageable. It is too late to find that out in such a confined place." The wrecks of ships lie thick along the shores of the strait.

Nevertheless, at dawn on July 14, 1885, the lip of the austral winter, on the *Beara Head*'s seventieth day at sea, it entered the northern entrance to the Strait of Le Maire—or rather, its captain thought it did. Visibility was about one mile, in rain, low cloud and forty knots of west wind. Maybe the ship was threading the needle of the strait. Perhaps it was too close to one side or the other: if the western side, they could bear off and clear an obstruction (if they saw it in time); if the eastern side, it might not be possible to haul their wind and beat clear to avoid the rocks or the iron-bound coast. Even if his ship started out straight, no navigator could predict the effects of the currents that swept the constricted water.

The crew deduced that they were approaching the strait by the fact that four of them were ordered aloft to the fore and main royal yards to "sing out if you see any goddamn thing," and by the engrossed concentration of the captain and mates on the poop. When the old man himself went up to the mizzen top to peer around the ship with his glass, it only confirmed his bold, or foolhardy, decision.

Benjamin was too tired to care. It even seemed to him that getting wrecked on Staten Island might not be such a bad thing. He would jump ashore, climb the rocks, find a sheltered nook or cave amongst them and sleep for a week. Then he'd think about food and getting back to the world. He had faith in the captain. But there was an air of anxious intensity throughout the ship that worried him; it was as if everyone aboard had taken a deep breath half an hour ago and still hadn't let it out.

"What about it, Jimmy?" said Anderson. "Is the bastard fuckin' mad?"

Russell the Yankee was standing amidships by the mainmast, scanning from side to side as the ship rushed on.

"Maybe, but he don't have an option," he said. "It's been a slow passage, even though this ship ain't a wagon—just bad luck. He's got to try to make up time. At least he don't back and fill. He's made a decision, and he's going on. I'm buggered if I'd do it! But all hands don't live aft, so here we go."

Urbanski and Kapellas were aloft, 165 feet up, one on each side of the mast on the main royal yard.

"What are you lookin' over here for, ya goddamn dago? I'm keepin' a lookout to port. You goddamn well look to starboard, for Christ's sake!"

"I just turn my head, Yankee. By heavens! If you look to port, why you lookin' at me? I'm to starboard. Just fuckin' look. This crazy bastard old man goin' to run us up on the rocks."

The Finn's opinion: This was more like it. This roastbeef captain was acting like a real sailor now. Run her hard through the strait and harder for the Horn. Get us out of here chop, chop.

The Dutchman's: Very bad decision. Too dangerous. If you can't see land, don't run the strait. That was the rule German officers followed, and it was the only reasonable thing to do in this godforsaken place.

Maguire, the shanghaied man: He had become a sailor, and had got used to the sea along the way, but this scared the bejesus out of him. Twelve miles wide, Russell had told him. Why would any man push a ship this size, at this speed, through a narrow gap when he wasn't sure where he was and had nothing, absolutely nothing, ashore that could help them if he misjudged? At least there were other ships about. Yesterday, they'd seen sails astern and to starboard, but now they were lost in this murk. Maybe there'd be a chance of rescue by another ship if they piled up.

The Elf: Good! This was right. There was just enough visibility to give them room to manoeuvre if they saw land; they could even haul their wind if necessary. Unless the old man was completely wrong, and they were headed directly for Staten Land and not the strait. If that was so, they didn't stand a chance in hell. But he must have shot the sun a couple of days ago—the Elf remembered some sunshine through the haze of dust and exhaustion as they shovelled the goddamned coal (which could have burned to hell, for all he cared). The captain couldn't make a big enough error in two days of dead reckoning to miss the strait altogether. On the other hand, the currents . . .

"Friend Ben," said Anderson as they stood by the bowsprit, peering ahead, "we've had the long trot under us, all right. Them eejits aft'll kill us all out at this rate."

"Friend Anderson, I couldn't agree with you more."

The cockney, whose release from irons seemed to be permanent, said nothing, only looked out over the sea with his usual loathing.

The mate: Driving for the strait with a short view and heavy weather was a risk, but a calculated one; he was beginning to like this paddy captain. He—MacNeill—would do the same. The only thing he didn't agree with was the goddamned old man going aloft himself. He should have stayed on the poop where he belonged. It didn't look good with him up the bloody mast, peering around like a worried old woman. Send some good seamen up there and hold yourself to the poop. Never let yourself look like you might be wrong. Decide and then do it; stand your ground by the weather shrouds and let the game play out.

The second mate: This risk was appalling—flinging the ship and the men at a goddamn hole in the wall at twelve knots in a gale! Now he was more sure than ever that it would be steamships for him from here on. Regular schedule, home every two months and let the bloody engineers worry about the engines.

The captain's wife: She was fearful—anyone could see how anxious her husband was—but she was exhilarated, too, by the ship's wild momentum as it thundered on through the cold mizzle and low, scudding cloud. The proximity of the fabled Horn was both frightening and thrilling. She was discovering unsuspected things about herself; a partiality to adventure, for example. How could she ever go back to the mannered confines of Bangor after living through this savage seascape? After watching the muscular, tattooed men swinging aloft like acrobats, their barbaric chants and shouts setting her hair on end? Even the mate's hard impenetrability impressed her—she could see the strength there as well, his being equal to anything; he was a man, like her husband, who would never give in, never say die, always find a way to meet any danger or difficulty. She suddenly decided to ask her husband to teach her navigation, sun sights and all the rest of his canny, subtle judgments that kept the ship situated on the sea's blank surface. Like everyone else aloft and on deck, she remained silent and staring, hoping not to see "any goddamn thing"—what they couldn't see couldn't kill them—but searching for it nevertheless.

The captain, soaked wet and stiff with cold, climbed down from his lookout on the mizzen top after two hours. He thought that they had to be through the strait. Even if he was off in his reckoning by ten miles, they would by now either have seen land (or hit it) or cleared Le Maire. He returned to the weather side of the poop and beckoned the mate, who had moved instinctively over to the leeward side, to join him. They were through, did he not agree? The mate did; it was damn fine work, he said. Did the mate also agree, however, that they should keep the lookouts aloft for another hour to be sure? He concurred with that too. After another hour, they could be sure.

At six bells of the forenoon watch, the captain ordered the upper- and lower-topsail and fore-course yards braced hard round, the jibs, staysails and spanker sheeted in and the barque brought up to sail full and by. Its course was south-southwest, towards the ice. The Horn was eighty miles away to windward. The seas were different—higher, their crests farther apart—more proof that they had cleared the strait and were drawing out of the relative shelter of Tierra del Fuego and the Wollaston Islands, of which Isla Hornos was the southernmost point of land, into the clear path of the untrammelled Southern Ocean swell that rolled round the world. Now all they had to do was make sixty miles southing and five hundred miles westing. It didn't sound like much: in ideal weather, they could do it in two days. They would pass from one great ocean into another, and then they would turn north towards the fifty-degree line and Valparaiso.

No one thought it would happen that way. There were no illusions aboard the *Beara Head* of a rounding either short or easy. It was winter. The barque was pounding to windward, the implacable west wind growing stronger already. The Southern Ocean current set against them. The rain was turning to sleet, and the temperature was dropping. Soon, the seamen would have to go aloft to shorten more sail. There were two men at the heavy wheel. Seventy days at sea. The struggle round the Horn was beginning, and the *Beara Head* joined battle.

CHAPTER EIGHT

The Uttermost Cape

Sailor or landsman, there is some sort of a Cape
Horn for all. Boys! beware of it; prepare for it in time.
Graybeards! thank God it has passed. And ye lucky livers,
to whom, by some rare fatality, your Cape Horns are placid
as Lake Lemans, flatter not yourselves that good luck is
judgment and discretion; for all the yolk in your eggs,
you might have foundered and gone down, had
the Spirit of the Cape said the word.
HERMAN MELVILLE, *White-Jacket*

Deep they lie in every sea,
Land's End to the Horn.
JOHN REED

I sailed past Cape Horn on the first of March, a cold, sunny day with scattered high cloud and a quartering wind. *Baltazar* passed by about a mile off the high cliffs, under full working sail in a ten-foot swell and slight sea. With its uneven rock face to the south and the treeless, green-grassed land sloping down from its highest point, the Horn looked like any mundane cliff in the Scottish Highlands. That made sense, because it's at about the same latitude south as the Highlands are north. The islands nearby—the Wollastons and all the lands south of the Beagle Channel—looked to me like replicas of the sea-lough shores of Scotland or the west of Ireland, without the sheep or people. On this bright, peaceful day, the fearsome cape was a pleasant and arresting landmark, a handy day-sail destination for a few photographs. We even considered anchoring off its eastern shore and taking the dinghy onto the beach to climb the hill to the Chilean lighthouse and weather station (manned by two lonely navy men). I had read of yacht crews doing that on calm days, but the swell we were sailing through was a little too high for a safe surf ashore. The other deterrent was the depression we had been tracking for a day. It was strengthening fast, and the Chilean navy had issued a warning about it just as we reached the Horn's eastern side.

Certainly, we had prepared for the worst, anchoring the previous night in one of our skipper Bertrand's favoured hurricane holes in the Wollastons: a bay more or less surrounded by the eastern end of Isla Hermite and the smaller islands of Maxwell, Saddle and Jerdán. Just as we entered it the day before, the squall line we had been monitoring for an hour hit with fifty-knot winds, heavy hail and then rain. We had unlashed the Zodiac dinghy, ready to take mooring lines ashore to a tree, and I spent twenty minutes lying on top of the rubber boat with one of the other crewmen to prevent it from blowing off the deck, all the while huddling away from the stinging ice balls. The next morning was fine, cold and clear, but we got ourselves and *Baltazar* snugged down as if we were going to sea for a month.

There wasn't a lot to do, because the fifty-foot-long steel boat was specially built and adapted for these waters. It had a keel that could be raised up into a cabin housing so the boat could take refuge in protected Antarctica bays, sliding over their shallow lips, which kept out the crushing, grinding ice; oversized mast, rigging and winches; a Plexiglas dome you could look out through when going on deck was very unpleasant or too dangerous; a small cockpit so boarding seas would not unduly affect the boat's stability; and two reels of floating nylon so the lines could be run off ashore quickly in the dinghy. It was necessary to move fast to get the boat secured in one of the little coves it had to duck into whenever the Horn weather made one of its quick, treacherous pounces. There were a few other sailboats in Ushuaia, and in Puerto Williams (across the Beagle Channel in Chilean territory), and they were all more or less like *Baltazar:* tough, heavy steel boats that could run fast and hide when that was possible, and stand up to brutal wind and waves when it wasn't.

Nevertheless, we went through the standard procedure, the mariner's customary prudence: harnesses and lifelines snapped on and double-checked; dinghy and other gear lashed down tight; hatches dogged; running rigging and sails inspected; the engine's sound approved. I was used to doing the same thing myself, even before starting out on a blustery fall crossing of Lake Ontario or a Caribbean trade-wind passage. The only dismaying aspect of our preparations was the skipper's wary watchfulness. Both Bertrand and his wife, Siv, were nervous—only a little, but it was obvious. They had rounded the cape a few dozen times, and they made two or three annual passages to Antarctica and the Falklands as well. They were veterans of this sea; they had been knocked down, almost turned turtle, coming north across the Drake Passage from an Antarctic voyage the previous year, and even that hadn't dissuaded them from sailing these beautiful, perilous waters. On our pleasant day, however, they were taking nothing for granted. They were suspicious of the shining sun, the gentle wind and sea. Having been hammered almost every other time they had done this, they assumed it would happen again. If they were going to be nervous, so was I.

Things change so fast here, said Bertrand. Just like that. *Tout de suite*. You have to be ready for anything.

As we cleared the lee of Isla Hermite and felt for the first time the full scend of the Southern Ocean swell, we looked to the west, where the weather came from, searching for the trap that would surely soon be sprung.

It wasn't, though. The cape came abeam, and we drank wine in celebration. I poured a little from my glass over the side and murmured the sailor's universal salutation off the Horn: "To the men who died here."

I added, "And to Uncle Benjamin, and to all my ancestors who came here."

Bertrand grew more relaxed and jovial as the Horn fell astern and we turned north towards the Paso al Mar del Sur, between Islas Herschel and Deceit, and our little bay of refuge. We had got easily round the Horn, although a real snorter was on the way.

The good weather was an anticlimax I had mixed feelings about. I had hoped to have the book material a Cape Stiff gale would have provided. That was why I had come: to look briefly and in reasonable safety at the Horn of legend and my sailor forefathers, and to experience its power. At the same time, I was happy enough to have the summery day. You never knew what could happen in storms in that mythic and feared sea area.

※

The reputation of the Horn as a place of tribulation for seamen is based on a few facts of physical geography that can be expressed in six words: compression, thin water, latitude, ice, mountains.

Compression: The two-thousand-mile width of the Southern Ocean, in which waves grow huge and roll unobstructed round the earth, is squeezed together off the tip of South America into a six-hundred-mile stretch between the Horn and Antarctica. This is still an ample body of water, but the wave energy of the wider sea spaces is, nevertheless, constricted. The Southern Ocean is two-thirds garrotted in the Drake Passage, and it fights back. The weather systems, too, are diverted or adulterated by the sudden intrusion of land. They become more complicated, pile up on top of each other, move in unexpected directions. The systems intensify; isobars crowd together, and winds jump and pop.

Thin water: The shallowing depths of the continental shelf off the Horn concentrate the compressed energy of waves even more. Seas build higher and steeper, and the effects of land alter their movement, creating cross seas or confused wave trains; rogues are born. It's similar to Biscay and the western approaches to the English Channel, or anywhere else ocean-wide seas come roaring into shallow water across thousands of miles of fetch: big waves turn mean and vicious.

Latitude: The Horn is farther south than ships usually sail voluntarily. Square-riggers running their easting down followed Maury's routeing advice and dipped down into the roaring forties and the strong westerlies that made good passage times and kept them in business. No vessel went deep into the furious fifties if it didn't have to. Yet that was what the Horn forced them to do—dive south into the ice and cold and stronger wind. To pass from one ocean to the other, there was no alternative.

"And ice, mast-high, came floating by, / As green as emerald," recalls the Ancient Mariner. In the Drake Passage in the vicinity of the Horn, there are usually some icebergs or growlers or bergy-bits drifting here and there. They are mostly impossible to see for any westbound vessels labouring south of the Horn in heavy weather and bad visibility (because of storms or fog)—conditions that exist more often than not—and it was a matter of luck whether a vessel collided with one. The dangers were multiplied for ships running their fast easting down, because impact at speed was usually fatal. It's impossible to say how many vessels were destroyed by bergs. By its nature, a collision was catastrophic, and it's certain that some of the ships posted missing were lost to ice, rather than storm seas. Near the Horn, vessels can expect pack ice anywhere below sixty degrees (240 miles south of the cape).

Mountains: the final complication. The Andes and the local mountains of the Darwin Cordillera funnel katabatic wind down their slopes and out to sea, hitting ships already sailing in storm-force wind twenty or fifty miles off the Horn with squalls well over hurricane strength. Large, well-organized and well-formed maritime weather systems are deformed and pushed around by the continental mountains, their winds mutated or augmented into shapes, directions and strengths that confound seamen.

All of this was bad enough for wind ships running across the Southern Ocean from Australia or from the West Coast back into the Atlantic—Yankee downeasters, limejuice wool clippers or Cape Horners carrying wheat, copper or nitrates. Running before strong wind and big seas had its hazards: broaching and capsizing (a quick death sentence), getting pooped (a wave boarding over the stern, sweeping the decks of houses, gear and men). Sailing at speeds of twelve to sixteen knots in bad visibility, a ship missed or hit icebergs by chance. Most square-riggers used to these waters had had close calls they'd survived: Conrad's *Torrens* ran into an iceberg head-on in 1896, three years after he had signed off; it mashed the bow and wiped out the bowsprit, but the ship made it to port. At least eastward-bound ships didn't spend long near the Horn. They swept by in a day or so, maybe close enough to see the rock, before turning northeast towards Staten Island and the weather complications of the Atlantic—unless they ran into one of the cape's strange calms or met with easterly gales or fields of ice. Then they got pinned off the Horn like any westward-bounder.

It was much worse going the other way, east to west, into the prevailing westerly wind and against the Southern Ocean current, which, near the Horn, could set a ship back ten to twenty-five miles in a day, wiping out any hard-won westing. A captain might measure his progress in tens of miles, or be satisfied if he managed merely to hold his own. Strictly speaking, seamen applied the term "rounding" or "doubling" the Horn only to an east-to-west passage. Running with the wind the other way was so much easier so much of the time that it didn't really qualify as a rounding—a vessel merely "passed" or "cleared" the cape or "left it to port." Ships like the *Beara Head*, westward-bound, were the ones that, sailors said, went to war with the Horn. Even more so when they rounded in winter: colder, windier, rougher, more snow and hail and frigid soaking rain, maybe more ice.

For all these reasons, the Horn acquired its legendary dimension for seamen: the bloody Horn, Cape Stiff. It was the alpha and omega of capes. No square-rigger man could claim to be a genuine deep-sea sailor until he had rounded the Horn, meaning the hard way, from east to west.

There are other dire promontories. The two other "stormy capes" are Leeuwin at the southwestern tip of Australia (more of a navigational marker of a roaring-forties passage, which ships cleared by a good margin as they ran to the east towards Australian ports on the south or east coasts) and the Cape of Good Hope, "the Cape" (which could be a terrible place when strong wind blew against the southwest-setting Agulhas current).

My father told me that his corvette was once the single warship escorting a small convoy of twelve merchantmen down the South African coast from Durban to Capetown in 1942 or 1943. There was a southwest gale blowing, and the ships struggled for two days through steep, high seas. One vessel fell into a deep trough and was unable to rise over the crest of the succeeding wave. It buried the ship under five thousand tons of falling water and kept it down, its buoyancy not enough to counteract the weight of the sea. The men never had a chance. However, the weather was too bad for the U-boat wolf pack to operate; it had picked off the ships of previous convoys like carnival ducks. It was considered an excellent outcome to lose only one vessel out of twelve, and only twenty-five men.

The first humans to see Cape Horn were probably a canoe-ful of near-naked Yahgan Indians. Together with three other tribes, they occupied the harsh lands of Tierra del Fuego. Like the Inuit at the other end of the continent, they astonish us with their ability to survive in a wild, barren, cold country. They ate shellfish and hunted guanaco (a form of large llama). They had flint weapons and made fire with fool's gold. They found the little sheltered coves and nooks in the land. When *Baltazar* anchored in some bay or off some quiet, sandy beach protected by trees or low hills, a little warmer than elsewhere, with grass and wildflowers, we inevitably saw signs of the old Yahgan life: the middens of shells and bones that had become small hills forming their own shields around the tiny settlements. *Baltazar* and the Yahgan had the same needs; we both found the scattered places—not as cold, more hospitable, out of the incessant wind—in which it was possible to live. The people are long gone, of course, done in by measles, smallpox, influenza or hunted down like game by white men on horseback with repeater rifles, a short, obscure chapter in the book of the Amerindian holocaust.

There's some evidence of Yahgan presence, perhaps temporary or sporadic, on the northern islands of the Wollaston group, but none on the southern islands, including Isla Hornos. However, it seems unlikely that these tough, far-ranging people would not have gone as far as they could go—to the very last piece of land of the continent, until, for the first time since their people had arrived in the Americas, they could see no more land but only the reach of the great Southern Ocean stretching down to the ice at the true end of the world.

Apart from the Yahgan, no one else saw the Horn for a long time. Francis Drake might have been the first European to do so. On his privateering voyage in 1578, he reached the Pacific after a fast seventeen-day passage through the Strait of Magellan, but he was driven back by a succession of gales into the open ocean to the south and then the east. In the desperate days that followed, his three unwieldy, unweatherly little ships struggled to survive.

Any nineteenth-century seaman beating around Cape Stiff would have understood the description of the place by the chaplain aboard the first Cape Horner: "The winds were such as if the bowels of the earth had set all at liberty, or as if all the clouds under heaven had been called together to lay their force upon that one place. The seas . . . were rolled up from the depths, even from the base of the rocks, as if they had been a scroll of parchment. . . . Truly it was more likely that the mountains should have been rent asunder from the top to the bottom, and cast headlong into the sea by these unnatural winds than that any of us would survive."

The *Marigold* and its twenty-nine men were lost. The *Elizabeth* became separated and eventually ran back through the Strait of Magellan and then to England. Only tough, tenacious Drake himself persevered in the *Golden Hinde*. He touched at land in several places to rest his exhausted crew, gather wood and hunt seals and penguins. At one point, Drake supposedly noted: "The Uttermost Cape or headland of all these islands stands near 56 degrees, without which there is no main island to be seen to the southwards but that the Atlantic Ocean and the South Sea meet in a most large and free scope." Drake called the island Elizabeth, after his queen.

It might have been the Horn, though according to dissident historians, it could have been the Islas Diego Ramirez, fifty-six miles south, or even Cape Brisbane on Henderson Island, fifty-eight miles west of the Horn. Drake's navigational tools were imprecise, and any of these are possible.

Without doubt, however, the privateering admiral had made the two most important discoveries of all: Tierra del Fuego was a group of islands off South America, not the northern tip of a great continent stretching far to the south from the shores of the Strait of Magellan; and south of Tierra del Fuego was a wide sea that allowed passage from ocean to ocean. Ships had an alternative to the narrow (and Spanish-dominated) waters of the strait. This southern sea became known as Drake Passage.

There's no question about the European who saw the Horn next (or first): he was Willem Schouten, a skilled Dutch navigator sailing for a group of merchants forming the Compagnie Australe under the direction of Isaac Le Maire. Schouten cleared the strait between Staten Island and the mainland, naming it after his patron. Beating hard for two days in heavy weather to make westing, he next sighted two islands to the southwest, naming them the Barneveldts, after the founder of the Dutch East India Company. (I saw them from *Baltazar* as we made our dash across the Bahia Nassau to and from the Wollastons.) On January 29, 1616, at 8 P.M., Schouten rounded what was obviously the southernmost point of land. He named it Cape Hoorn, after his home town in the Netherlands. (The cape's name had nothing to do with the shape of a horn—the English and Dutch words are accidental homonyms, what the French call *faux amis*—although when I looked at its profile from the southeast, I thought you could make a case for a resemblance.) Schouten was out by 110 miles in his calculation of the latitude of Cape Hoorn (its longitude was anyone's guess). Understandable, since his ship was rolling its guts out in the usual gale as he used his primitive instruments.

This early, brittle European contact with the Horn turned into a story of Dutch and English freebooters, privateers and outright pirates motivated by an ache for Spanish New World and Far Eastern gold, and of Spanish flotillas trying to stop them. National and imperial interests came along for the ride. Later, neighbourhood gentrification took place as the search for loot gave way to voyages of exploration and scientific research. One by one, the small, frail expeditions made their way south towards the Horn—for them, the perilous yet open gate to the Pacific.

First, there were the Spanish Nodales brothers, who sailed with two caravels through Le Maire and close by the Horn, correcting its latitude to within two miles, an extraordinary feat done aboard a ship with instruments as crude as those of Schouten. They clawed their way west and then northwest against heavy gales with snow, reaching the western entrance of the Strait of Magellan, which they sailed through, thus completing the first circumnavigation of Tierra del Fuego. The Islas Diego Ramirez are named after the brothers, who had demonstrated that Spanish interests in the south were indeed at risk from seaborne raiders.

The Nodales included in their report of the voyage vivid details of the terrible gales and cold they had encountered. The Spanish did nothing, counting instead on the weather to deter opportunistic Dutchmen or Englishmen. With a few exceptions, that's how things worked out for the next eighty years.

Two Dutch raiding fleets, in 1624 and 1642, were rendered ineffectual by weather and disease. Then the Englishman John Narborough, unusual in that he was a captain in the regular navy and not a freebooter or, more typical for those days, a combination of the two, sailed through the Strait of Magellan in 1669, carrying out some useful hydrographic work, but nothing else. The English buccaneer Bartholomew Sharp, looking for prizes, made for the Horn down the west coast of South America ten years later in a captured Spanish warship. His was probably the first British passage past the Horn, although from west to east. He conned his ship precariously through a sea "gleaming with icebergs." Later, the pirate captain John Cook, with the future circumnavigator William Dampier aboard as a seaman, rounded the Horn, claiming afterwards he had caught glimpses of a mass of land to the south, a *terra australis*. Other captains would search for this land for more than a century, with Dampier himself being the first. He rounded the Horn in 1699 and made three adventure-filled circumnavigations in all. After the first, he published a book, *A New Voyage Round the World*, which became a best-seller in its time (an early European example of the extreme-adventure genre, in the old tradition of *The Odyssey*) and made the hybrid buccaneer-naval captain rich and famous. On a later voyage, the irascible Dampier, whose crews appear to have been in a permanent state of mutiny against him, marooned his equally prickly sailing master, Alexander Selkirk, on one of the Juan Fernandez Islands. The castaway, the inspiration for *Robinson Crusoe*, was rescued four and a half years later by Woodes Rogers, another Horn-rounding English privateer (with Dampier on board as pilot; there's no record of how he and Selkirk got along during the remainder of the passage). French pirates made an appearance, then more English ships and the Dutch once more—all in the perennial search for Spanish prizes.

One of the English vessels was the *Speedwell*, captained by George Shelvocke, notorious even in that violent time for his ruthlessness. During the *Speedwell*'s attempt to round the Horn, it was driven far south by storms, below sixty-one degrees latitude. The little ship was nearly overwhelmed many times, and the crew believed they were dead men. One of them, more disconsolate than the rest, believed that a black albatross that had followed the vessel for several days was an ill omen and shot the bird, hoping fair winds would result. The ship did survive, and eventually it made its way north again. The *Speedwell*'s ordeal and the slaughtered albatross inspired Coleridge's poem seventy-six years later.

Shelvocke left one more legacy: a list, prepared for the Admiralty, describing what happened to the eighty-seven men (of a crew of 115) he lost during his three-year circumnavigation. Thirty-seven men deserted, with only a handful getting back to England; seventeen died aboard ship at sea; two drowned; three were killed by enemy action; eighteen were taken prisoner by the enemy; four were murdered; six were discharged or marooned by the captain.

The next Horn-bound ships belonged to a Spanish squadron. Of its five vessels, one disappeared and two were wrecked in the usual Horn weather. Only one made it round the cape. The Spaniards had been dispatched to intercept the English captain, George Anson, who in 1740 started out with a fleet of eight ships with the goal of capturing the legendary Spanish Manilla Galleon, with its cargo of precious metal. He eventually made a circumnavigation, with half of his nearly two thousand men dying in the process, most of them in the rounding of the Horn and most to scurvy, one of the disease's more successful interventions. Another seven hundred or so deserted or came back on ships that gave up and headed home. Longitude confounded part of the fleet: two of Anson's largest vessels, separated from the others in the constant gales off the Horn, finally turned north when their dead reckoning put them far enough west; they sailed north for many days before realizing that they were still on the wrong side of the continent, a navigational error of three hundred miles or so. By then, it was too late to follow their flagship.

"I had my topsails reefed for 58 days," Anson wrote. "My men are falling down every day with scurvy. . . . I have not men able to keep the decks or sufficient to take in a topsail, and every day some six or eight men are buried." The ships lost men overboard, too, as the terrible gales persisted. Then the worst storm of all hit the surviving ships. A young midshipman aboard the *Wager*, John Byron (known as "Mad Jack" or "Foul-Weather Jack," the grandfather of the poet), described the impact of the first ferocious squall: "In its first onset we received a furious shock from a sea which broke upon our larboard quarter, and rushed into the ship like a deluge; our rigging too suffered extremely from the blow . . . so that to ease the stress upon the masts and shrouds we lowered both our main and foreyards and furled all our sails; and in this posture we lay to for three days." It was no use: the *Wager* was partially dismasted and went onto the rocks. Byron survived, but not many others.

When the four ships that got round the Horn made a rendezvous at the Juan Fernandez Islands, one of them was so damaged and rotten that Anson ordered it burnt. On the other ships, corpses lay all over the decks, "it being impossible to conceive the stench and filthiness which men lay in or the condition that the ship was in between decks," wrote one of the captains. Anson had lost to scurvy 280 men of the complement of 531 aboard his own ship, the *Centurion*. The *Tryal* had only five men fit for work. Almost three-quarters of the *Gloucester*'s crew was dead. Rats fed on the corpses and gnawed the fingers and toes of the sick.

Nothing stopped these rapacious men. A few months later, with only one ship and seventy-one men left, Anson attacked and captured the Manilla Galleon and its cargo, which was worth more than £400,000 in the currency of the time (about £20 million today).

What really astonishes about all the voyages of the seventeenth and eighteenth centuries is that any of these slow, clumsy, unweatherly, hemp-rigged little wooden tubs actually survived their months rounding the Horn, beating into the gales and seas of the Southern Ocean. That they did indicates two things: the ships, if not handy, were tough and well-built, and the men who sailed them were the original iron men. They had determination, endurance and the ability to absorb the psychological effects of losing ships and having men die around them for weeks on end. Death to disease and plague was common enough ashore, and people were inured to it, but putting up with this attrition while simultaneously fighting the gales of the Horn and preparing to engage the human enemy was something only the most robust, and greedy, men could do.

Anson's voyage was the last large-scale expedition of government-sponsored piracy and plunder. No voyage had lost more men or a greater percentage of those who started out. His experience was the catalyst for the work of James Lind and his treatise on the causes of scurvy and its cures. The next voyages round the Horn were made by hydrographers, scientists and explorers. The advent of the Enlightenment in Europe encouraged such projects. And the search began for the unknown continent, *terra australis incognita*, which was supposedly glimpsed by the privateer John Cook, and which many geographers thought had to be there to balance out the earth's land masses and keep the planet in stable rotation and orbit. At the same time, circumnavigating sailors brought back stories about lands that did exist: the happy islands of the Pacific. Seamen on the beach, and aboard their diseased and straitened ships, hungered for the sweet plenty and easy women of paradise.

The Horn became a much busier place. Ships had three purposes in passing through the Strait of Magellan or rounding the cape: to chart the surrounding waters of the Strait and the tortuous channels and gnarled islands of Tierra del Fuego; to search for the *terra australis incognita;* and to pass, in the course of circumnavigations, into the Pacific Ocean for all the purposes of trade and exploration there.

First came John Byron, the only officer to survive Anson's expedition, with the *Dolphin* and the *Tamar.* The *Dolphin* was an Admiralty experiment: its hull was sheathed with copper to find out if that would stop, or at least slow down, fouling. (It worked.) Byron had to have been traumatized by the ghastly suffering and dying aboard Anson's ships twenty-five years earlier, and by his own bare survival in the wreck of the *Wager.* He was obsessed with preventing scurvy, and he tried everything he could to keep his men healthy and safe, even to the point of ignoring many of his orders—to explore the West Coast of North America and to search for the Northwest Passage, for example—so that he could stay close to land for fresh food and avoid undue danger. He made the fastest circumnavigation to that date—twenty-two months—and he lost only six men from each ship. He was popular with his men, if not with the Admiralty.

Near the Horn, Byron wrote: "It is certainly the most disagreeable sailing in the world, forever blowing and that with such violence that nothing can withstand it, and the sea runs so high that it works and tears a ship to pieces."

Byron sailed so fast (and with unusual competence) and skipped so many of his orders for exploration that another expedition was assembled and set off as soon as ships could be fitted out. Samuel Wallace and Philip Carteret eventually passed through the Strait of Magellan.

Even though beating through the strait, with its williwaws, narrow waters, shoals and fast currents, was a difficult feat, it still had to have seemed a better proposition than clawing round the Horn in the open water of the Southern Ocean. Carteret's vessel, the *Swallow*, was so slow and cumbersome, however, that it took the little fleet 115 days to clear the strait. (Magellan had done it in thirty-seven days, even though he didn't know where he was going, and Drake in seventeen.) Once in the Pacific, the two ships were separated, either by weather or because Wallis could no longer bear to wait for the ponderous *Swallow* to keep up with the *Dolphin* (he had inherited Byron's ship). In the course of his subsequent circumnavigation, Wallis discovered the labyrinth of reefs and atolls of the Tuamotu Archipelago and, most important, the island of Tahiti, the very heart of the image of paradise. Lumbering along behind in the *Swallow*, Carteret stumbled upon Pitcairn Island and searched through sixty degrees of longitude for the unknown continent, encountering nothing but scurvy.

The French mounted what could be described as the first genuine scientific expedition under Louis Antoine de Bougainville, a soldier but also a member of the Royal Society in London (he wrote a work on integral calculus). Like his predecessors, de Bougainville used the Strait of Magellan, reaching the Pacific in 1768. Byron had written: "I would prefer it twenty times over to the going round Cape Horn," and the Frenchman concurred. Nevertheless, de Bougainville was criticized in both England and France for avoiding the Horn; it disqualified him from claiming the same mariner's status as Drake, Schouten or Anson. In his defence, de Bougainville cited scurvy. Like Byron, and all commanders now, he feared the disease that could destroy an expedition faster than any cape. During his transit of the strait, he landed often for fresh food. He wrote: "I make no doubt but the scurvy would make more havoc among a crew who should come into the South Seas by way of Cape Horn than among those who should enter the same seas through the Straits of Magellan. When we left it we had no sick person on board."

One footnote to de Bougainville's voyage is that it carried the first woman to circumnavigate. She was a servant for one of the botanists on board, and she managed to pass herself off as a man until the ship reached Tahiti. The islanders, who it seemed were more observant about matters of gender, recognized the *faux* man for what she was, and she finished the voyage as a woman.

Next, it was the turn of the Great Navigator himself. One purpose of James Cook's first voyage was to explore the Southern Ocean passage around Cape Horn. The Admiralty speculated that, so long as it was accomplished in the southern-hemisphere summer, rounding the Horn might be a better option than taking months to pilot a ship through the strait; it ordered Cook to test the proposition. With his usual understated panache, Cook made his fast thirty-three-day rounding, surveying the coast and establishing the Horn's longitude as he went. Cook was lucky with the weather too. He didn't run into the ferocious gales his predecessors had experienced; he even had a few days of a Cape Horn calm, during which the voyage's scientific patron, Joseph Banks, was able to row a boat off from the ship to shoot birds for study and food. One of them was a grey albatross with a ten-foot wingspan, but no harm, supernatural or otherwise, fell upon the *Endeavour*.

Cook was equally lucky on his second passage past the Horn in 1774, this time sailing from west to east. He ghosted by the cape in a gentle westerly and spent Christmas anchored in a sound on the southwest coast, sending parties ashore to hunt and explore. The crew ate local geese for their holiday dinner. After landing on Staten Island, however, the ship was driven east into the Atlantic by strong gales, as the truer Horn reasserted itself.

Cook was the first seaman to experience the Horn as anticlimax. He noted after his first (east-west) passage that he had not had to reef his topsails once since clearing the Strait of Le Maire, "a circumstance that perhaps never happened before to any ship in those seas so much dreaded for hard gales of wind, insomuch that the doubling of Cape Horn is thought by some to be a mighty thing." The Horn was never easy, and no one could count on Cook's own relatively tranquil passage, but the seaman who rounded in summer—from west to east, if possible, to take advantage of the westerlies—might do well enough most of the time. If he had luck too; a seaman always needed luck.

A series of expeditions over the next forty years took the route round the Horn. The Comte de La Pérouse, with two ships, rounded the cape from east to west (in spite of Cook's advice to do it the other way) in 1785. Later in his circumnavigation, La Pérouse and his flotilla disappeared after sailing from the English settlement at Botany Bay, in Australia.

A circumnavigator who needs no introduction, William Bligh, successfully beat round the Horn in 1788 but was immediately driven back by strong gales. The *Bounty* fought hard for a month, but then, with his crew worn out and the little ship battered, Bligh turned and ran off to the east, reaching the Pacific by running his long easting down south of the Cape and Australia. The long struggle at the Horn, and the longer eastward passage, laid the foundations for the legendary, ruinous mutiny.

The first Russian, Ivan Krusenstern in 1803, reinforced the view of the Cape Horn route as the accepted, normal pathway from Europe to the Pacific. His voyage also demonstrated that things were getting better for seamen as the knowledge of how to prevent scurvy circulated through the maritime world. Krusenstern didn't lose a single man from either of his ships during his three-year round-the-world voyage (demonstrating a very high level of seamanship too), a remarkable change from the carnage aboard Anson's ships sixty years earlier. A second Russian, Thaddeus von Bellingshausen, who hero-worshipped Cook and aspired to complete the Englishman's work of exploration and survey, was probably the first to sight Antarctica, in 1820 (although there's doubt about this; an anonymous English or American sealer, hunting deep into the pack ice, could have made an earlier sighting). Von Bellingshausen made for Tahiti, the sailor's valhalla, for rest and relaxation, but the place was changed by disease and missionaries. The exotic, savage island had been transformed into a sedate Christian community. Nubile brown women no longer offered themselves to deprived seamen. Instead, they dressed in yellow-and-white robes and went to church.

France had settled down after the convulsions of revolution and Napoleon, and Jules D'Urville, who had been commissioned to look for the remains of La Pérouse, made a circumnavigation between 1822 and 1825, taking the unfashionable, slower route through the Strait of Magellan. On a later voyage, his men landed on the coast of Antarctica, the first to do so; they brought back earth and rock samples as if they were returning from the moon.

The most famous of all was a young working passenger aboard the *Beagle*, Charles Darwin. His later fame obscured the maritime achievement of the ship, which, on a previous voyage between 1826 and 1829, had begun to chart the channels, shoals and rocks of the waters of Tierra del Fuego. The vessel's captain, Pringle Stokes, worn out by the strain of the job, killed himself while near the Horn. Surveying was particularly nerve-racking because it was often carried out in execrable weather conditions and, of necessity, close inshore, where every channel, bay and cove was a new experience and unknown hazards abounded. Stokes was replaced by the twenty-three-year-old Robert Fitzroy.

The *Beagle* sailed under Fitzroy again in 1832, now with Darwin aboard. The young naturalist explored Patagonia and the Andes. Fitzroy, four years younger than the youthful Darwin, was a religious man and believed in the literal meaning of the first chapter of Genesis. When Darwin eventually published his great book, Fitzroy was indignant at its unorthodoxy. Later, like Stokes, his predecessor aboard the *Beagle*, Fitzroy committed suicide.

In the seas around the Horn, the *Beagle* suffered the usual trials. Near the Islas Diego Ramirez, it was nearly overwhelmed by three huge waves that caused a broach and near capsize. Later, the vessel was driven by stress of weather into a small bay, where it picked up five men, deserters from an American whaler. Even though they had lived off the hard land of Tierra del Fuego for a year, they were in better condition than the *Beagle*'s crew. Fitzroy continued his surveying, and he did it so well that modern charts, updated by Chile, are essentially copies of his work. Darwin was not a happy sailor in those turbulent waters. When he saw Cape Horn, he wrote: "the sight . . . is enough to make a landsman dream for a week about death, peril, and shipwreck."

In spite of Darwin's dire associations, the Horn was turning into a necessary sea-mark, as more and more ships found they had business in its waters. By the 1830s, the cape seemed a little tamer and less alien, a more manageable risk than the awful promontory that had terrorized and mauled the little caravels, carricks, frigates and brigs of the sixteenth to eighteenth centuries. The French now often refer to the

Horn as *le mythique Cap Horn*. It has always been that, but to a lessening degree. Demystification began in the course of Cook's voyages, when he ghosted the coast of Tierra del Fuego in sunshine and landed for Christmas goose, and continued in Darwin's time. The Horn was still bad enough—as Darwin's own description shows—but getting round it was no longer a desperate, reckless venture, justified only by the prospect of a lifetime's supply of Spanish gold. Now doubling the Horn was an acceptable commercial risk—insurers would underwrite it—in the course of conducting trade and carrying goods. In an accelerating rush, the commodities of the Pacific were hunted, gathered or extracted: whale oil, copper ore, guano, sandalwood, tea, wool, lumber, grain and gold. All of them were loaded on ships that came round the Horn to get them and went round again to take them home.

Whalers had probed the hunting grounds of the Southern Ocean for years—British ships began whaling in the South Pacific in the late 1780s, then Yankee vessels joined the hunt, with New Bedford and Nantucket ships establishing their dominance from the beginning. The numbers of whales seemed inexhaustible, and indeed, they remained so while the hunting technology was restricted to close-in combat with the harpoon and the lance from small, frail boats, men wounded and dead in each year of every voyage. The modestly self-sufficient whalers ranged the oceans for three or four years, sometimes as long as five. For Melville, the whale ship was "the pioneer in ferreting out the remotest and least known parts of the earth. She has explored seas and archipelagoes which had no chart, where no Cook or Vancouver had ever sailed." Scores of anonymous Nantucket captains "were as great, and greater than your Cook and your Krusenstern."

Melville claimed that it was whalers who truly cracked open the Cape Horn route, and he had a point. The Chilean coast between eight and thirty degrees south latitude was a prime whaling ground where ships hunted by the dozens, having rounded the Horn to get there. The Spanish tried to maintain their habitual restrictions on access and trade along the west coast of South America, but they couldn't resist the pressures of business and nationalism. "It was the whaleman who first broke through the jealous policy of the Spanish Crown, touching those colonies," Melville wrote. True or not, they certainly participated in the process; nationalism completed it. In the early 1800s, Chile and

Peru fought for, and won, independence from Spain and immediately opened up the coast for trade.

Welsh merchants began shipping top-quality coal out to Chile (whose own deposits were copious but of poor quality) and bringing back copper ore, which had been mined there for centuries by the Indians and then the Spanish. This was the beginning of a two-way transport that would endure until the last days of sail: Welsh or English coal out, New World resources home (the commercial symmetry of the *Beara Head*'s voyage). The copper-ore carriers were small barques. They had to be sturdy because these bulk cargoes were a new thing for ships to carry; they were hard on the hulls during loading—all those small pellets and rocks pitting and hammering the wood—and seamen soon discovered that coal or copper could shift catastrophically in one bad squall or rogue sea, imposing near-fatal strains on the many tiny joins of wooden hulls. The Welsh ore carriers became famous for their toughness during the two-way roundings of the Horn, and for the hard-case crews who drove them round in all seasons.

Other resources became cargoes for ships on the Horn route: guano, nitrates (saltpetre) for fertilizer and explosives (with which Europeans would eventually blow each other up), Hawaiian sandalwood and hides from California.

We know a great deal about the hide trade because it was the purpose of Dana's voyage to the West Coast in 1834. On its rounding of the Horn, the *Pilgrim* stood to the southwest on the starboard tack, making good westing before turning north; captains had learned already that this was the best tactic to adopt to avoid the easterly currents and the rocky shore. The brig made a fast passage of nine days from Staten Island before it headed north, although it lost a man in the process. (As the *Pilgrim* made its westing, another little brig, the *Beagle*, was casting about the channels and islands of Tierra del Fuego, surveying as it went, with its passenger, Darwin, aboard, his data piling up, intimations forming.) In lighter weather one day, the *Pilgrim* met another vessel, the whaler *New England*. Its captain rowed over and spent four hours visiting Dana's ship. He apparently thought nothing of rowing back to his own vessel, by then almost eight miles astern— he was a whaleman, after all, and used to rowing small boats on the open ocean, even off Cape Horn.

Once on the West Coast, Dana became so disgusted with his captain, the shouter and flogger of men, that he used his Boston family connections to get aboard another hide drogher, the *Alert*, and he came home on it. This was the frigid, ice-bound passage Dana described in *Two Years before the Mast*, as they worked their way round the Horn in perverse easterly gales. Without a chronometer, the vessel had no idea of its longitude apart from the dubious intuition of dead reckoning. When they closed with land on their third attempt to clear the cape, they thought it was the Horn, but it turned out to be Staten Island instead, 110 miles to the northeast. Before arriving in Boston, the *Alert* had to beg potatoes and onions from another ship to treat crewmen who were down with scurvy.

Gold had driven most of the first voyagers to round the Horn, and gold was the lure again. It was discovered in California in 1848, and in the next few years, thousands of ships made the passage round the cape. The volume of traffic was enormous, although it didn't last long: 760 vessels from North American ports alone in 1848–49; fourteen hundred the next year, including more than a thousand American ships, from New York, Boston, New Bedford, Nantucket, Philadelphia, Baltimore, New Orleans and dozens of other ports. Crews jumped ship along with the passengers in San Francisco, stranding hundreds of vessels. It was a short-lived rush; by 1851, most would-be miners crossed the isthmus of Panama by train, taking their chances with yellow fever and malaria rather than Cape Stiff. Nevertheless, the increased traffic further tagged the Horn as a viable, necessary route for commercial shipping.

Gold was struck in Australia in 1850. A stream of ships that had dropped their loads of avid adventurers there headed back east round the Horn for England or the American East Coast, to pick up fresh cargoes of men. At the same time, British and European demand for wool and tea spurred yet more Southern Ocean passages south of the cape towards home. In both cases, speed was of the essence—miners wanted to get digging; first home with the tea got the highest price— and the slow, tubby, bluff-bowed old-time vessels were increasingly replaced by lean, fast ships with fine lines and tons of canvas aloft. The Horn was the logical route. The driving captain and the ramrod, bucko mate came into their own as men who had the nerve to carry sail that almost drove ships under, if that was what it took to make fast passages. They had to regiment and hound seamen to do it; discipline became more strict and the fo'c's'le a harder place, more like a man-of-war. Clipper ships like the *Cutty Sark*, the *Thermopylae* and the *Flying Cloud* were the epitome of the search for speed under sail. They were the standard until industrial and economic requirements later in the century produced bigger and bigger wooden vessels, then iron and steel ships and four-poster barques "built by the mile and cut off by the fathom," the wind ship's dying form.

Bigger and stronger ships reduced losses and the mortality rates of seamen using the Horn route. Iron hulls and wire rigging enabled a vessel to stand up to battering Cape Stiff weather that would have broken a wooden ship. Running into ice was not as likely to be ruinous. Hundreds of square-riggers rounded the Horn in each direction every year, and most of them made it to port. Nevertheless, the attrition of men and ships continued, and it could be heavy. In an average year, a Cape Horn sailor had a one-in-twenty chance of dying during a voyage. In bad years, the chance doubled or tripled.

Square-riggers were lost in a variety of ways. At one end of the spectrum of cause and effect were the obvious and least destructive circumstances—for example, the four-masted barque *Pindos* was gutted on rocks and lost, but the twenty-seven men of the crew were saved. More seriously, a ship may get stranded and become a total loss, suffering some casualties. Worse, a vessel may founder in the open ocean or run into ice in the south, but survivors are able to take to the boats and win through to somewhere. (Even one man will do. Ishmael, for example: "The devious-cruising *Rachel*, . . . in her retracing search after her missing children, found only another orphan." The last man living from the *Pequod* can tell the story.)

Or a vessel disappears. There are no survivors, but later, searchers or passing ships find wreckage floating at sea or pasted into the crevices of a rocky shore. The ship is classified as "lost with all hands," and even though the exact circumstances of the end remain veiled, the location and condition of the wreckage may provide some clues or propose a reasonable explanation (the figurehead of the *Loch Vennachar* washed up on Kangaroo Island, off South Australia; a sole survivor from the Tasmanian barque *Brier Holme* staggering ashore near Hobart; a few fittings on the west coast of Ireland that might have been from the *Bay of Bengal*, bound from Cardiff towards Taltal, Chile).

It happened so easily and so often to the far-ranging wind ship, silent and incommunicado. And the vessel didn't have to be sailing the edges of the world. Ships disappeared in narrow waters as often as in the intractable wilderness of the roaring forties. Because they were snugged down and secured for sea, they often gave up little or no wreckage or gear if they went down in open ocean. Still, some lee shores became repositories for the bits and pieces of the ships and their crews: the west coast of Ireland, Biscay, the southern coast of Chile and the islands in the South Atlantic to the east of the Horn.

> *It tosses up our losses, the torn seine,*
> *The shattered lobsterpot, the broken oar*
> *And the gear of foreign dead men. . . .*

When Ernest Shackleton and his five companions stumbled ashore on the windward coast of South Georgia after their epic open-boat voyage to seek rescue for their marooned crew—their desperate mountain-and-glacier climb still ahead of them—Frank Worsley, once a square-rigger seaman, reported the assorted flotsam of a hundred Cape Horners scrambled along the rocks and sand of the remote island: topmasts, figureheads, handrails, broken skylights, gratings, casks, the ribs of ships, the bones of men.

There was a procedure ashore when a ship failed to show up. First, it was posted as "overdue" in Lloyd's Shipping Intelligence. This was alarming for the owner and for the relatives of the people aboard (although it triggered a flurry of commerce, as some optimistic insurers reinsured the overdue ship at a hefty premium, counting on its eventual appearance). No one was surprised when a ship became overdue. It could be anything: bad luck in the doldrums, unusual headwinds, more gales than usual off Cape Stiff, a foul bottom slowing down an already unwieldy iron hull, a stranding on some non-fatal shore. Everyone waited. In three weeks, maybe four—a little longer if the ship was a known wagon—Lloyd's changed its posting to "missing." All hope was not immediately lost. Missing ships turned up sometimes, the victims of mutiny or the captain's decision to run off east and get to the Americas by way of the Cape and the Indian and Pacific oceans. But hope dwindled, shrank and tailed away, day by day. Maybe survivors would show up somewhere eventually, more dead than alive in their lifeboats, and a cause could be ascribed. Maybe the ship remained "missing." It had disappeared and taken the riddle of its end with it forever. The dreadful enigma is complete.

Land-dwellers, and seamen, were intensely curious about the circumstances in which ships went missing. Anyone who wanted to travel anywhere across a sea or ocean had to go by ship, of course, and sometimes, they had to brave the worst sea conditions in the most isolated parts of the planet: New York to San Francisco round the Horn (before the railway); Britain to Australia, or back, through the Southern Ocean. Naturally, people wondered about the particular cause of a ship's abrupt disappearance from sight and sound—ice? Waves? Rocks? What was the sequence of events? How did the crew respond (assuming they had a chance to do anything)? Maybe it happened quickly, leaving not much time to react. And always, the morbid fascination with what it had been like for those aboard in the final hours, or minutes, of their lives, when they knew they would die. What had they done? How had they behaved? With dignified resignation? Fighting to the end to save the ship? Or with panic, a desperate, hopeless struggle to keep water out of their lungs or the cold from killing them?

People must have thought about missing ships the way we contemplate downed airplanes today. We ask ourselves about the causes of a crash, and about what it was like for the people on board during the last few minutes or seconds. They're the same questions that sea travellers dragged up out of their own well of nightmare and fear.

"In the word 'missing' there is a horrible depth of doubt and speculation," wrote Conrad. He got a hint of what could happen one day in 1887, when he was first mate aboard the Scottish barque *Highland Forest*, on passage from Amsterdam to Java. In the chaotic seas left after a three-day Southern Ocean gale, the vessel rolled so heavily that some piece of gear or other aloft broke away. It was serious enough that the mate had to go up to see to repairs. Conrad took the carpenter and a couple of hands with him. Sometimes they had to drop everything and hang on with arms and legs as the barque threatened to roll them into the sea. The gale had driven the vessel much farther south than they had intended to go. Suddenly, the carpenter saw something ahead in the "empty wilderness of black and white hills." The men stared at it for some time before they realized that it was an iceberg, as low in the water as a raft, melted down but still big enough to sink a ship. With no time to get down to deck, Conrad shouted a warning from aloft and the barque was steered just clear. Later, on deck, the captain (whose name was McWhir, later to appear in *Typhoon*) mused to his mate as they looked astern at the ice: "But for the turn of that wheel just in time, there would have been another case of a 'missing' ship." (During the same heavy-weather passage, a seaman died and a flying spar struck Conrad, injuring his leg; he spent six weeks in hospital in Singapore and had to resign his berth.)

Dana had a similar near hit. On the *Alert*'s second attempt to round the Horn on its homeward-bound passage from west to east—supposedly the easy way, with following winds and current—the brig was beating south against persistent aberrant easterlies, hoping to get round the big icefield that had stymied its first try. In the middle of the watch on a misty afternoon, all hands were called out by "a loud and fearful voice." As they came on deck, Dana's watch saw that the helm was already hard up, the after yards were shaking and the ship was beginning to wear round. They tailed on to the braces, driven by the captain's orders "as though for life or death." The lines were stiff with ice in the intense cold, and the vessel turned slowly. As it stood off on the other tack, Dana finally saw the danger: an ice-island looming out of the mist, and around it, pack ice, dim and rolling in the swell. "In a few minutes more, had it not been for the sharp look-out of the watch, we should have been fairly upon the ice, and left our ship's old bones adrift in the Southern Ocean."

One of the worst of the bad years around Cape Horn was 1905, when the majority of the ships making the passage were big, first-class iron or steel vessels (like the *Beara Head*), almost as durable as wind ships could be. The weather near the cape that year, especially during the winter months, was execrable, and ships foundered and ran up on the rocks and disappeared off the face of the sea by the scores. To take only a small portion of the shipping traffic going one way round the Horn: 130 wind ships left European ports between May and July (sixty-two British, thirty-four French, twenty-seven German, a scattering of four other nationalities), bound for ports on the west coast of the Americas. Of these, fifty-two arrived at their destinations intact and more or less on time, four were wrecked, twenty-two put into port with storm damage incurred off the Horn. Of those ships, four British and one Danish were posted as total losses. By the end of July, fifty-three ships were still unaccounted for. Most of the overdue and silent vessels eventually turned up at their ports of destination after unusually long and arduous voyages. Four vessels, the *Glenburn*, the *Alcinous*, the *Bay of Bengal* and the *Principality*, fell into the bottomless and mysterious hole of the "missing." These numbers do not include ships sailing in other months or leaving from North American ports, or those running before the gales from west to east across the Southern Ocean from Australia. Casualties in the latter group were bound to be relatively lighter, since they spent much less time off the pitch of the Horn. Nevertheless, the captain of the *Invergargill*, bound to Queenstown, Ireland, from Sydney, said that the gales off the Horn in 1905 were the worst he had experienced in thirty years as a captain.

The ship *Garsdale* was dismasted and abandoned off the Horn; its entire crew was rescued by the French ship *Bérangère*, which stood by in atrocious weather and put its own men in acute jeopardy to save the twenty-three Englishmen. The French sailors shared their meagre food and clothing with the saved men. No one knew about the heroic rescue until the *Bérangère* arrived in Greenoch, Scotland, many weeks later. The *Bidston Hill* went down off Staten Island, and eighteen men were lost. The ship *Agnes* burned off the Horn; a few men survived and were picked up by an American vessel. Winds well above hurricane force literally shook the masts out of the *Alcyon* four miles off the Islas Diego Ramirez, on the supposedly safer homeward-bound route. The crew made for the islands in their two boats. One disappeared in the attempt. The other reached land but was destroyed in the surf. Two men survived out of the entire crew. They lived on albatross and melted snow until they were spotted and rescued by a Norwegian ship.

It wasn't strange that they were seen; ships kept watch as a matter of course for shattered survivors from their brotherhood. The veteran Cape Horn captain James Learmont always ordered a close lookout, especially onto the shore of Staten Island, whenever he ran through the Strait of Le Maire; sailors from foundered or wrecked ships were often there, waving or lighting smoky fires, a passing ship their only hope before exposure and starvation did them in.

The British Isles was a big full-rigged ship with a reputation as something of a brute that lost men on each passage. A vessel its size rigged as a three-masted ship (square-rigged on all three masts) had gear so heavy that it guaranteed exhaustion and killed men. It was an advertisement for the efficacy of the four-masted barque rig: a greater number of smaller sails made a vessel easier to handle, especially with one mast's worth of fore-and-aft canvas. The British Isles left Port Talbot, Wales, in June 1905 with a cargo of coal for the Chilean nitrate port of Pisagua. Twenty seamen signed on for the passage. According to the main narrative of the voyage, in the heavy weather and bitter cold off the Horn, the ship lost three men overboard; three others died from injuries (caused by boarding seas or breaking gear), while two were permanently and three partially disabled by injuries and frostbite. The captain amputated one man's gangrenous leg with the cook's knives. The ship took seventy-one days to round the Horn from fifty south to fifty south. It was long given up for lost when its surviving men and boys worked it into the roadstead in Pisagua and let go the anchor.

Almost all wind-ship voyages were anonymous, witnessed only in the laconic, bitten-off entries of the logs. For ships that were wrecked or burnt or foundered, even those records are absent, and the story of the vessel's eventual end died with the survivors. Of the ships that went missing, we know nothing at all, no maritime version of a "black box" to play back the courage or panic of the last moments. Their men died without a sound or a sign. In the case of the British Isles, however, we know a great deal about its 1905 passage because two books describe it in detail. Even so, there are uncertainties and ambiguities (a signifier of sea-voyage narrative false memory syndrome). There's some dispute, for example, about the number of men lost. The figure of six is given by one of the apprentices aboard, W. H. S. Jones, who later became a master mariner and wrote about the voyage in The Cape

Horn Breed. The captain, James Barker, a superb seaman but an unrelenting driver, also wrote a book, *The Log of a Limejuicer,* in which he described four deaths. Three was the more likely number, according to the seaman-writer Alan Villiers, who studied the logs and other documents. Not that it mattered much in the end, said Villiers: "God knows three were enough to lose from any ship even on a Cape Horn winter's rounding, especially going one at a time."

The year of 1905 was vile but not unique. In 1907, for example, six big ships disappeared off the Horn and were posted as missing, taking at least 150 men with them. Some of the losses had known causes. The *Glencairn* and the *Indore* were wrecked on Staten Island; the *Tillie E. Starbuck* had to be abandoned. The *Metropolis* was knocked down, and its crew resorted to the desperate measures of cutting away its fore and main topmasts and jettisoning 250 tons of cargo before it hobbled into Port Stanley. The *Gladora* was damaged and towed into Puerto Williams, on the Beagle Channel, for repairs. The Italian barque *Volturno* was dismasted and stranded; two of its crew drowned. The Welsh barque *Denbigh Castle* gave up trying to round the Horn and ran off to the east, arriving, like a ghost ship, in Freemantle, 223 days out and presumed long lost.

Between 1904 and 1908, twenty-six big British Cape Horners were posted missing. They carried fifty-nine boy apprentices, 312 able seamen and a larger number of ordinary seamen, idlers and officers.

Apart from such disasters, the low-level attrition continued, ship by ship, man by man, right up to the end of sail. A sample Lloyd's news item from the 1930s in its entirety: "Hamburg: *Padua,* s.v. [sailing vessel] from Pisagua. Weather damage. Lost four men."

It has never been a cape like any other; it is even distinct from its familial "stormy capes," Leeuwin and Good Hope. Random natural forces created the Horn as a promontory, a natural formation of land jutting down into the high southern latitudes. If that was all it was, the cape, and the sea and weather conditions surrounding it, would be a mere hostile curiosity, a thing for humans to avoid, like Antarctica or the deep ocean. But humans could not avoid it. For a hundred reasons of commerce, science and war, ships had to get round it, and the Horn became Cape Stiff, not only the sailor's desire and ambition, but also his nemesis and nightmare. Because the seas there are the worst in the world and the weather unequalled in its severity, the encounter with the Horn became the archetype of the ancient struggle between man and the sea. Nothing was worse than rounding the Horn, and nothing distinguished the sailor more than doing it. Melville wrote that it is the place that "takes the conceit out of fresh-water sailors, and steeps in a still salter brine the saltest. Woe betide the tyro; the foolhardy, Heaven preserve!"

At first, the Horn beat down or destroyed any man who tackled it. Surviving a rounding attempt was not even likely. Later, the chances improved—to 50 per cent, then to 60 and so on. By the nineteenth century, rounding the Horn had become merely very dangerous, something a seaman would live through more likely than not, although it was never anything like a sure thing. "At the present day, the horrors of the Cape have somewhat abated," wrote Melville in the 1840s. Somewhat.

The Horn became the synecdoche of the wilderness of all the stormy seas below forty degrees south. Seamen like Melville and Dana felt able to take the chance that they would survive sailing round the Horn, and then, because they were also writers, to use the cape as a literary device, a symbol of the intractable world in which humans struggle and suffer (Melville's "Spirit of the Cape," which disposes of human will and aspiration as it sees fit). There is a Cape Horn for every man, says Melville. Like Ithaca, the Horn is a reason for the voyage and its true end. After it rounds Cape Stiff, the ship stands on, making for Valparaiso or Pisagua or San Francisco to offload its cargo, but that's merely a coda.

CHAPTER NINE

The Bloody Horn

The true sea—the sea that plays with men
until their hearts are broken, and wears stout ships to death.
JOSEPH CONRAD, <u>The Mirror of the Sea</u>

And now the STORM-BLAST came, and he
Was tyrannous and strong:
He struck with his o'ertaking wings,
And chased us south along.

With sloping masts and dipping prow,
As who pursued with yell and blow
Still treads the shadow of his foe,
And forward bends his head,
The ship drove fast, loud roared the blast,
And southward aye we fled.
SAMUEL TAYLOR COLERIDGE, "The Rime of the Ancient Mariner"

Deck log, Beara Head, *barque, Liverpool to Valparaiso, July 15,
1885: 56° 05´ S., 65° 52´ W. Wind WSW 9, 10. Blowing a full gale.
Sky threatening appearance. Continuing under lower topsails and fore
course. Very heavy squalls blowing at hurricane force with snow and
sleet throughout the day. 2 P.M., took in fore course as wind increased to
near-hurricane strength from the west. Headreached to the south-
southwest at two to three knots. Barometer continuing to fall. Mizzen*

*topsail, spanker topsail and inner jib split and blown to pieces. Very
high sea filling the decks.*

The snorter came down on them like the wolf on the fold, the entrance of the villain right on cue. The *Beara Head* had run through the
Strait of Le Maire in a blind brazen rush with a gale-force wind on its
beam. That had been a regular Atlantic gale, a known beast with a
measurable, predictable size and power. It would blow its allotted
time and then fall away to a breeze and a gentling sea. Or so it had
seemed to the men on deck and aloft, straining to see land and hoping
not to. South of the strait, in the bellows of the Drake Passage,
Antarctica two fast days' sail south, abruptly open to the full-bore on-
slaught of the Southern Ocean, the wind and waves were like a meta-
morphosed breed. Everything about them was more intimidating:
sight, sound, strength, duration. It was like coming out of the foothills
into the mountains, or turning a sheltered corner into a full blast. And
behind the immediate sensations was the suggestion that there might
be no pause, that this great natural force might exert itself against
them without end. That was what shrivelled a seaman's guts. In other
oceans, the storms came and went, but there was a chance in between
to recover and rearm. In this great Southern Ocean, storms might be
separated by the barest of margins, one depression riding on the back
of the other, another one behind it and so on. Snow, sleet, hail, cold,
long darkness of winter nights, icebergs if the ship was forced to flee
far enough southward. This war against the Horn began unfairly: the
Southern Ocean wind fresh and invigorated by winter; the ship, seventy days out, its gear worn down and the men worn out by too little
sleep, food like swill, incessant hard labour. Seamen could do only
one thing: be of good and stout heart to pit against this sea's "tiger
heart."

Deck log, Beara Head, *barque, Liverpool to Valparaiso, July 16,
1885: 56° 30´ S., 65° 45´ W. Wind WSW 9, 10, 12. Continued blowing a very strong gale from the WSW with heavy squalls throughout the
day. Intervals between squalls from 5 minutes to 20 minutes. Barometer low but steady. Continuing snow and sleet. Headreaching to the
SSW under fore and main lower topsails and lower staysail. Sky a*

wild and threatening appearance. Spanker gaff carried away. Main royal blown out of gaskets and lost. Heavy seas over the bow and filling decks. Port gig shifted on skids and stove in. Saw two barques to northeast.

Getting a ship round Cape Horn could be the most difficult job a captain would ever do at sea. Some roundings were short and sweet; others were longer but no worse than a gale in lower latitudes; still others were on a par with an Indian Ocean cyclone, say, or a series of storm systems sweeping over a ship on the Australia run through the roaring forties of the southern Pacific and south of the Horn to home. But if the weather was especially bad, with prolonged westerly gales in the vicinity of the Horn, then even by the extravagant standards of seamen's suffering, rounding the cape was the worst thing of all.

Considering how hard a battle it could be, the rules of engagement with Cape Stiff were terse. They were most succinctly stated by the British captain James Learmont, who set the amazing, and lucky, record time for a rounding in the barque *Brenhilda* in 1902—five days, one hour, from 50 south to 50 south. It was much the same as watching the cyclonic depressions that race across the Southern Ocean. "You've learned the signs for shifts of wind—the slight clearing in the southwestern sky, a movement in rising cloud, then the swift, sudden shift. It's the same off the Horn except the wind is madder there, the shifts faster, nights longer, sea higher, ice nearer—you know the odds." A captain must watch the weather; always be ready to put her round; never be afraid to wear ship the moment it will pay (every time the other tack will give her a better course); get sail off in time but put it back on too, every chance he gets. And look after the crew. "You'll get no sleep. You'll get so wet so long your skin will come off with your socks, if you get time to take them off. But, with luck, you'll get past Cape Horn, and, by the grace of God, you won't kill anybody."

The Admiralty sailing directions were equally laconic, if humourless. First, there's a hopeful note: even though they're mid-winter months in the southern hemisphere, June and July can be the best time for rounding the Horn because there's more chance then that the gales will be in the eastern quadrant, rather than the western. However, the days are short and the weather cold. Then come the "instructions," which boil down to this: Head west whenever you can. Sometimes it's a good idea to head south first; a ship has a better chance of finding easterly winds at 60 south than farther north. And there's no point in trying to hug the cape, scold the Admiralty seers, even with a fair wind, because the east-setting Southern Ocean current is strongest there. Better to stay south, at least as far as the Islas Diego Ramirez. Sailors call them the Rammerees.

So much for the theory.

Deck log, Beara Head, *barque, Liverpool to Valparaiso, July 17, 1885: 56° 42´ S., 65° 40´ W. Wind WSW 10. Heavy squalls continued until forenoon. Heavy seas coming over bow. 4 P.M., mizzen upper topgallant yard carried away and secured with great difficulty. Towards nightfall, squalls increased again, some to hurricane force. Barometer beginning to rise. Driven off farther to east. Heavy snow at times. Saw barque to the north. Number of albatross and other birds about.*

He had never seen a seascape like it and could not have imagined it. Benjamin wedged himself into the railings of the windward poop steps and looked out over a sea that writhed and heaved with such monstrous, uncontrollable energy that he was amazed that the barque continued to live within it. He looked up at wave crests like waterfalls toppling down eighty-foot-high hills, then down into troughs that seemed bottomless, the sea there merely a darker part of the deep hole he teetered above. This was not his first gale, but it was his first hurricane wind, and his first time in the mythic Southern Ocean.

Going aloft to secure the mizzen upper topgallant yard took every scrap of courage and dreg of energy he had. The wooden spar had broken away from its port-side restraining lines, and it swung back and forth across the width of the ship like a scythe cutting air thick with salt spray and snow. It was fifty feet long and a foot in diameter. Both watches went aloft to grapple with it. The yard was beyond control, and it crushed no one and smashed no one into the sea only by pure luck and the desperate agility of the men. To everyone's surprise, it was the second mate who did the job. He slipped a running bowline over one end as it lunged by and managed to take two lightning turns around the mast with the line's standing part. Then, somehow, he hung on and absorbed the shock of the yard's swing to leeward. This slowed it down enough for half a dozen hands, and then a dozen, to get hold of it, lashing it tight to the topgallant mast and shrouds.

Benjamin played his part in this sharp skirmish: a skinny paddy who had gone to sea for adventure, maybe transportation to the New World, 120 feet in the air, clinging with bloody hands to wire and rope that thrummed and sang in the wild wind, the deck below near invisible through the blowing snow of the sea-blizzard, its atonal scream and shriek in the labouring barque's rigging blocking out all other sound, his shipmates' shouts reduced to enigmatic lip movements. The only other sound Benjamin could hear was himself humming and whistling—to himself—the same few tunes over and over, Christmas carols: "Good King Wenceslas," "Deck the Halls," "The First Noël." It was something he'd done for as long as he could remember. He'd whistled under his breath as a boy, directing his troops against the neighbouring Fenian gangs, and later, when the little coasting smacks ran into a forty-knot wind and he had gone forward onto the slippery, dancing foredeck to muzzle a sail on some black night. "Once in Royal David's City." He sang often aboard the *Beara Head*. When he was on deck watching the sea, waiting for something to happen, or lying in his bunk wondering what it would feel like—the inevitable beginning of the inevitable end—the insistent music in his head hardly stopped. He hummed and whistled the tune of some old familiar song as a kind of audible talisman, familiar and comforting when he was afraid.

Everyone aboard the barque carried the burden of fear and wonder bred by storms and the drawn-out passage, emotions felt most intensely now in the precinct of the Horn. Each responded in a different way.

She had never imagined she would feel like this. The captain's wife stood just out of the wind in the companionway scuttle, looking forward through horizontal snow and spray, the bowsprit a ghost of a spar in the distance. She watched the seamen climbing the windward ratlines towards the careering yard that her husband said might tear the sticks out of her if they couldn't get it secured. The men receded into the murk above her, moving slowly and deliberately, their soul-and-body lashings unable to restrain oilskins that flapped and rode up over their heads and arms. At times, the seamen froze as the breakaway yard whipped from side to side above them; from where she stood, it seemed to miss them by inches. The light nearly left her eyes, she told her husband later. Clamped onto the rigging, the men looked like a cluster of black insects with odd, flickering wings. As she had so often before, she pitied them their ordeals. She had seen what they ate—as different from her own diet as a vassal's from his lord's. When they came up onto the poop to handle the jigger sails or the mizzen braces, she had seen their bloody hands, the suppurating saltwater boils on wrists and necks. Even the burliest of the men had become lean; their eyes stared with fatigue. But not fear. That had been a surprise. They always looked to her like men who had a job to do, and who were getting on with it. Even when the *pampero* hit and she thought the ship was lost, the seamen had worked with steady determination and without a hint of despair that she had seen. During the darkest, wildest hours of the gravy-eye watch, when this Cape Horn gale always seemed at its worst, she heard the many-voiced barbaric chorus singing out on the poop above her as they jerked the lines, following the unheard chant of the shantyman, his single voice lost in the storm.

She had even changed her mind, a little, about the mate. He was still impossible to understand, or to make contact with. His obduracy was impregnable. He filled the cabin with his muscled bulk and wide head, his wind- and sun-polished face. She heard him all over the ship, driving the men with that great voice, like a doomsday weapon, insulting, haranguing, threatening. He had driven that poor little pressed clerk to kill himself; she had no doubt of that. He was a dangerous man, his violence constrained only by the ship's discipline—her husband's unquestionable authority—and perhaps, she had to admit, by the mate's own sardonic rationality. She was never sure that he wasn't acting a part. Maybe ashore he dropped his role and became a human being with whom one could have a conversation. . . . No, he wouldn't change. He was an exile; a life spent at sea in the stringent routines of sail had created a hard man aboard ship and a monster ashore. In spite of all that, she saw his strength and unfailing competence. After the *pampero*, he had been magnificent. When she heard his booming voice, it was comforting beyond measure, the strong human sound in the chaos of wind and sea. She admired her husband even more for his control over this Caleban than for his own undeniable skill with the ship and its endless gear.

The only harsh words she and her husband had exchanged in the unprivate close quarters of the aft saloon and their own small conjugal cabin had been about the seamen. He treated them like slaves, she said. They are slaves, he said, or, softening a little, much like slaves. Aboard ship, they belonged to him; his word was the law. They did what he said when he said; he would punish any man who did not. That was the way things had to be—because the sea itself showed no mercy, had always to be watched, and seamen primed to jump at the instant. But that didn't mean they couldn't be treated with some courtesy, given better food, she said, half complaint, half question. They signed the articles voluntarily, and they understood only a whip of a voice and the whip itself. The food was what the Board of Trade prescribed—no more, no less. Some of the men were all right—the two Yankees in the port watch, for example, or the Scowegians in his own—good men who knew what to do and took the initiative. Others didn't have a whole brain among them; he had to drive them. Anyway, they were all merely his means of carrying out his commission from the owners to get the ship and its cargo to Valparaiso; they weren't expendable exactly, but he'd damn well expend them if he had to. She could neither agree with nor win this argument. Her sympathy for these sorely tried men grew day by day as the Cape Stiff weather ground them down.

After Jagger and the seamen had secured the yard and climbed down—she was amazed, as she had been so often before, that they were all still there—she stood for a long time, shivering under the scuttle, watching the sea. She felt the biggest surprise of all: she did not want to be anywhere else on earth. Who ashore could ever imagine a sight like this ocean in turmoil? The barque's plunge and buck into the face of the gale seemed to her like a valiant and stubborn battle against great odds. *They that go down to the sea in ships . . . These see the works of the Lord, and his wonders in the deep.* She found it difficult to imagine that she could ever again endure the constrained ceremonies of the drawing room, making polite conversation with the ladies, pouring more tea for the vicar, Sunday promenades along the front in Bangor. The sea then had been a mere backdrop, pretty or sullen, to her walking and talking.

She would never be able to ignore it like that again. Now she wouldn't be able to keep her eyes off it. She would watch the patterns of waves in Ballyholme Bay, which was open to the sea, searching for signs of what would happen next. Watching the sky too. Everything there meant something. You could read the clouds—their colour and shape, their complex layers—like a book of prophecy, limited in time to be sure, but infallible if you knew the portents. The wind would never sound the same to her as she lay in a snug bed ashore. She would listen for its tone and pitch. Thirty knots? Forty with gusts to fifty-five? She could tell the weight of wind now by its sound. On dark nights, when winter gales rattled and banged their way across the land, she would feel the sailor's ambivalence: Thank God I'm here in this unmoving house and not aboard some rolling, threatened barque, but I wonder what the heart of this gale is like. What is happening out there? What sail would she be carrying now? Shortened down to topsails and fore-reaching? Or running off before it with topgallants as well? Decks awash, the seamen gathered in their small number by the poop, watching and waiting. The sense of a grand accomplishment afterwards, as the wind slackened, waves dropped, sun broke through; the feel of life renewed and sweeter by far than it had ever been before.

She saw something else too aboard the *Beara Head* as its battle to round the Horn got well underway. The officers continued to direct the ship's campaign with their calm and methodical competence—her tough, beloved husband; the fearsome mate; even the second mate, whose confidence and strength she could see growing by the day. Yet there was much less of the bullying and hazing, the shouting she detested. Sometimes, its absence had astonished her. When the crew was aloft, struggling to restrain the yard, she saw the mate on the poop, head craned back, never taking his eyes off them but saying not a word. More than anything else, his uncharacteristic silence demonstrated for her the desperate danger of the men aloft. When they climbed down, the mate didn't even berate them in his usual foul argot (what a new language she had learned on this voyage!) for taking so long; he dismissed the watch, and that was that.

The men were changing too. The Yankee sailor, Urbanski, and the grinning Greek, Kapellas, had called off their incessant and bitter bickering—like frustrated lovers, she thought with surprise one day. The cockney, who had been a month in irons and always looked at her with hatred (although he looked at the entire world that way), seemed calm now—no loud, obscene rants. She could see it amongst all of them: a binding together, a mutual solicitude. Perhaps that was what always happened down here near this cape—the crew nourishing itself in brotherhood in proportion to the rise in power and malice of the sea.

The westerly gale blew itself out after three and a half days. More accurately, it blew past the *Beara Head* towards the east, still strong and fierce, full of life. Without the obstructions of land, the vast whirl of the storm's low-pressure system could stay alive and maintain its heft over the long stretch of the southern Atlantic until it encountered other vessels on the shipping tracks that swung to the south of the Cape of Good Hope. It would terrorize them too as they surged and surfed towards Australia, running their easting down in the roaring forties and the furious fifties. The Southern Ocean waves from the same gale, rolling round the world, could test and destroy ships ten thousand miles apart. It was nothing personal, nothing like an animal predator that got the taste of human blood and then sought it out, a new-minted man-killer. A gale was dispassionate and aloof. It existed and moved and tore the bejesus out of ships for a time, then churned on. They survived or they didn't. The storm, and the Southern Ocean that nourished its long life, was as unconcerned with human existence as ancient rocks or stars.

By the evening of its seventy-fourth day at sea, the *Beara Head* had made a little cherished westing—a score of miles. Mostly, the barque was heading farther south. It had to because the wind was still out of the west, and the waves left over from the gale remained high and more sloppy and disorganized without the ordering strong wind. They came at the ship in an endless succession of serried watery walls thirty-five feet high, breaking over the bulwarks and filling the decks, battering the starboard bow, driving it off to leeward, forcing the vessel back off its course and to the south. The captain was content enough. They had sidestepped the new gale that had threatened them from the northwest—he had read the signs in cirrus and cumulus and a watery sundog. Sixty south was where he wanted to go. Maybe down there he'd find one of the Admiralty's easterly July winds. He would be happy with even a wind-slant north or south of west so that he could drive along for a few days, clawing his way through some longitude. He could still make a reasonably quick passage round; the recent gale might be their last setback.

Deck log, Beara Head, barque, Liverpool to Valparaiso, July 20, 1885: 57° 57´ S., 66° 46´ W. Wind W. 6, 7, 6. Fresh breeze and squally. Continuing to make some westing. Seas still running high. Heavy masses of cloud to the north. Very cold. Saw two large icebergs and smaller ice pieces to the south. Ship in company. Port boat repaired. 120 miles east of longitude of Horn.

The air was so cold that the sea felt warm. Benjamin saw the ice to the south. Its sight was merely the latest thing to scare him in two and a half months of novel heart-stoppers. He certainly wouldn't be dozing off standing up anytime soon while on lookout on the frigid, water-logged fo'c's'le deck. Gales were one thing, known quantities you could see and fight against. They ended, and you had a reprieve, however brief, until the next one. But he had a sense of ice as a silent, constant presence—a passive, sneaky ambusher of ships. Ice ripped the guts out of a wooden vessel. He knew that an iron hull had a better chance of surviving a collision. The old hands had told him about ships that had lived through such an event and limped into a port of refuge, their bows smashed and blunted like an old prizefighter's nose. But there were all those ships that just went missing, as if they'd sailed off the edge of the Finn's flat earth. Who knew what happened to them? Iron ships as well. Maybe some of them had hit ice; at just the right angle of impact, it could perhaps slice into the hull, ripping it open, spilling coal or grain into the sea as the sea rushed in faster than any pumps could handle.

The cold was beginning to affect the crew. The only time they were anything other than wet and cold was after a few hours in a bunk, all-standing under a blanket. Body heat, the only source of heat on board except the doctor's stove, steamed a man warm and his clothes halfway dry. Until he went on deck again. Then the sheets of spray and the deck-sweeping waves soaked and chilled him down again in two minutes. Frostbite had already appeared. After the desperate work aloft during the gale—in reality, it had been a blizzard—and the two days of sail handling and salvage that followed, both Maguire and the

Elf had developed spots of pale, cold skin and, here and there, blisters on their hands and faces. Some of the skin was turning black. Not good, said Paddy. That was how gangrene started, and then the only solution was to amputate.

"Are ya goin' to cut off me fuckin' head, then?" said the Elf, his right cheek hard and blackened.

The frostbite was a bad sign, thought Benjamin. It meant debilitated seamen. And the ship was still heading south into the heart of the southern winter. The air and sea seemed to get colder by the hour. A gale down here would be an ordeal he dreaded already.

They weren't alone. About two or three miles on their leeward side, he could see a three-masted ship on a parallel course. It was a fascinating reflection of what he supposed the *Beara Head* looked like: a wind-beaten, weather-stained, rust-streaked ship; seas breaking aboard continuously; its hull and half its tophamper dipping out of sight in the troughs; a third of the hull's length exposed as the bow reared and pitched into the Southern Ocean's big swells, beating hard across Drake Passage towards the great southern continent, hunting for a chink in the wind through which it could drive a few miles of precious westing. Benjamin wasn't comforted by the presence of the other vessel. On the contrary, its spectral profile and ungainly progress seemed to emphasize how far away they were from the world and its sanctuaries.

Deck log, Beara Head, *barque, Liverpool to Valparaiso, July 23, 1885: 59° 07′ S., 67° 20′ W. Continued blowing a strong gale from the west. As I did not wish to run farther to the Southward, I wore ship onto the port tack under lower topsails. Barometer continuing to fall. Wind W. 10, WSW 11, W. 10. Squalls with higher wind. Heavy seas filling the decks. Lower main and mizzen topgallants split and lost before the hands could bring them in. Heavy dark clouds with the sun at times showing out very bright.*

The gale that Benjamin feared had come. There were no easterly winds at sixty degrees this year. First, there was snow. Then it stopped, and the wind blew hard and constant under low, scudding cloud, the sun showing through from time to time and a half-moon and stars at night. They got all sail in, except the lower topsails, their Southern Ocean storm rig, before conditions got too bad. The captain wore ship onto the port tack when thick snow was still falling but before the real weight of the wind slammed into them. The change of course to the north was a relief for everyone. It was illogical to find comfort in it—the ice was still lurking around them—but they felt that they were sailing towards relatively safe water. Some imagined they could feel the sea warmer within a few hours of wearing ship. Beating north, soaked and frozen, into the ice-laced seas of a Southern Ocean snorter seemed a much better proposition than beating south into it. The seamen's knowledge of the ship's position was rudimentary; the old hands could only put it somewhere east and south (maybe well south, by the feel of the sea and air) of the Rammerees. Neither the captain nor the mate shared navigational inspiration with the men. Like soldiers in the line, the seamen wheeled north or south, or sat tight, depending on their orders, but no one was about to confide in them the whys or how-fars of it all. In fact, the *Beara Head* had done better than they thought: when the new gale hit, the barque had worked its way west to almost the exact longitude of the Horn, although the cape now lay more than two hundred miles to the north.

Jagger, the second mate, had the watch, and even though the wind howled unabated at storm force, the captain was below, leaving the deck to him. Not that there was much to do except keep an eye on the lower topsails for signs they might blow out of their bolt ropes, make sure the helmsmen—the wheel needed two men at a time—held the old hooker up to the wind, spot an iceberg before they barged into it and try not to bloody well freeze to death. The men of his watch had found their usual nooks and hiding places below the poop, where they were at least out of the wind, although not the worst of the boarding seas. Ten days since they'd passed through Le Maire, and all they'd managed was this jaunt south into the ice and then back north again, making westing to the Horn and now getting driven back. The ship had lost close to fifty hard-won miles since this gale started up. In a while, they'd be damn well back at Staten Island again.

Still, Jagger felt cheerful. Things had changed for him over those ten days; he had changed too. It was odd how one act could have such consequences; even though it took only a minute, thirty seconds, it could alter your life. Snaring that topgallant yard had been one of those things. In desperation and terror, he had lunged up out of the footrope, unsupported by anything except his thighs braced against the yard and the steady weight of the wind on his body, and flung the bowline at the runaway spar as it swung past his head. Without knowing whether he'd got the loop on or not, he made a dive for the mast and whipped the line twice around it. The sudden jerk as the flying yard brought up astonished him with its force and the instant pain in his shoulders and back. He almost slid off the footrope and might have done so if not for someone—he still wasn't sure who—grasping his oilskin jacket and yanking him back from the abyss. It was all he could do to hang on while the seamen threw themselves at the broken spar and tied it off. Someone else helped him climb down, his shoulders burning, muscles trembling. Back on the poop, before he could report a word, the captain said: "Well done, Mr. Jagger!"

With warmth and a smile! By heaven, he'd remember that moment and those four words for the rest of his life. And he would remember the mate's reaction too. MacNeill looked at him, looked him in the eye, without expression. That was all. But it was the first time Jagger could remember on the entire passage that the mate had looked at him as if he was actually *there*.

Three days so far and no sign of slackening off. It had to be a goddamned monster of a storm. A thousand miles across, maybe two, and moving slowly. The captain's optimism for a fast rounding drained away with each day of this interminable gale. He'd followed the damned instructions—what else was he supposed to do his first time at Cape Stiff?—and gone south looking for an easterly, or even a bloody shift one way or the other in this endless westerly, just a few degrees to give him a favoured tack. Risked the ship down close to sixty degrees, poking around in the ice, freezing everyone's balls off; worked her west to the Horn's longitude and for his sins, nothing but a living gale dead out of the west. He hadn't been able to shoot a sight in four days, but he knew they were getting pushed back to the east—by the current, by the leeway made by these galumpus slab-sided carriers and by the seas blattering his hull every ten seconds and flooding his decks with five hundred tons of water. What ship could beat to weather when it was half submerged half the time? Even this iron ship couldn't keep out all the water, and yesterday in the first dogwatch, for the first time on the passage, he had had to keep the watch at the pumps for an hour, much longer than the usual ten or fifteen minutes needed before they sucked dry. Water got below down ventilators, and enough of it got down the aft companionway and through his own goddamn cabin for that matter. But that much in the well meant that some must be getting through leaking plate joints and around rivets too. He hadn't heard of an iron ship actually breaking up under long stress of weather, or leaking so badly that the pumps couldn't keep up. Wooden ships, yes. He was damn glad he wasn't down here in one of them; he'd be bloody well taking a turn at the pumps himself by now. With iron ships, though, it might just mean that when it happened, it happened catastrophically. You never knew about all those ships that disappeared—what specific affliction did them in. For now, he had no choice: headreach to weather and avoid losing too much ground— "Keep 'er up, goddamn it!" he shouted at the helmsmen, who were sweating with exertion in the frigid wind and spray—if these gawks could be induced not to steer the goddamn ship like women. He'd been through worse wind than this, a cyclone on a Calcutta jute ship when he was a young second mate, almost as green as that damn Jagger (although the boy did well with that spar aloft; he had some promise after all). That time, the wind had stripped away the sails, masts, boats— and half a dozen men to boot. But the old man had kept his nerve, used the wreckage as a sea anchor streamed off the bow, and they'd survived. The old girl looked like a hulk. Still, they got her to the Hooghly under jury-rig, only a month late and just two weeks after being posted overdue. But that had been a warm wind, and they knew they had to endure it for only a day, maybe less. This wind down here was much worse: not as powerful, of course—nothing equalled your tropical cyclone for force—but it was icy cold, and that sapped the sand out of the men, and worst of all, it went on and on. You couldn't predict when it would end, and you knew that when it did, another one was right on its heels to pick up the slack, pronto.

Sooner or later, though, the wind had to veer or back, and he'd take his chance: wear ship in a flash, day or night, if that gave him the best angle; crack on the topgallants if it blew anything less than a hurricane. And then he'd make some goddamn westing. He wasn't one of those useless old poreens, captains who thought it was still the sixties and they'd never built the canal, and the steam engine was still a toy to play with round the coast. They thought they could take forever with their cargo, lie to under lowers whenever it blew up past a full gale, and wait for a shift and a blue sky so they'd skite by the bloody Horn with dry decks. Sometimes it took them a month or more before they found weather that suited them. Then there were those captains who ran off to the east and got to the West Coast by way of a circumnavigation of the goddamn world. What sense did that make? Even if there was only a short gap between gales, the wind had to change direction as the next storm came on, and eventually, it had to give you a lift. The only problem—the injustice of it all—was that this constant west wind made him look like a doddler, not the driver he was. The bloody owners never understood. Even if they'd been seamen themselves, they forgot what it was like out here for their captains. Like generals at the rear, eating and drinking themselves into a self-righteous stupor and damning their soldiers for not charging dead into the guns, like those poor bastards in the Russian War. Well, there was nothing for it but to hang on, wait for the shift and then drive hard. Now he'd go below and see how wifey was. Time for her navigation lesson. What a pleasant surprise that she had taken to this life so well, and to this wild ocean.

Deck log, Beara Head, barque, Liverpool to Valparaiso, July 27, 1885: 56° 37´ S., 66° 40´ W. Wind W. 9, 7, WSW 7. Got upper topsails, topgallants, fore course, lower staysails, jibs and spanker and topsail set and I am able to make NNW on a true course. Nevertheless, I have been pushed north and east. We are only 20 mi. NW of our position 9 days ago having covered 300 mi. during that time. Sky clearing but seas remain high and confused. Barometer steady but high mares' tail cloud. Barque in company. Number of porpoises about.

The high cirrus cloud, the mares' tails whose presence the captain noted with apprehension, didn't, this time, presage another spell of bad weather. The barometer remained steady, and the ominous clouds dissipated. The sun shone occasionally, and the decks of the *Beara*

Head were not awash for the first time in two weeks. In the cold winter sunlight, the pale, frostbitten, boil-infested seamen dried out donkey's breakfasts and clothes so stiff with salt they stood up by themselves. When dry, they had to be beaten with fists into lumps of cloth soft enough to conform to the shape of the body. The men picked lice off each other. The cold weather at least made the bedbugs torpid and gave the sailors peace while they slept. However, it drove the lice to seek the warm, furry crevices and folds of the human body. The deckhouse door and ports were opened to release the thick fug of smoke and stench of unwashed men. The doctor made lobscouse. "Salt pork tastes better to a man in fifty south than a banquet to an epicure," said one old, literate sailor. That was merely making the best of a bad thing. Lobscouse, a hash of salt horse, pounded hard tack (and weevil) and a few soft, mouldy onions, was a seaman's delicacy, an exotic break in the consumption of monotonous food that belonged in a trough. With his clean face and apron and unending politeness, the doctor had also managed to beg a small quantity of marmalade from the aft-cabin table, and he served it with hard tack for breakfast the next morning. The seamen received this largesse with expansive gratitude—a luxurious, self-satisfied stretching of gut and limbs in the sudden and unexpected kindness of Cape Stiff.

On the poop, the captain's wife walked in her heavy shawl, turning her face to the sun whenever it appeared, thanking God for this warmth and near-fair wind. Surely their struggles were over. They had paid the piper, had they not? This wild and savage Southern Ocean had frightened yet fascinated her, but the vast Pacific lay ahead. She eagerly awaited the chance to glimpse its secrets and surprises.

The barque beat into diminishing seas under near-full sail. Most gratifying, the wind had finally given them the slant they needed, going south of west, allowing the barque to make if not westing, at least some respectable nor-westing. The captain by the jigger weather shrouds noted their progress with satisfaction, his anxiety and frustration ebbing away. On July 28, he shot good sights and found the ship a mere fifteen miles south-southeast of the Horn—close enough to see if it hadn't been for haze. To avoid the strengthening Southern Ocean current closer inshore, he wore ship—the seas were still too big to chance tacking, especially with a foul bottom slowing them down—and went south again, really a little east of south, to gain more of an offing. The next day, at about the same latitude as the Islas Diego Ramirez, he wore ship and the barque resumed its encouraging progress to the northwest.

Later that same day, eighty-five days out from Liverpool, the *Beara Head* passed the longitude of the Horn. It was one of the few occasions aboard ship when the captain informed the fo'c's'le just where in the watery world they were—he did the same thing, for obvious reasons, when they crossed the line. Getting by Cape Stiff was significant enough that even the rabble forward deserved to know. There was a feeling of accomplishment and pleasure aboard the ship, but not one of celebration. The captain's wife might well have been mistaken when she thought they'd paid their dues. It happened often that ships got past the Horn, only to be driven back east of it again. Eighty degrees longitude was a long way west—430 miles, to be precise—and that was what they had to aim for before turning north. Only then would they have a reasonable margin of safety to leeward, between the barque's thin iron hull and the rocks of Chile's iron-bound coast. For now, however, the sun shone from time to time, the sea was blue, the wind steady and fair enough. Soon they would have to wear ship and tack to the south again, but if the weather held for two or three more days, they would have room to manoeuvre when things got bad again.

Sleek black-and-white porpoises by the score skimmed and sped about the ship. Sometimes they jumped high and clear of the water. In the vast sphere of sea and sky around them, that was the only worrying sign. Seamen liked and admired these animals, their sleek agility and intelligent regard, but not when they leaped. Everyone knew the old and infallible saw: "When the sea-hog jumps, look to your pumps."

Deck log, Beara Head, *barque, Liverpool to Valparaiso, August 1, 1885: 56° 14´ S., 68° 13´ W. Wind WSW 8, 9, W. 9, 10. Blowing a strong gale since early morning. Heavy seas with breaking cross-seas filling the decks. Sky with low cloud moving fast. Barometer falling. Lying to under lower topsails and two staysails. Lost spanker topsail and main and mizzen lower topgallants. Two barques eastbound to the north.*

When he found out they were to the west of Cape Stiff, Benjamin was elated. He'd done it! Rounded the Horn! He could wear the gold earring, piss into the wind, swagger through any Sailortown in the world.

Too good to be true. That's what Russell said. Benjamin had hoped the Yankee was just being pessimistic for superstition's sake: don't count your chickens; don't jinx things with a fool's bright-eyed babble about having done it; don't bid the hateful Horn adieu until you're at eighty west. Maybe Russell was just a killjoy. He'd be quare confounded when the ship kept beating its smooth way west; he'd eat his Yankee words. Damn him for trying to spoil the moment!

Russell was merely a realist. First the bloody porpoises couldn't stay in the damned water where they belonged, then the sky showed its signs too. The wispy mares' tails appeared, but that didn't mean anything—they had come, and then gone, a few days ago. But later, the mottled cirrus, which looked like fish scales, crept in, and the sun, before it disappeared, shone hazy and weak, dogged by a rainbow ring. The scend of swells rolling in from the west grew by the hour. Benjamin had been more than long enough at sea to know what the signs meant. In the aft cabin, the old man was no doubt watching the glass go down like a mine-slag bucket. Benjamin wasn't afraid of the weather to come. He now believed the ship could withstand anything in this stormy Southern Ocean; gear might break and sails might blow out, but the hull and hatches would hold—unless they ran into ice or a rogue sea. The ship would tuck its head under its wing, lie to, give a little ground and live to fight another day. It was the weariness, the deep-in-the-bone tiredness, that was the worst. That and the pain of oilskins rubbing his skin raw until it bled, and his hands like open wounds, uncooked meat, ten fingers and six fingernails. That wasn't bloody well natural, or nice either.

The *Beara Head* lay hove to on the port tack for two days. The captain had given up any ideas of sailing south—or anywhere else, for that matter—looking for fair wind. The old copper-trade captain James Giles told him once: "There's the enemy; lick him if you can. If you cannot, steer away south and turn his flanks." He'd tried that and got another damned westerly gale for his trouble. Now he would stand and fight here, in the shallow waters between the Horn and the Rammerees, short-tacking and biding his time.

As stubborn and tenacious as the barque was, it could not avoid losing ground. The heavy overcast prevented any sun sights, but the captain's estimated position on August 3, taking into account leeway and current, put them back east of the Horn. On August 4, worried about how close the ship was drifting to the Wollastons, he wore ship onto the starboard tack. The wind had settled into a routine: it blew at storm force or higher without a pause for half a day, then slackened off for a few hours before coming at them again at fifty or sixty knots. Not enough of an interval to crack on some sail. A captain with enough men might have done so to gain a few miles before having to shorten down again. Those were the instructions: don't waste even the time between bells if you can make westing—it's a war of half-hours and cables of progress. But with a short-handed crew, it was just not worthwhile to set sail in these so-called lulls, when the wind continued to blow at forty or forty-five knots anyway. The canvas could be loosed and trimmed in a hurry; the problem was getting it in again an hour or two later. On the wrong side of the Horn again, he had to save the strength of his handful of haggard labourers for many hard watches to come. And it was an even smaller handful now: two men out of each watch incapacitated by frostbite, a man injured during the *pampero* still not fit and an able seaman in the starboard watch with two broken fingers. A boarding sea had swept him halfway out of a washport. As his mates dragged him clear, the door swung shut, smashing the bones. The captain set them, arranging the pieces more or less in line and lashing on a splint, as the seaman, with half a pint of rum in him, grunted stoically. When he saw the watches muster at the poop at eight bells, the captain was always struck, and often appalled, by how few they were: sixteen or seventeen men and four boys to handle a twenty-five-hundred-ton ship. It was ludicrous, outrageous, but it was the way things were in sail those days.

Snow, and sometimes sleet and hail, fell intermittently throughout the gale, and for a few hours one night, thunder and lightning crashed and flashed around the straining barque, suddenly illuminating the falling snow like bright sunlight. Sea conditions were the worst yet, the waves steeper and higher in the shallow waters of the continental shelf, which was only forty or fifty fathoms deep in places. Men aloft looking down could not believe the ship wasn't foundering. Had the hatch covers given way while they'd been up there? For minutes at a time, all they could see were the raised poop and fo'c's'le decks and the deck-house tops; everything else was covered in green water and swirling foam.

"What the Jaysus would she look like if she was going down?" Paddy asked rhetorically.

Deck log, Beara Head, barque, Liverpool to Valparaiso, August 4, 1885: 10 A.M. While we were hove to, a tremendous sea swept the decks, carrying away all boats, stoving in the starboard bulwark from the fo'c's'le deck to the mainmast. The hands' deck house and the galley and half-deck, as well as the aft cabin, were flooded out and damaged. Main and aft hatch covers were ripped partway off and much water below. Lost fore and main lower topsails. Fore and main topgallant masts sprung and yards damaged. Two men were swept overboard, but one was recovered by Mr. MacNeill. As it was not possible to launch boats (there not being any boats), the other man is lost. Managed to secure hatch covers and set new fore topsail. Pumping steadily. I decided to run for the lee of Staten Island to effect repairs. Wind, W. 10, 11 with squalls above hurricane force.

Ten seconds.

The mate heard it coming before he saw it, a deep, rumbling roar that boomed over the screaming racket of the gale. It reminded him of something. What? The sound was like a huge waterfall, except this one was bearing down on them at twenty-five knots. The falls on the Shubenacadie River! He'd been taken there once as a child to visit an uncle. The cataract's steady, unending thunder, its relentless energy—where did so much water come from?—had frightened him. He couldn't wait to tug his mother away from the abyss towards a nearby building, a comforting human thing to enclose him from the fearsome rush of nature.

Seven seconds.

"Bear up! Hard! Hard!" he yelled to the helmsman, leaping to the wheel himself, throwing his body onto the spokes.

Five seconds.

"Montyreevo the watch! Lay aloft!" he shouted to the men at the break of the poop.

They had to hear it too. They had to get off the deck into the rigging. Whatever was coming at them was a goddamn big rogue. Anything on deck, anyone, was a goner.

Three seconds.

The sonofabitch wasn't coming round fast enough! There wasn't enough time. They had more of a chance head-on. But this pig of a wagon wouldn't turn!

"Turn, you fuckin' bitch!"

Two seconds.

He could see it now as it emerged from the mist and low cloud surrounding the ship. It was a big rogue, all right, a black wall rearing over the bowsprit. The top third—thirty feet? forty?—was breaking white foam, tumbling down the wave face like the waterfall he had remembered. As high as the topgallants! He saw that it would hit the barque at an angle, about two points on the starboard bow. They would broach. They might capsize. If ever a ship this big could capsize, here was the boyo to do the job.

One second.

"Hang on!"

The helmsman jumped for the jigger shrouds and scuttled aloft.

No seconds.

He watched the sea enfold the ship, as if the *Beara Head* was disappearing into a monster's maw. The bow rose to meet it and then began to fall off to one side, the beginning of a broach. He waited for another second, straining at the wheel as the wave rolled aft. Then he jumped too, an adrenaline leap to the weather ratlines and up twenty feet like a bounced ball, wrapping his arms and legs around the shrouds as the barque rolled over to leeward and the sea submerged him.

Benjamin heard the wave from the break of the poop. He stared forward towards the sound. It couldn't be a steamer bearing down on them. There were none down here, especially not in this weather. They went through the straits, through Magellan. Calm water there, and the williwaws didn't bother them. Not another vessel. It had to be a wave, then. Holy God, it sounded like the hammers of hell! He heard the mate's yell, disembodied through the wind and the fury of this water. He ran to the mizzen rigging and began to climb, running up like the nimble tar he had become, aware of others scrambling around him. At thirty-five feet above the deck, he looked forward just as the bow rose up to the immense sea and disappeared into its black water.

❊

The mate, halfway to the jigger top, was still under water, holding his breath for what seemed like a minute. Had to be less time than that, though. The water drained away from around him, and he took deep breaths, aware of the coldness of the wind after the sea's warmer embrace. Curiosity—that was what he felt. He was intensely curious to see what the ship would look like and how long it would take for her to go down. Would she roll right over, leaving them nothing to cling to? Maybe she wouldn't surface at all, driven under by the tons of water that had dropped down on her. Or would she linger upright until the weight of water in the hold—he had no doubt the hatches had been breached—overcame the hull's buoyancy and dragged it down, the trapped air bubbling out, each seaman (the ones who weren't already in the sea) making his own decision about how he was going to die (look for flotsam to grab hold of and die slowly of the cold or drown right away)? He had always been curious about what happened on those ships that just disappeared, all the missing—hundreds, thousands of them, with all their men. What was it like at the end? It was the same thing for the men, of course; they drowned, and that happened in the same way for each man. It was the ships he was interested in, the angles and attitudes they assumed as they rolled over or slipped down by bow or stern, how long they took, what it sounded like. Now he'd find out.

He felt the *Beara Head* come upright again, or more or less so. Looking down, he saw that the decks were visible, although the barque listed to port even more than it had after the *pampero*. The goddamn coal had to have shifted again. Of course it had. How could it not when they damned near turned turtle? The vessel was lying beam-on to the wind and seas—a large wave broke over the windward side as he watched. She was upright, after all; she'd go down slowly, then. As the wave drained away, he could see that the main and mizzen hatch covers were askew, several detached yards hung from aloft, the iron plate of the deck houses was buckled inwards. He saw the shape of a man in the water just off the leeward side. They were all as good as dead men, but this one was trying to get back aboard, to live a little longer. The mate shinnied down to the deck, leaned over the bulwark and grabbed hold of the man's long grey-brown hair. It was the shantyman. The mate pulled him out of the water a little until he could grasp Paddy's oilskins and haul him aboard, one strong arm and hand taking the weight of fourteen stone of seaman and waterlogged clothes.

"Mr. MacNeill!" The captain's voice.

"Aye aye, sir. Here, sir!"

"Mr. MacNeill, see to the hatches if you please. And put some men to the pumps."

Stupid bastard, the mate said to himself. He didn't want to be running fore and aft and thwartships, buggering around with useless details, and miss seeing exactly how the *Beara Head* would die, solve the mystery of the missing ships once and for all. He wanted to see how it would happen.

But as always, he obeyed.

"Aye aye, sir."

The captain was at the wheel; he looked as if he was steering the barque out of the Mersey on a sunny afternoon. His wife stood in the listing companionway scuttle, staring forward in shock at the ruined ship. The little dog squatted beside her, motionless for once.

"And then, Mr. MacNeill, we'll need sail forward, I believe. Get her turned and we'll run off."

"Aye aye, sir."

Stupid bastard! The mate went looking for men to delay the ship's inevitable death.

Benjamin, together with most of the port watch, had climbed high enough so that he was above the water that engulfed the ship. He watched the vessel disappear beneath him, and not just the lower decks this time, but the entire ship—poop, fo'c's'le deck, deck houses, everything except the masts. For ten seconds or so, they were perched on one of four tilting spars that rose up, incongruously and mysteriously, in the middle of the ocean. Benjamin looked out over the sea.

He knew they were close to the Horn. Could they get there? Dry out and build a fire? Eat shellfish? Wait for rescue? No, that was foolish. If they were a cable off the rock, it wouldn't make any difference in this sea. He looked back down, and to his surprise, the ship, dim in the faint winter light, reappeared from under the water, slowly and with reluctance, it seemed, but more of it was there every second.

"That was a big 'un." Russell shouted the words, but calmly, as if he was judging the best one out of a herd of placid farm animals.

"Sonofabitch!"

"Ohoooo! Boys, now that was a breaker!"

"Smoke an' Oakum!"

"*Satan!*"

"Who'd sell a farm?"

"Friend Ben, I've decided to retire from the profession of seaman," Anderson shouted into his ear.

He was laughing, Benjamin saw with surprise.

The watching, marvelling men saw someone slide down to the leeward side of the main deck and haul a man out of the sea. They recognized the mate by his bulk and the strength he displayed, and then they saw that Paddy was the man brought back from the dead. They hadn't missed their shantyman until then. They heard the captain giving his orders—making it sound as if they had a chance, Benjamin thought. Maybe they weren't foundering. He heard the mate's ready, loud reply.

Who else was overboard? No one he could see. Then the starboard watch began shouting: "Man overboard!"

One of the Scowegian able seamen.

They climbed down, straining to see the water round the ship in case the man was nearby. In the gloom, it was impossible to make out anything more than half a cable from the rail in the white foam. They had to get a boat in the water, Benjamin thought. He realized it might be impossible. He had heard of ships sailing on when a man went into the sea because conditions wouldn't allow a small boat to be lowered, let alone stay afloat if it did get launched. But they weren't sailing anywhere; the *Beara Head* was dead in the water. They could get a boat down. He ran to the port-side skids. Goddamn it! The boat wasn't there.

"Where's the bloody boat?" he screamed.

"All gone, you stupid goddamn paddy!" yelled the Finn. "Boats are all gone."

Benjamin stood despairing. No lifeboats. If the ship went down, they were lost. Even though he knew the boats could not have survived in this wind and sea, he felt bereft, doomed. Then the mate's familiar voice, the background noise of their lives for three months, broke the rogue sea's spell of paralysis and anguish.

"If a man's in the water, he's damned well scuppered. Now give an ear, ladies! Port watch, lay onto the pumps! Mr. Jagger and you, Johnston," he was addressing the carpenter, "take your watch and get these hatch covers secured. Russell, Lundy, you, Finn, and you, Maguire, lay aloft and set the fore topsail. Haul up a goddamn jib while you're at it. Sing out when you're ready! Chop, chop! Jump to it!"

They jumped to it. The string of orders got them moving again. They had lost a man, thought Benjamin. Only one? Were there more in the water they hadn't noticed missing yet? But maybe the ship was going to stay afloat. The captain and the mate seemed to think so. He ran with the others, quick and sure, up the foremast ratlines and out onto the footrope of the lower topsail yard.

As he gave his orders, the mate began to change his mind. His confident directions to the seamen had the effect of resuscitating his own belief in their survival. He thought he had been about to die, and he'd been ready for it—no regrets, just that passionate curiosity. It took a while for a man to come back from that serene edge, to summon up the energy to begin struggling again. Things weren't as bad as they had seemed at first. The ship had been very lucky. It was afloat—the hatches had mostly held, against all odds—and if they could stop more water from getting below and pump out what was already there, they might be all right, full and by. He was reluctant to admit it only because he wouldn't find out what happened when a ship went down with all her souls aboard. Not this time.

The wind scoured the broached barque unabated. Big seas broke continuously over the high windward side, each one striking the iron hull with the sound of cannon fire. Another rogue would finish off the ship and its crew; the mate would get his wistful wish after all. Such seas were not uncommon in a long gale like this one in the shallow waters off Cape Stiff. But the odds were in the ship's favour that it had survived the worst. In the meantime, the *Beara Head*'s seamen worked for the ship and their lives—secured hatch covers, got sail set forward, pumped and pumped—beginning the barque's resurrection.

> *Deck log,* Beara Head, *barque, Liverpool to Valparaiso, August 5, 1885: Wind W. to WSW 10, 11 with stronger gusts in the frequent squalls. I am running off to the NE under fore and main lower topsails, for Staten Island about 150 mi. distant. Listing twenty-five degrees to port as the cargo has shifted once again. Heavy seas but the ship is running as well as can be expected. Barometer low but steady. At dawn, sky had a wild, threatening appearance. 8 A.M., lost topsail but set another one.*

With two sails set forward, the barque payed off and got its stern to the wind, much as it had after being laid low by the *pampero* off the River. The wind was a point or two on the port quarter, and the vessel ran hard before the Southern Ocean greybeards. Only the most experienced seamen could handle the ship in these conditions, and they spelled each other off every half-hour in twos. Another broach, while running at speed, would probably be the end. The captain stood beside the helmsmen, looking aft at the seas rearing up behind them and giving instructions to bear off or head up a little to take the blow at the best angle. The helmsmen never looked back, a standing order on some ships in heavy weather. No one wanted them getting edgy at the sight of a thousand tons of water apparently about to bash their brains out.

The mate oversaw the ship's repairs, or such that could be done under conditions that were still extreme and hazardous. The sail cloth that the crew had secured over the hatches as temporary protection was taken off, and new tarpaulin and reinforcing planks were laid down. This had to be done quickly and often under water, as seas continued to break heavily aboard. More water got below, and the pumping went on without cease. It was work to break hearts and backs. The pumps were on the most exposed part of the main deck. The seamen had to lash themselves to the mainmast fife rail to avoid getting washed away, and they worked in water that was often up to their waists, even necks, frigid enough but less so than the cold wind. Not a man aboard could remember what it felt like to be dry. Paddy, recovered from his swim, sang a shanty for the pumps:

> In Amsterdam there lived a maid
> Mark well what I do say!
> In Amsterdam there lived a maid,
> An' she was mistress of her trade,
> We'll go no more a rovin' with you, fair maid.

The pumping outlasted the shantyman. "A bad leak is an inhuman thing," said Conrad's Marlow. It forces unremitting work on men. They labour to preserve themselves but soon feel, nevertheless, as if they are already "dead and gone to a hell for sailors." However, the *Beara Head* was not a wooden ship with a limitless capacity for absorbing sea water, and by the end of the dogwatches on the next day, after the crew had pumped almost continuously in four-man shifts for eighteen hours, the mate sounded the hold and, satisfied with the water level, gave the welcome order: "Belay the pumps!"

During the same day, the crew secured the wreckage of yards and gear stripped off by the rogue sea. The mate himself went aloft to inspect the masts. He discovered that the wooden fore and main topgallants had sprung—there were long splits in each—and would need repair. Half a dozen sails had been ripped out of their gaskets by the rogue and, restrained by only their clewlines and buntlines, sliced to ribbons by the wind. The men collected the pieces with care and presented them to the sailmaker, who was now sewing uninterruptedly, with a man and two of the boys to help him. The mate went out onto the bowsprit to look for damage there. He disappeared under water several times, the bowsprit driving into the back of waves ahead as the stern rose high to meet the next comber. He was tethered onto a line with three men on the other end, but his own strength kept him attached to the spar each time, arms and legs wrapped around steel wire and manilla lines, holding his breath until the stern slid down into the next trough and the bowsprit popped out of the sea.

"She'll hold," he said when he crawled back onto the fo'c's'le deck.

A wooden ship would almost certainly have gone down, perhaps an older iron vessel too. But the wire rope kept the rig in place, the Belfast rivets turned out to be "Protestanter" than the sea and the hull plates held firm.

By noon on the second day after the rogue sea's assault, the look-outs saw Staten Island off the port bow. Before sunset, the barque ran around the eastern end of the island—staying to seaward of the heavy tide rips that extended five or more miles out—and hauled its wind to sail north into the lee of the land. A dozen other vessels jilled about there—British barques and ships, a French four-poster and a Yankee schooner—hove to or standing on and off, waiting for better times. All of them had suffered in this bad year of Cape Stiff and were making repairs of one sort or another. The barque joined the wounded fleet.

Twenty-three days after clearing the Strait of Le Maire and beginning its attempt to round the Horn, the *Beara Head* was fifty miles farther away from its goal than when it had started. It had suffered serious damage; its beaten-up crew included half a dozen men incapacitated by injury and frostbite, and they had lost a seaman. They would wait in the tenuous lee of this godforsaken island, salve their wounds as best they could, make repairs and then sail west and try again.

The extent of their misery is difficult to imagine. The *Beara Head*'s crew lived such a penurious life at the best of times that it was hard to see how the rogue wave could make things worse, but it had. They had been poor men; now they were paupers. Most of their bedding—the noisome and infested straw of the donkey's breakfasts—had been washed away, together with a number of blankets. Some men would sleep on bare bunkboards under scraps of sailcloth. Sea chests had been lashed securely to prevent them tumbling about in heavy weather and kneecapping a man, and they had remained intact. However, sea water had broached many of them and washed away the meagre collection of ragged clothes inside. Three of the port watch had the dungarees, jumpers and oilskins they stood in and nothing more. Even worse, the galley had been swept by the wave, damaging the stove. It could be repaired, but the carpenter would have to forge an iron part, and he had more urgent things to attend to. In the meantime, there was no hot coffee. Already, the seamen felt the lack of this more than anything else.

The sea had destroyed stores too, enough of them that the crew would be on short rations for the remainder of the passage, creating an even greater gap between calories consumed and expended. The officers were only a little better off: their mattresses, clothes and blankets were intact. However, sea water had contaminated a good part of the aft cabin's store of food. At least this deprivation swept away the divisions of privilege. The captain ordered the food pooled and divided amongst the afterguard and the fo'c's'le so each man aboard would receive the same rations. No more plain salt junk for the men while the officers ate oatmeal and marmalade. Now they would feel the pinch and cramp of hunger together.

The captain's wife surveyed the crew when it mustered at the poop for orders. What a ragabrash they were, she thought. To think that everything depended on this small troop of thin, exhausted men in their worn and torn oilskins. She felt despair. How could they ever get the ship round this accursed cape? Suddenly, she was struck by the contrast between the great, battered ship—with its lofty masts, acres of sails, heavy, cumbersome gear—and the puny, diminished seamen. It seemed impossible that they would ever work their will on this mass of iron and canvas. And the ship was farther away from the Pacific than when they started. She counted the men of both watches: fifteen seamen; four boys more dead than alive; the idlers, who would be too busy repairing damage for a good long time to sail the ship; the two mates. One poor soul lost; the man had just disappeared. What a terrible way to die! Six of the crew out of action, the two most serious frostbite cases lying even now in the tiny cabin next to hers. She tended them with warm water—although there'd be no more of that for a while—and some ointment her husband had prescribed from his "Captain's book on how to keep seamen alive a little longer if you don't kill them first," as she had teasingly dubbed it. Maybe they should run off to the east, she thought. Her husband had always derided the idea, but perhaps he should reconsider. . . . No, they didn't have enough food to do that now. They'd starve before they arrived anywhere, and it was un-thinkable to stop in port for food for a hungry crew. Her husband would be a laughingstock; he'd be commanding a Donaghadee fishing smack next. No, they would have to persevere; there was no alternative to rounding this damned Horn (she allowed herself the blasphemy). She had loved the first days in this wild ocean, its grandeur and power, its freedom. But now she dreaded the renewal of the battle, which had worn down even the sturdy fascination of her gaze.

It took five days to make repairs and to shovel the coal back to star-board and trim it. Work went on by oil lamp after the early dark and be-fore the late dawn of the short winter days. The fire had turned out to be a blessing. When they were shovelling the cargo around the hold to get at its source, the captain had ordered more shifting boards installed to further isolate the hot coal from the rest. He and the mate agreed: the extra planks had made the difference between the relatively small shift that occurred when the sea overran them and a catastrophic slump to leeward of the whole cargo, like a landslide pinning the barque down right over on her side, maybe even making the difference in whether or not she turned turtle.

Getting the coal back in place was dog-work, fit for the scrawny ap-prentices, the unskilled idlers—doctor and steward—and most of the ordinary seamen of both watches, Benjamin included. From the port watch, he, Anderson and Maguire shovelled coal, Maguire content enough to be on the business end of a spade yet again—it comforted him, the rhythm reminding him of home, its long days but certain life.

The mates, the carpenter, the sailmaker, the able seamen and such experienced ordinary seamen as the antagonists-turned-chums Kapel-las and Urbanski did the skilled work. They called on the filthy coal whippers when they needed help with the heavy lifting. They rerigged the detached yards and rove new lifts, braces and sheets. They fished the fore and main topgallant masts (that is, they lashed a spare spar to the sprung or cracked portion for reinforcement, like a splint). They made new hatch covers, which were ready for fitting when, as Grey put it, "the slack-'olders below've finished playing with the goddamn coal." (The cockney was a permanent part of the crew again. They needed every man's muscles and skill. His truce with Russell continued.) Finally, the repair gang inspected every piece of iron and line aloft for wear and weakness, replacing the most slightly suspect. The ship was going back to war, and everything had to be ready.

It would have been a good week's work for a well-fed crew of experienced riggers in port, with dockside cranes to lend a hand, let alone a gang of half-starved, frozen, weary seamen a hundred days out.

Deck log, Beara Head, barque, Liverpool to Valparaiso, August 12, 1885: 54° 40 ´S., 63° 55´W. Wind E. 3, 4, 5. Having completed repairs and trimmed the cargo, and the wind going light to the east, I am clearing Staten Island and will make to the west once again. Sky partly overcast. Seas are moderate but a high swell continues to make in from the southwest. Barometer is rising slowly. A number of vessels in company.

On their ninety-ninth day at sea, they set full sail and made to the southwest once again. Eighty degrees longitude was six hundred miles away, but the barque might cover two or even three times that distance if the wind was contrary.

At first, it looked as if their luck had changed. In the cold and low overcast, the wind held east, and even increased, so they were able to make some progress against the persistent opposing swell throughout the first day and night and the second day. No more Irish luck, said Paddy. The so-called luck of the Irish was, in reality, all bad—that most distressful country. That's what they'd been getting. Now maybe it was time for some other nationality's real serendipity. They would have been happier with an easterly gale. Even if it did make their lives miserable to crash headlong into an adverse swell, at least they'd make some serious westing. Their slow, if relaxing, headway was keeping them too long to the east of Cape Stiff. The fair wind wouldn't last forever; they needed to get past the Horn before the west wind inevitably resumed. But the easterly stayed light, and on the third day after their departure from the lee of Staten Island, it fell away completely. The ship wallowed and rolled like a proverbial drunken pig (although who the hell ever fed good grog to pigs anyway, Paddy wanted to know) in the hollows and hills of the unending Southern Ocean swell. The motion threatened to jerk the masts clean out. The watches went aloft to furl the square sails so they didn't slat and flap themselves to pieces, and the men spent half their time hanging onto yards and jackstays and various lines as tight as barnacles on iron to avoid getting flipped into the sea like stones flung from catapults.

> *Pitch, pitch, Goddamn your soul*
> *The more you pitch the less you'll roll*
> *Or roll, roll, you sonofabitch*
> *The more you roll, the less you'll pitch.*

The *Beara Head* had stumbled into that most aggravating of weather phenomena, a Cape Horn calm.

Sometimes it happened that there was no wind around Cape Stiff, and that could be worse than a "reg'lar snorter." As if to prove that there's no end to the variety of things that can do a ship in, some were wrecked by calms in the vicinity of the Horn. It has happened in the conventional way: a vessel running through Le Maire—or close to Isla Hornos or one of the other islands of the Wollaston group, or to some slab of rock farther west—tries to weather it without having to wear ship and tack off in some direction well off its desired course; the wind suddenly falls away; the undiminished swell or the set of the current takes over; the ship is driven ashore and smashed apart; it might take less than an hour, half an hour.

Needless to say, calms near Cape Horn drove captains mad. At the one place on earth they thought they wouldn't have to endure lack of wind, a prolonged calm was a gross and enraging insult. They might, as Alan Villiers observed, "turn into surly, cross-tempered morons, raving about the poop and finding fault unnecessarily, or jumping on their sou'westers in a fury." Villiers heard about one captain in the grip of a week-long calm—in winter!—who, in his despair, broached the ship's whisky supply and drank his way through a couple of cases of the stuff. By the time he was finished and the wind came back, he had delirium tremens and had to be lashed to his bunk until they passed and he could resume his now compromised command.

Usually, however, a Cape Horn calm was short-lived—it might last a few hours, or even less than an hour—and then it was the foreteller of a renewed westerly gale. Or sometimes, an easterly gale. With Cape

Stiff, you never knew. During Dana's difficult passage round the Horn, a rare fine west wind fell away to a calm. Less than an hour later, however, the *Alert* lay hove to under a close-reefed main topsail (Dana sailed when square sails were still reefed—that is, left set but reduced in area) in "the fiercest storm that we had yet felt," and it was from the east, the direction in which they were heading. They drifted about like this for the next eight days, the wind falling calm for an hour or two, usually around noon, with a few tantalizing puffs of fair west wind before the easterly gale came roaring back down on them. During all that time, a cold, drenching rain fell. It had replaced the snow they had experienced earlier. Snow was bad enough—it had to be very cold for it to fall—but at least it didn't soak a man's clothes clear through, said Dana. "For genuine discomfort, give me rain with freezing weather."

Deck log, Beara Head, *barque*, Liverpool to Valparaiso, August 15, 1885: 56° 03´ S., 66° 23´ W. Wind calm, slight var. breeze, W. 4, 5. 30 mi. ESE of the Horn. Heavy rolling in long, high swells from southwest. Heavy masses of cloud to the north and west. Three ships and two barques in company. 8 P.M., breeze finally filled in from the west. I set all sail again and brought ship to wind on the starboard tack. Very cold with some small icebergs about. Barometer falling.

He never thought he'd yearn for wind, even a dead-ahead west wind, while rounding the Horn, but for Benjamin, and everyone else aboard, the calm, combined with the continued big swells, was almost intolerable. It was impossible to keep the barque's bow facing the waves; it tended to fall off and lie broadside-on to them, rolling so hard that the yardarms plunged into the sea with a pendulum's regularity. Maybe the bloody coal will shift again, he thought. They were heeling over almost as far as they had when the great rogue sea swept them. The last thing he wanted to do was shovel more coal. He'd loaded it in Liverpool, trimmed it off the River after the *pampero*, shovelled it out onto the deck to put out the fire, shovelled it back in when the fire was out and trimmed it again after the big rogue. And he knew he'd have to unload it, shovelful by shovelful, when they finally ended this goddamn endless expedition and actually made it into Valparaiso. He'd never look at a coal fire or a coal scuttle the same way again.

All this time, the current was setting them to the east. They must have got close to the Horn's longitude before the wind fell away, and now here they were, drifting back again. Benjamin was getting hungry already on the reduced rations. The small luxuries of oatmeal, marmalade and mouldy onions from the aft cabin were no substitute for a full belly. He never thought he'd miss smelly salt horse, but he would rather have enough of that than a dainty meal of the captain's condiments. This stalled progress would only make their grub situation worse. A calm off the Horn? No one had ever told him of such a thing.

"It happens," said Russell. "But we ain't on our beam ends yet. You were getting fat as a porpoise anyway, there, young Ben—too much first-chop salt junk. This'll be good for you. Get you pretty for them little brown Paradise gals."

The port watch stood at the mizzen mast fife rail, hanging onto it, in fact, against the ship's death roll or clinging onto rails and lifelines nearby. The starboard watch was below, trying to sleep, the men wedged into their bunks to avoid getting pitched out onto the deck like topers from a tavern. The crew continued to lead a "Cape Horn life." There was no chipping or painting. All they had to do was handle the ship, and if, for the moment, the mate could think of nothing to be done in that department, then they were free to chat and smoke. (Even the no-smoking-on-watch rule had been abrogated for the duration of the hostilities.)

"Well, how long will we have to sit here with our brains rolling out of our lugs?" Benjamin persisted.

Grey laughed.

"Not long, little paddy. Just take a butcher's 'ook. Big fuckin' swells. That cloud to the west. You'll be dancin' on the fuckin' footropes again in a watch or two."

"I'll like it better nor this," said Benjamin.

He didn't.

By four bells of the forenoon watch the next day, he'd been on duty for ten hours straight. He had lost count of the number of times he'd scrambled up the ratlines and out onto the footropes of the lower topgallants and upper topsails, furling and passing the gaskets, casting the gaskets loose; down on deck, hauling yards up and then lowering them away again, hauling clewlines and buntlines taut, then letting them go; out on the bowsprit, muzzling the topmast staysail and inner jib, then hauling them up again. It was one damned thing after another every ten minutes. He had hummed his way through every Christmas carol he knew.

The wind had filled in from the west, just as Grey predicted. From the first welcome breaths, cold on Benjamin's face, to a full gale of fifty knots took less than an hour, and the work became a steady parade from belaying-pin to belaying-pin, and yard to yard, fisting the sails into lumpy shapes they could lash down. The wind was fitful in velocity. It fell away to thirty-five or forty knots, then suddenly increased to fifty-five or sixty in violent squalls that raced across their path with heavy snow, sleet and hail driving horizontally into their raw faces. Nevertheless, the captain and his ramrod, the mate, worked the ship as if they had a full complement of fresh sailors aboard to scurry up and down all day long—as if they'd been out ten days instead of more than a hundred, and port was in the offing, not twenty-five hundred miles away.

By the end of the afternoon watch, the wind had finally stabilized into a howling gale right out of the west, with the usual frequent lulls and squalls, which left no time for sail tweaking. It was back to the familiar routine of head-reaching on one tack or the other under lower topsails; plunging into the endless lines of Cape Stiff greybeards, decks awash; the watch crowding at the break of the poop, trying to avoid both the glacial wind and the frigid seas breaking aboard waist-deep. The mate sent the port watch below after it had been sixteen hours on deck. Benjamin flopped and crawled, all-standing, onto his hard, bare bunkboard, hungry-gutted, too tired to piss, and fell asleep. As he did so, he thought, These new tactics of the captain's will bloody well kill us all before we get to eighty west, never mind bloody Valparaiso.

❋

The captain knew he might push them beyond endurance, but what choice did he have? He'd mollycoddled them, tried to pace them so they'd last the distance. If he judged it so the whole goddamn passel of them dropped dead from exhaustion the moment they anchored in

Valparaiso—no, the moment after they'd off-loaded the coal in Valparaiso—that would be fine well with him. But the goddamn eternal westerlies and the rogue had put the end to that. So he'd changed his battle plan. Like any good field general, he had to know when to abandon tactics that weren't working and try something else. In the fog and smoke of the battle of the Horn, he had made his decision: he would push, push and push. Every chance he got, he'd make sail and make westing; when wind and sea became too much, he'd tuck his head under his wing, but never for too long. At the slightest sign of easing, up they'd go, by heaven, to loose some canvas. These were merely the standard instructions; he knew that. He had fudged them because he was short-handed. That was all over now. Now he'd heave ahead every chance he got, and he'd see what these fo'c's'le hallions had in them.

> *Deck log,* Beara Head, *barque, Liverpool to Valparaiso, August 17, 1885: 56° 24′ S., 66° 50′ W. Wind W. 9, 10. Ship to wind on starboard tack under lower topsails. 11 A.M., I tried setting the main and mizzen upper topsails, but they blew out immediately. A man was lost overboard when he fell from the main topsail yardarm. Heavy seas filling the decks. Masses of cloud about and frequent squalls with snow and sleet. At sunset, about 5-minute intervals between the squalls.*

It was the Elf. While helping to furl the remains of the main upper topsail, which had blown into rags and shreds within ten minutes of being set, the shanghaied seaman Cavers, nicknamed the Elf for his short stature and slight, although deceptively strong, frame, was knocked off the yardarm by a flick of one of the larger canvas remnants. He fell into the sea without a sound that anyone heard. His watchmates spotted his head and upper torso almost immediately and kept him in sight for about a minute, shouting all the while that a man was overboard. The Elf floated off the side of the barque's hull, no more than twenty feet distant, but the vessel was sailing away from him at approximately three knots. He made no effort to swim, either because he was injured or because he didn't know how. Then a large sea washed over him and he disappeared from sight. A lookout was kept for several minutes by all the watch on deck and several hands who came up from below, including all of the idlers, as well as the captain, the mates and the captain's wife, but they saw nothing. Visibility was very bad, with snow and heavy spume blowing hard across the sea's surface. The captain gave no order to heave to for the purpose of attempting a rescue because the *Beara Head* had no boats to launch, all four having been stove in and swept away by the rogue sea the vessel had encountered two weeks previously. In the captain's judgment, it would have been fruitless to launch a boat even if one had been available, because the risk to its oarsmen would have outweighed any chance of finding the man in the water. Five minutes or so after the Elf's fall, the mate ordered the watch aloft once again—they had descended to the deck in their agitation—to finish furling the blown-out topsail.

Later in the day, the Elf's belongings were auctioned off. His only possessions of any value were his knife, which the Greek, Kapellas, bid for and won, and his donkey's breakfast, which was awarded to the Yankee, Urbanski. Anderson, the paddy, won the card draw to decide who would move into the Elf's upper bunk, which was drier than the lower berths occupied by the green hands. With the Dutchman and Maguire incapacitated by frostbite, there were now only seven seamen in the port watch fit for duty.

By noon the next day, the *Beara Head*'s 105th day out and thirty-four days since it had cleared the Strait of Le Maire, the gale backed to the southwest and eased slightly. This was an encouraging sign. Their previous experience had been wind that blew for days on end without so much as a pause or a shift. It had been near impossible to make westing; even if the captain had decided to press his men hard, there had been little opportunity. This weather was behaving more conventionally; it might leave a gap between itself and the next snorter, which was no doubt hunting them down even now. They could get the upper topsails and a few staysails set and make some miles.

They did more than that. As soon as the wind shifted, before it was clear that it would also slacken a little, the captain wore ship onto the port tack and headed northwest. He ordered topsails and lower top gallants set. Even though it was still blowing more than a whole gale, the shift in direction was the important thing; it provided a chance, whose duration no one knew, to get to the west, maybe to weather the Horn for the second time. Under its increased sail area—more than was prudent, but prudence had got them nowhere—the barque made heavy going of it, pitching into the forty- and fifty-footers, burying its bow back to the foremast, throwing spray past the poop, driving as hard as any ship could be driven. The crew of a wooden ship would have been pumping constantly under this strain, becoming too worn out for any gale-force sail-handling. Only an iron ship could withstand the pummelling the captain now imposed on the *Beara Head*.

Deck log, Beara Head, *barque, Liverpool to Valparaiso, August 19, 1885: 56° 05´ S., 67° 47´ W. Wind, SW 8, 9, 8. Heavy squalls with snow from time to time, and seas remain very high. The ship's motion is violent and I am watching the fished topgallant masts carefully. Barometer low but steady. 6 A.M., sky very threatening and furled topgallants and lowered staysails. 8:30 A.M., set same sails once again but lost mizzen lower topgallant immediately. 3 P.M., estimate passed the longitude of the Horn. I am carrying on to the northwest to weather Isla Hermite before wearing ship. Many porpoise and albatross about. At 5 P.M., the sun shone out bright and the wind eased a point more to the south.*

No celebrations for Benjamin this time. He had no difficulty assuming the old hands' wary indifference. Doubling the Horn was one thing; keeping the damned rock at your arse-end was something else. According to the second mate, Cape Stiff wasn't more than fifteen miles distant when they crossed its longitude, and it was less than twenty miles to the east even now. It was odd, Benjamin thought, that you could fight so long and so hard just to get past this infernal rock, without ever seeing it. It surprised him, now that he thought about it, that this greatest symbol of a seaman's hard life could, like a wraith, lie always just beyond the wind ship's constricted, cloudy horizon, even as the ship gave its all to leave it astern. And lost men in the process. It was an unseen enemy, lurking, never looming, just out of range, made more powerful and terrifying by its invisibility. Some of the able seamen had been around the Horn two dozen times and had glimpsed it only once or twice. And that when they had been aboard homeward-bounders running free to the east, when the ship was easily manoeuvrable and it was safer to pass close by. Still, he wouldn't worry too much about it. In fact, you could argue that not seeing land down here in this murk was a good thing after all; if you could see it, you were too damned close.

✳

The slight wind shift to the south was a gift from God. It enabled the *Beara Head* to weather Isla Hermite, the island immediately west and a little north of Isla Hornos. And it gave the barque the opportunity to steer clear of Falso Cabo de Hornos, the False Cape at the tip of Peninsula Hardy, thirty miles northwest of the true Horn. As its name implied, it had been the death of many an eastward-bound square-rigger that, running hard on a week's worth of dead reckoning, thought it was the real thing—there is a superficial resemblance in the profile of forbidding bare rock—hauled its wind and turned to the northeast, towards where it thought Staten Island lay. In fact, such vessels were heading into the narrow waters of Mantello Pass and soon found themelves irretrievably in the surf, and then on the rocks, of the northern Wollastons.

By late in the afternoon watch of the following day, however, the captain decided that he had chanced his arm enough. He had managed an uncertain sun sight the previous day, and he thought he still had some sea room to the north and west. But he was uncertain about the set of the local currents, hived off from the Southern Ocean stream, as he closed with the coast. His dead reckoning put the group of the Islas Morton no more than fifteen miles off his bow, and Islas Henderson and Sanderson even closer than that. More worrying were the notations on his chart several miles out from the islands themselves: "Breakers," a word spread, vaguely and with menace, over several miles of sea; *"Arrecife Peligroso"*—dangerous reef—said another, with an *x* marking the spot. These were the perils that the survey expeditions had discovered over the past two hundred years or so. Who knew what they had missed? The barque might be closing in even now on some unmarked, unknown *peligroso* reef. He had no desire to leave his legacy in the Southern Ocean: *Arrecife McMillan*.

Deck log, Beara Head, *barque, Liverpool to Valparaiso, August 21, 1885: 56° 30′ S., 68° 30′ W. Wind, SW 8, 9, 8. At 8:30* A.M., *nar-*

*rowly avoided collision with homeward-bound barque. I was unable
to determine its name or hailing port. Islas Diego Ramirez close
aboard and I wore ship just before noon. Carrying on with lower topsails, main topmast staysail and jigger staysail. Lost fore topsail while
wearing ship and bent and set new one. Barometer continuing to fall.
The sky changes its appearance very often. Heavy seas. Two new frostbite cases.*

As if they didn't have enough to worry about with a month of continuous gales and rogue seas and ice and avoiding every rock in creation
and the Rammerees somewhere close off their bow, they nearly got
scuppered by a bloody homeward-bounder running blithe and blind to
the east. It wasn't that surprising when you considered it, Benjamin
thought, because there had to be fifty vessels in the sea area round
about the Horn, either clawing their westing or running their easting
down. It was a wonder they weren't bashing into each other every
night of the week. And every day too, for that matter. With the spray
and mist, what lookout could tell what the hell was coming from windward a cable-length away? That was a tenth of a mile, and a ship running hard would cover the distance in—what?—thirty seconds or so.
Not enough time to do anything except yell, "Mother!" which is what
men scream when they think they're about to die.

Benjamin was on deck at one bell of the forenoon watch, the wind
howling unabated, seas worse than ever, the old man still driving hard
(although, for once, not on the poop), the watch wet and frozen, hud-
dled under the overhang, wretched in the scant light of the winter's
late dawn, when going aloft or hauling lines waist-deep in icy water
was the most repellant prospect imaginable. He heard the lookout's
frantic shout:

"Ship ahoy! Ship! Dead abeam to windward! She's on top of us!
Jesus jumpin' Christ!"

Even as he began to react, Benjamin found himself thinking how
surprising it was that Anderson's distinctive Belfast speech patterns
were so clear even through the man's near-panicked warning. He
thought, An Ulsterman guldering away at the bow. He wondered if he
sounded like that when he yelled in fear. Of course he did.

During these incongruous linguistic ruminations, Benjamin's instinctive and illogical first physical reaction was to run to the weather railing, the exact spot where, if Anderson's snap judgment was correct, the other vessel would make its imminent, catastrophic impact. He peered through the sleet and spume, but all he could see was the dark slab-side and tumbling crest of the next wave in the infinite sequence of waves that—more irrelevant musings to avoid thinking about what was at hand—he had no doubt spanned the whole huge orb of the Southern Ocean.

He heard two things almost simultaneously: the mate's booming voice ordering the helmsmen to bear off (it sounded as if he had switched into his highest mode of vocal amplification) and the sound of the other ship, the watery, spilling scree of its bow wave, the rising crescendo of its passage through the wind and of the wind through its rigging. Benjamin and the other members of his depleted watch yelled and screamed into the wind's teeth like fools: "Look out!" and "Head up!" and "Ahoy!" and "Ye fuckin' eejit!" and "Ya goddamn sonofabitch!" and "Ya damned lubber!"

And so on. Not a syllable of which was of the slightest use to anyone on either ship, but this was one of those few occasions when seamen could do nothing at all to avoid or mitigate disaster. There was no time for anything but shouting. It was up to the helmsmen on both ships—had the other vessel even seen them?—and the mechanics of relative trajectories, turning radii, velocity, blind luck.

It was a magnificent sight! Even in the midst of what he supposed might be his last few minutes on earth, or at sea, Benjamin was astounded by his fast and passing glimpse of the eastbound vessel—a four-poster like the *Beara Head*, as it happened. Of course, he had seen many ships at sea, some up close (but in gentle weather), some in storms (but at a distance); he had seen them come and go near harbours, but then they were under shortened sail and in narrow waters. He had never seen anything like this ghost-barque that was thundering down on him at fourteen knots, accelerating down the long-fetch breakers, with pale spray and even solid water, like dark sheets, flying out from its bow and quarters. Its outline appeared from the misty stew of the gale's spume and sleety rain, maybe a quarter-cable away, already into its turn to port—it had seen them—its bowsprit rising sky-high out of the wave's trough and its rudder biting into the face of the sea behind, giving it turning power, wrenching the onrushing ship to one side. As it swung by the *Beara Head*'s stern, bowsprit clearing the transom and spanker boom by twenty feet—no more than that, the mate swore later—Benjamin could see along the length of the other vessel the shapes of its bellied topsails and topgallants, bare masts thrusting above and below the straining sails, starboard topside awash, seas rushing across its deck and pouring out the washports. They were close enough that he could see the outlines of oilskinned figures staring back, motionless like him in the deadly thrall of the moment, and hear the loud curses and oaths of a voice like the mate's voice, a vocal doppelgänger, bellowing away, like all mates, on the other barque's poop. Even before he knew it would miss them, Benjamin felt the hair on his neck stand up, shivers run over his skin. He breathed "Dear God!" in awe and wonder at the sight of the shadowy, lofty wind-driven ship soaring past him and past the Horn, surely man's most beautiful machine, a sight to rouse the heart in the midst of this wilderness.

All of them, the *Beara Head* and its entire company, nearly wiped out. The homeward-bounder's crew gone as well. Benjamin had no doubt that if the unknown barque had plowed into their beam at fourteen or fifteen knots in the midst of a full gale, neither ship could have survived—the certain result if it had been dark. Two more to add to the thousands posted overdue, then missing.

As he thought about it, Benjamin concluded that the near collision was the first of two events completing his transformation into a deep-water seaman. It was the lesson of the few seconds that night when sixty men (a woman, maybe two, as well) either survived or didn't, the helmsmen's frantic efforts sufficient to make the difference or not. A seaman had to realize that he could fight and struggle for the ship, and therefore for himself, but that all that courage and nerve might not mean a thing if his time had come. By luck or some kind of design—who knew what or whose?—in the end, the sea decided who died, and when.

The other event was the Elf's death. Before his watchmate had gone into the sea, Benjamin had a paradoxical sense of the dangers that were inescapable in his new occupation: they were palpable, yet abstract. He could feel the imminence of violent death every time he went up the ratlines in heavy weather or leaped clear of a deck-sweeping sea. Who would not? Nevertheless, the possibility remained somehow theoretical. True, the Clerk had died, but by his own pathetic hand; he was a weak man, literally out of his depth a cable from shore. Benjamin hadn't known much about the Scowegian seaman swept overboard by the rogue sea, and somehow that seemed to be an easier death to avoid than a fall from aloft—keep your wits about you and hang on or jump for it; it wasn't difficult. The Elf was different. He was a capable seaman, the nimblest and handiest man of them all aloft. If he could go into the hole, anyone could. Benjamin could. He hadn't seen him fall, but he had glimpsed his head in the water. He had climbed down to the deck and lost sight of the doomed man, and then he hadn't been able to spot him again in the roiling sea. It was the lack of drama or ceremony that was most unnerving: the sudden, quick, voiceless fall; the pointless attempt to keep him in view; the immediate reversion to the ship's business in the climb back aloft to finish furling the few scraps of canvas the Elf had died to save; the quick distribution of his belongings; the obliteration of his life on the ship; the disappearance of a man.

Deck log, Beara Head, *barque, Liverpool to Valparaiso, August 23, 1885: 56° 43´ S., 69° 52´ W. Wind, SW 10, 11. Strong squalls above hurricane strength about every half-hour, with snow and hail. Very heavy seas filling the decks. Carrying on under fore and mizzen lower topsails on starboard tack. Sky constantly wild and threatening in appearance. Some small icebergs about. A barque to the north. One new frostbite case.*

Benjamin heard the scuttlebutt from the aft cabin that they had got to 110 miles past the Horn when the gales stopped them again. That was about three hundred miles from eighty west. But the news was good too: the troublesome cape was finally being spoken about in the past tense. They had made enough westing, and if they could hold their own, they would be able to turn north when the weather eased; the Chilean coast with its offlying archipelagos was already curving towards the northwest at this longitude. It couldn't happen soon enough, Benjamin thought. He was at the bitter end of his rope; he didn't know which side of himself was up. It was all damned near impossible: this never-ending toil in endless storms on a passage without end. The ship was disintegrating around them, everything in a foother—halyards, sheets, braces and lifts parting or chafing through; blocks exploding; ratlines and footropes collapsing. He hadn't had more than an hour at a stretch below in days. Up aloft furling rags of canvas to be stitched together into new sails whose scraps he'd be furling next watch. The two topgallant masts liable to come down on his bloody head any minute. Only seven men in each watch fit for duty. He was just not upsides of it all. He could count his ribs; the only thing worse than salt horse and weevily hard tack was not having enough of it. He didn't feel hungry any more, though, just weak, as if his arse weighed ten stone. The songs he hummed ran through his brain in an obsessive loop. At times, as exhausted as he was, the pain of frostbite, bloody hands and saltwater boils kept him awake during his one hour below in every eight. He promised himself what all sailors promise in storms and forget later: When—if—I get ashore again, I will become a good man, and I will never set foot on a bloody barque or ship or smack or hooker again, God strike me dead if I do.

Deck log, Beara Head, *barque, Liverpool to Valparaiso, August 27, 1885: 56° 40´S., 68° 57´W. Wind, S. 6, SE 6. Carrying on under all sail except for the fore upper topgallant and royal because of the damage to the mast. Seas are high but subsiding. Continuing to repair sails and gear. Sun for most of the day but very cold and icebergs about. A ship in company.*

The storm was like a hero's final test before winning the prize or like the last gate out of hell: stronger than those preceding it and harder for wearying men to breach, but still the final obstacle. The barque popped out the other side of the long-lived snorter into cold sunshine, steel blue water, small, scattered icebergs, and a wind that was strong enough to drive the ship into the big leftover seas and, for the first time in no one could remember how long, abaft the beam. A wind fair and favourable.

Deck log, Beara Head, *barque, Liverpool to Valparaiso, August 30, 1885: 55° 13´S., 79° 52´W. Wind, SSE 6, 5, SE 7. Clear and warmer. Wore ship to the north.*

The terseness of the log entry was the best of signs, a reflection of well-being, progress, the return to the ship's normal, beneficent routine. Long screeds by the captain meant that hell was breaking loose: the weather was kicking up or things were going wrong (sails splitting, masts springing, men dying).

Thus on its 117th day at sea, the *Beara Head* made its westing to eighty degrees longitude and wore to the north. The rounding of the Horn would be officially complete when they crossed fifty south, but that was a formality. The turn north was the momentous manoeuvre, the tack of tacks, the beginning of their return to the world.

Sun not warm but bright, wind strong but fair, seas high but not dead ahead, gear and sails fragile but holding, food allowance short but . . . well, you couldn't have everything—these were the constituents of happiness aboard the *Beara Head*. The barque was under all sail, making eight or nine knots across the long, rolling Southern Ocean swell. A weight was lifted, tension dissolved. They breathed easier, relaxed; the world seemed bright again, and they forgot their bargains with God.

The captain thought he'd prepared himself for the worst, but he hadn't really. Forty-six days from Le Maire, fifty since they'd crossed fifty south (a numerical coincidence, nothing more). Two men lost. The grace of God that Captain Learmont called down, which would keep men from dying, had failed the *Beara Head* this time. Of course, it could have been ten worses: losing the ship altogether—she could easily have gone down when the rogue blattered her—or stranding her on some goddamn out-by rock, or a dismasting and jury-rig into Stanley, where the bloody sheep-shaggers rob you blind for repairs. Doubling Cape Stiff didn't get much worse than this, if you were in the category of ships that got round it at all.

But look: he'd done it! He'd got the damned wagon round! And he hadn't made a mistake in the doing of it either. He'd gone south as per the fuckin' instructions and found nothing but ice and more westerlies. He'd gone north and the wave had near killed him. He'd run the Barney's brig for Staten Land and made repairs. Then he'd gone right back at it and had stood his bloody ground between the Horn and the ice, driven those fo'c's'le hallions hard through his chosen instrument, the mate (a savage, dangerous man whom he'd also had to keep under control, by the way), and ground out his goddamn westing cable by cable. If any captain deserved to bloody well wear ship to the north after getting by Cape Stiff, he bloody well did. And he'd kept the cargo intact too. They'd given him three thousand tons of coal, and that's just about what he'd give back to them in Valparaiso. That was the nut of it.

Another time round for the mate, his fifteenth. Not the worst, but one of them. One difference this time, though: he'd never had a moment at sea in his life like the few minutes after the rogue submerged the ship before his eyes. Thought he'd seen everything. But those moments, which seemed endless when he thought back on them (which he did often—that was unlike him too), when he had thought he was about to die and realized he wasn't afraid, merely curious, that chunk of time was something he would always remember.

He'd been at sea since he was fourteen, a cabin boy beaten and cornholed on a Bluenose schooner. He had never considered doing anything other than working ships. But that rogue had shaken him, not because it damned near killed him, but because . . . he wasn't sure. It just made a man think about other things he might do with his life. But ashore? He hadn't been more than ten miles inland since he was six years old—hardly more than a few cables, for that matter. The land was alien; he didn't understand its people or its customs. He'd noticed how people looked at him if he ventured away from the shoreside streets of bars, boarding houses and brothels: as if he were an animal or a being fallen to earth from a world in space, the ones he navigated by, Betelgeuse or Orion or Altair. Enough of all this! Ships and Sailortowns were his home ground. He was a seaman, and he'd have to stay at sea. Simple as that. No doubt, in the end, he would read the script he came so close to deciphering on this passage: the story of how a ship and its men went down.

A hard go this time, thought Russell, the Yankee; like his last ship. Maybe he was getting too old for the Horn. It had been bad, but he still liked the other things: being the best man in the watch; leading the others; the admiration and ready response that he was accorded by other seamen—except for bastards like the cockney. But they were aboard every ship; you couldn't give them the slip. He'd never been able to get the hang of the navigation; that was what had held him back from getting a ticket aft. Anyway, it was time for a change. Maybe he'd jump ship on the West Coast and sign on to a Yankee next time, if he could find one. The blamed limejuicers were everywhere these days, crowding the American ports. A Yankee schooner, then; do a little coasting for a while. The food would be better. That was the problem with limejuicers: they starved you to goddamn death. On the other hand, the Yankee bucko mates tried to kill you quicker, with their belaying-pin soup. There was his choice. He'd think about it. No question he'd stay afloat, though; it still gave his heart a lift to look at a good ship with sweet lines on her. Prettiest things on God's earth.

The Finn was goddamned if he'd sign on to a limejuicer again. These roastbeefs didn't know how to drive. A bloody month and a half to round Cape Stiff with half a goddamn watch. Back to a Norwegian ship or a Dutchman next time. Even one of the bounty ships, although he hated Frenchies too. Still, they had good gear and a proper crew of men. This passage on this bloody wagon, he'd done the work of two seamen—if they'd been Finns; three for these fuckin' roastbeefs and paddies. First passage he hadn't been able to finish his model. Now he'd have to work fast to get it ready for Vallipo if he was going to make what he needed for pisco and some of those little brown gals.

For Grey, the cockney, this was it. His last passage! No question, this time. He'd jump in Vallipo. Maybe he'd wait for the goddamn Yankee arse-licker in one of the joints on Gaff Tops'l Street. The local peelers—the Gilantes—never got up that far; their horses couldn't climb the goddamn cliff. The mate's fore-and-after was bound to come in there for a drink, and when he did, he'd stick the son of a whore. Then he'd get drunk as an admiral and stay that way until his money ran out, and then . . . wherever he went, he'd go by land—a topside seat on a stage up the coast. It didn't matter. This time, he would stay ashore for good. No backing and filling. This time, he meant it.

How Paddy the shantyman loved the Vale of Paradise!

> *In the roundin' of Cape Horn, me lads,*
> *Fine nights and pleasant days;*
> *An' the very next place we anchored in*
> *Was Valparaiso Bay,*
> *Where all them Spanish girls come down,*
> *I solemnly do swear,*

They're far above them Yankee girls,
Wid their dark and curly hair.

Then another ship, or even stay aboard this one, since it might head for Frisco, a port he'd always liked too. Or maybe he'd light out inland for a while, to the mines.

I'll cross the Chile mountains,
Shallow, shallow Brown,
To pump the silver fountains,
Shallow, shallow Brown.

The *Beara Head* had been a good go, though. Short allowance and the usual bollocks round the Horn. Nearly scuppered there, but for the mate. Still, they were good shipmates. Besides, everyone liked a shantyman, and he liked being liked. A man felt squared away in a cozy fo'c's'le with good men who liked him.

Urbanski, the other Yankee, was goddamned if that Greek didn't get up to his tricks the minute they got clear of Cape Stiff. He'd been fine then, but now he'd tacked back with his sly smiles and his smooth, brown skin. He thought he was a copper-bottomed little bastard, all right. The plan: go ashore to the *puerto* and get both sheets aft on booze. Then maybe he'd think about knocking the little dago galley west sometime before they sailed again, just to make sure he was ashore in the hospital or the *calabozo* when they hauled anchor.

Life was so strange, thought Maguire, the shanghaied man. One minute you were living your life: digging and carrying for twelve hours a day; off for a pint, then home to the old woman and childer; one day on a whim going into a new place in Sailortown, meeting a woman, probably a hag-whore, but she looked comely enough after whatever was in the drink. The next minute, you're coming to with a head like a sack of spuds on the iron deck of a bloody ship already careening down the Mersey, with the city falling away behind you. Now here he was, four months later, a reg'lar old tar on a Cape Horner. He could go aloft, and he knew the ropes, and he could steer this slab-sider straight and true in a cross-sea in half a gale, and he could do a tolerable splice in a line or even a piece of iron wire in a pinch. And he'd just rounded the

Horn! When he got back, he'd drink on his stories for free, for a few weeks anyway. No longer than that, since shanghaied men weren't exactly uncommon in his neck of the woods. If he went back at all, of course. He didn't miss the family like he used to. Didn't see them much anyway: back late from the pub of an evening, which just left Sundays. Maybe he'd stay at sea for a while. He had a feeling it might be hard to go back to shovelling sand for a living. He'd decide after he'd seen Vallipo, maybe Frisco. The thought of these waiting new worlds excited him.

Jagger, the sonnywhacks, was a real second mate now, by heaven, and not a papa's boy at all. He was a different man after the incident of the runaway spar and the rounding of old Cape Stiff. Or maybe he was a man now and he'd been a boy before. Reaching north out of this cold, sunlit sea, away from the ice and storms, a new idea had entrenched itself in his mind: he would stay in sail. Steam was no place for a real sailor, a real man. His heroism in lassoing that flailing yard had begun the change, and running his watch round the Horn (although the old man was on deck day and night, but that was normal) had confirmed that he belonged on the poop. The captain was as solicitous as ever, but things were different with everyone else: no more of those slight, telling pauses between "Aye aye," and "sir"; no sarcastic emphasis on "Mr." when the mate addressed him. The captain's wife no longer patronized him with her questions; now she treated him with more of the respectful distance she accorded the mate. That was appropriate. He was getting his sea time; soon, he'd be able to get his mate's ticket, and then maybe he'd become one of those twenty-seven-year-old captains driving his own four-poster to Rangoon or Singapore, Cally-o or Sydney. He'd certainly get a command before that bastard of a mate, who was too much of a goddamned barbarian for any owner to trust his ship to.

What the captain's wife wouldn't do for a conversation with another woman! In Valparaiso, there would be a harbour full of ships; she could count on a dozen other captains' wives being there. The luxurious prospect of socializing with anyone other than the shy Jagger and the taciturn and terrifying MacNeill filled her with almost uncontainable anticipation. It was even difficult to talk to her own beloved husband these days. He had been exhausted for weeks by lack of sleep and anxiety and had been in no mood for conversation. His days were filled with navigation and worry. She had feared for his health when he spent days and nights on the poop, soaked through and freezing. At least seamen got their blood going when they worked; her husband merely stood, getting wetter and colder. Sometimes when he came below, she had had to take his oilskins off; his hands wouldn't work, and he shivered for half an hour while she wrapped him in blankets and put her arms around him, her warmth radiating into him, his dampness seeping out to her. More than that, though, it was hard to have a conversation with him simply because he was a different man at sea. She had heard of captains who were civilized and gentle husbands and fathers by the hearth and changed into Bonapartes or czars, raging, profane tyrants, at sea. Her husband wasn't like that, but he obviously found it difficult to change from the hard, loud, oath-spouting driver on the poop to a loving, gentle, attentive husband, all in the course of descending the dozen steps down the companionway to the cabin. If he expected instant obedience on deck, it was hard for him not to expect it below as well. She understood that; it was just the way of it, and after all, obedience was expected of women no less than seamen.

She would be happy simply to step off the *Beara Head*, whose poop deck and cabin she had been confined to for nearly four months, climbing down to the main deck a handful of times, but even then never forward of the mizzen mast. The middle and fore parts of the ship were mysterious, forbidden ground to her. Yet she knew it was the prospect of talk with women that thrilled her (as well as a hiatus in the wearying stream of oaths and blasphemy aboard ship that she had never really become used to), not the prospect of leaving the sea.

She had been reading some poems: "The waves are full of whispers wild and sweet; / They call to me." To her too. She had brought stories by women about the sea. In "The Story of Avis," she read: "The rhythm of the tide . . . It seemed to her a great song without words, full of uncaptured meanings, deep with unuttered impulse." And in "The Awakening": "the voice of the sea is seductive; never ceasing, whispering, clamouring, murmuring, inviting the soul." These words resonated with her: the infinite sea made her feel closer to God, to His magnificent and sublime creation. But the days and nights in the turbulent, unbridled Southern Ocean had invited her soul to a more earthly awakening that both excited and saddened her: the adventure and freedom of life at sea, sailing round a wide world that would be always limited and curtailed for her because she was a woman, and a man's wife.

Benjamin, the ancestral paddy, hadn't thought he would feel this way: uncertain, a conflict in his heart. (He would have said "ambivalent" if he'd known the word.) True, he was happy as a pig in shite to be heading north out of these hellish high latitudes. The warm sun would soon heal his burning boils and dry out his sodden, cracked and bleeding hands, and in a week or so, they would arrive in port and have enough to eat, fresh food too. He was satisfied with himself, with how he had behaved through the past fifty hard days. He had stood his watches and gone aloft and done everything that was required of him. So had everyone else, of course, but at least he hadn't been the one to break down or get the Cape Horn fever, even though he thought he'd come bloody close to it some of those terrible black nights.

He had dreamed for so long of going to sea and becoming a deep-sea sailor, a Cape Horner. It had been his goal in life since he was old enough to walk down to the little walled port by himself to watch the coasters work their way in and out of Carrick, and farther off in the loch, to see the great tall ships towing into dock in Belfast or out into the Irish Sea, the world ahead of them. The excitement he felt when he watched them had to have been in his blood. But he hadn't forgotten his promise to God in the worst of the last gale: to stay away from ships. Cape Stiff had been almost too much for him. He wasn't sure at all that he could go through it again. It wasn't necessary, of course; he could ship out across the Pacific to the East, whose fragrant, alien ports he'd always dreamed of, or to Australia, where the sun shone ever hot and the vast, wild land out back of the coast, whether you went there or not, was a permanent memento of breathing space and possibility. Somehow, though, avoiding the Horn, with all the meaning it had for sailors, would spoil everything else. It would be a kind of desertion. As the old shanty said, the Horn was part of the sailor's way. You couldn't pick and choose so as to avoid it. If you were a deep-sea sailor, you had to face Cape Stiff. He wasn't sure he wanted to do that again. He and Anderson had always talked about jumping ship and exploring America. They had no schedule or ties, no obligations; they could roughneck ashore for a while and then, as experienced seamen, ship out anytime, for anywhere. It was just that he didn't want to desert ashore because he was afraid of the Southern Ocean. But if he was going to do it anyway . . . Och! It was all a pig's breakfast.

He had certainly learned a thing or two, though. About sailorizing, of course, but about himself as well. He was a different man; he had found out what he was sure all seamen discovered: the secrets of himself. The knowledge of what he was and wasn't capable of doing and enduring would stay with him no matter what else he did. If he could keep his head clear and his guts strong in the face of a Cape Stiff snorter, no one and nothing could bamboozle or intimidate him again. In rounding the Horn, he had got round himself as well.

Deck log, Beara Head, *barque, Liverpool to Valparaiso, September 1, 1885: 49° 54´ S., 80° 31´ W. Wind SW 7, W 6. Heavy SW swell. Steady breeze. Crossed 50 N.*

CHAPTER TEN

Not a Voyage Complete

When you start on your journey to Ithaca,
then pray that the road is long,
full of adventure, full of knowledge.
CONSTANTINE CAVAFY, "Ithaca"

It's a long, long time and a very long time,
It's a long time ago.
"Belay!" said the mate.
C. FOX SMITH, "What the Old Man Said"

On its 128th day at sea, the *Beara Head* entered into a vale of paradise.

The barque ghosted towards the anchorage of the bay off Valparaiso near noon, under topsails, jib and spanker. Its foul hull cut its speed in half and made its turns reluctant and sluggish. When it was a mile off the shore, boatloads of apprentices pulled out from half a dozen of the seventy or so anchored ships. The boys swarmed aboard the barque. Fresh and nimble aloft and brawny on deck, they did most of the work of anchoring, clewing up and furling sails.

The *Beara Head* had been posted overdue and probably lost. A wind ship coming back from the dead was an occasion for joy and philanthropy, even if it did happen all the time. The British consul came out with the pilot to welcome in the prodigal and deliver the news that five other vessels were overdue and presumed lost off the Horn in the persistent winter gales. (Three of them would turn up, in Pisagua and

Iquique, nitrate ports farther north, but two are still listed as "missing since 1885," as if they still might come in off the sea 117 years late.)

With the anchor down and the barque settled back into its slot, the apprentices sped back to their own vessels and the *Beara Head* was left alone, its crew disoriented by this sudden incursion of kind strangers. The lack of motion was a strange and disconcerting sensation for its crew, and they staggered a little, weaved and stumbled as they walked the deck, their brains responding too slowly to the absence of stimuli.

Ashore, the buildings crammed the pocket of flat land below the hills that stretched away in a haze of heat and sand towards the interior. The hot, coastal desert wind blew over the ships; dust began to collect in the crevices of gear and in the folds of the sails, furled tight in their harbour stow. The seamen were dazzled by the smell and gritty feel of land. On one side was the town, with the lubber section of the Almendral. On the other, the cliffs of the sailors' old *puerto*, Sailortown and Gaff Tops'l Street, where they would find what they craved. Valparaiso was a good place for booze, women and the knife, the first two of these paradise enough for sailors.

Benjamin was shovelling coal again. He knew it had been coming. He'd moved the bloody stuff around so often, he was sure he recognized some of the individual lumps; their odd angular shapes and contours, like faces or animals or clouds, were familiar. From six in the morning for twelve hours, the *Beara Head*'s entire crew, except for the three men still down with frostbite and broken bones, stood in the hold and filled sacks that were hauled up and over the side five at a time and into lighters that took them ashore. Some other poor bastards shifted them there. The dust was the same as in Liverpool. In fact, there was even more of it, maybe because the coal had been broken up by all its travails: the ship's wrenching and tossing, the manhandling it had got over and over as they had shovelled it in and out of the hold when it burned, and twice backwards and forwards in the hold when it shifted. Liverpool had been warm when they shovelled, the South Atlantic and the Horn cool; in Vallipo, they worked under an iron deck beneath a thirty-three-degree latitude sun, with little ventilation in a wind that, if they had been able to feel it, was cool only in relation to the forty degrees of heat they sweated in. When they came on deck for dinner and at the end of the day, they shivered in the sweltering air. Later, at night, a cold wind blew in off the Andes and the only place to stay warm was under blankets, for the men who still had them; those without put on shirts and jumpers as if they were back at sea in fifty south.

At the end of the first day, three men jumped ship, climbing over the side in the dead of the gravy-eye and into a crimp's waiting boat. Two were from the starboard watch and one was Jan the Finn. He took his few clothes, the near-finished model ship, his knife and spike but not his old rotted and hogged sea chest. He also gave up his pay, the few pounds he had left after his advance and slop-chest tobacco purchases. Several of his watchmates heard him drop out of his bunk and slither out the half-open deck-house door, but no one spoke. Better to leave a deserting man alone, especially a grumpy Finn with a slick blade. Anyway, it was standard procedure to jump; the only surprise for the listeners was that it was one man and not half the watch. Benjamin slept through the departure. He realized that men had gone only when the diminished gangs mustered at the hatches the next morning.

With the desertions, three men sick and the deaths at sea, seventeen seamen and boys remained to offload the stipulated one hundred tons per day. It was a minor mathematical chore to calculate that they would be shovelling for thirty days, assuming no more desertions. They wouldn't be able to keep up the pace with fewer men. The next night, though, two more seamen jumped, one of them Kapellas. Like the Finn, he stepped off the *Beara Head* with whatever he could carry over his shoulder, light and unencumbered (the articles were an irrelevant piece of paper that meant nothing in the dusty *pueblos* of the West Coast), into the waiting boat and ashore to drink and beachcomb, evading the gendarmes (the captain would make a token effort to redeem his seamen) until he ran out the crimp's credit and found himself in yet another fo'c's'le with a new mate to hound and drive him, maybe another Yankee watchmate to dance the strange dance with.

Fifteen bodies left to discharge the coal, but the desertions stopped there. The captain doubled the anchor watch and had the mates stand deck watches each night. It wasn't necessary. The men who were going had gone; the remainder were content to work out their time and wait for a legal spree ashore.

A routine was established. The fifteen seamen and boys laboured and sweated all day in the iron oven of the hold, filling and sewing shut the sacks and taking turns at the cooler work of hoisting them up and over the side; a half-hour break was allowed for dinner, which now included fresh bread, vegetables and meat (voluptuous with their crisp or silky-soft textures, their outlandish tastes motley surprises, but oddly insipid without the bite of salt, the unyielding hardness of the ship's biscuit). In spite of the new food, the seamen grew skinnier, their flesh liquefied by the heat.

After twelve hours' work in the hold, the apprentices were told off to row the captain and his wife to one of the other ships in the anchorage for a social evening of talk and drink—the captain relieved, for now, of his sea strains and the need for shouting, and restored if not to his drawing-room charm, at least to his aft-cabin cordiality; his wife ecstatic at the feel of other decks and of the shore itself some days. She walked in blissful conversation with other wives round the Almendral, avoiding the cliff paths to the *puerto*, where day and night the shouts and music rang out of sailors, deserters, miners and beachcombers from half the world's countries and empires, in their wild sprees and fandangoes with booze and whores. On morning walks past the *calabozo*, the women saw morose sailors, the unlucky few who'd been picked up drunk and disorderly the night before, grooming the Gilantes's horses as punishment.

The *Beara Head*'s idlers were excused from the filthy business under the hatches, and they worked to repair the barque's storm damage and its degradation after a voyage of four months. Most of the number-one canvas had been shredded or destroyed; the sailmaker would sew a new set. The bulwarks and deck houses had been bent or buckled, the masts sprung and temporarily fished, wood trim and teak deck planks stripped off, blocks "froze," gear of all sorts demolished; the carpenter would pound and cut them back into shape, fashion new masts and yards. When the crew had finished with the coal, they would send down the jury-rigged spars and step (set up) the new ones. The steward properly cleaned out the aft cabin and saloon, which had been gutted by the rogue sea, and served port to the captains and wives who visited the *Beara Head* when its turn came up on the social round. The doctor luxuriated in his galley, now restored to order and cleanliness, and in preparing food that wasn't ripe with wriggling, scuttling protein.

At six in the evening, the seamen sluiced off the coal dust, ate and talked around the foremast or sang with Paddy, watching the lights on the so near, so far shore. When the cold night wind rose, they went below and slept (every man with a blanket now, courtesy of the slop chest and half a month's pay).

Benjamin and Anderson made a decision. They would jump ship. Not here in this little shithole, where they would soon have to go back to sea or dig ore out of an inland pit. They'd done a lifetime's shovelling. When they got to a white man's town, maybe Frisco or across the Pacific to Sydney, they'd jump for it together. There were rumours that it would be Frisco for grain. That would be all right; America, the New World, would be good enough for them.

Many years later, when he was dying of tuberculosis, Benjamin would wonder whether the weeks of sucking coal dust into his lungs had sprung the fatal weakness, begun his death right then and there—in Liverpool, and when he shovelled the coal in three oceans (Vallipo the worst because it went on so long).

The heat and dust were unrelenting. The mate rode them with the same fervour he had brought to the job at sea. He had changed a little during the battle with the Horn, softened towards them, given small signs of his concern. It didn't last. Those glimmers of solicitude began to dissipate as the wind went fair; by eighty west, when they squared the yards and began to make their northing, he was the old driving bucko.

Discharging the cargo took forty-one days, the pace slowed by the first days' desertions and the compounding exhaustion of the seamen. As they got down towards the bottom of the pile, the lighters began bringing ballast out to the *Beara Head* to maintain its stability. The men had to shovel both ways: coal out, slippery rocks and gravel in. The ballast was dust-free, but it stank of shit and the foreshore it had been dug from. At first, they needed only three hundred tons or so as harbour ballast because it seemed there was a chance of a nitrate cargo in Valparaiso. That would mean a passage back around the Horn to England, the only choice for the seamen to renew their acquaintance with Cape Stiff and spend three or four more months at sea or jump ship right here in dago-land. Benjamin and Anderson were undecided. Then the lighters began to arrive with more ballast rock, enough to keep the barque upright in a gale of wind; it was clear they were going back to sea. Word filtered forward from aft: no nitrates, but grain. They would sail for Frisco after all.

✳

Before they sailed, they went ashore.

For the first time in 170 days, Benjamin and the remaining men of the port watch stepped onto land. It was one watch ashore at a time, the captain's attempt to keep repair work going aboard the *Beara Head*. The shantyman, the cockney and the two Yankees had been here before and knew where to go: up the narrow cliff-paths to the Tops, the rows of dives, bars, whorehouses and fandango joints along Shit Street and the Street of the Nail. Benjamin, Anderson and Maguire followed them. So did the Dutchman, who had seen twenty Sailortowns, but not this one. They drank pisco and grappa, vino, German lager and *aguadiente* in its sausage-like goat skins (good for smuggling back aboard wrapped around a leg or a waist). They drank at the Shakespere, run by the crimp Tommy Jenkins; the Liverpool Ship, run by Paddy Byrne; the Cross Flags and the House of All Nations.

As they tavern-crawled, the able seamen pointed out to the greenhorns the famous crimps and shanghai artists of Vallipo, the men to stay away from if they wanted to avoid waking up the next morning on a Chileño whaler or some Yankee hell-ship bound for God knew where and for how long. There was Nigger Thompson, who had lost a finger as a mercenary for one side or the other, maybe both, in the Chileño-Peru War. And there was Dutch Charlie of Shit Street, who killed men who owed him money. Paddy knew the shanty about Jimmy the Wop, owner of the English Bar:

> *But Jimmy the Wop he knew a thing or two, boys,*
> *An' soon he'd shipped me outward bound again;*
> *On a limey fer the Chinchas fer guanner, boys,*
> *An' soon wuz I aroarin' this refrain . . .*

They would also be well-advised, Russell told them, to keep to windward of *los gringos locos*, the British and Yankee deserters who manned many of the ships in the Chilean navy (the Peruvian too). There were gunboats in the port now, and when these hard men hit the beach, said Russell, it was way for a sailor and stand from under. He once saw a gang of liquored-up, bloody *gringos locos* down on the waterfront stand up to a mounted charge by a dozen Gilantes. The seamen faced down the cavalry, dragged the gendarmes off their horses, beat the tar out of them, took away their swords. When they began slitting the throats of the horses, their officers came ashore to restore order; they had to shoot three of their own *gringos*. They lashed the others to the guns aboard ship and flogged them; you could still whip men in the dago navy. After that, though, when the *gringos* came ashore, the Gilantes always disappeared.

Men who had downed nothing more than a few tots of rum over the previous six months got drunk fast, even the hardened topers. Benjamin's memories of Sailortown in paradise, at first rich and detailed, dwindled away to isolated and confused fragments: more drink; two men fighting with knives—the cockney and someone else? he wasn't sure— near-naked women dancing; a kaleidoscope of men of all colours— bronzed white, shades of brown and black, yellowish—flashes of tattoos; a bloody, sweaty chest; shaved heads; men's hair braided in queues; more knives; loud fandango music; shouting and singing in a dozen languages; the stink of vomit and dung; the smell of heavy-sweet perfume; more intense colours—a whore's shawl, bright-shirted Indians, a blue sky giving way to velvet black—some seaman crying for a scuppered chum; bits and pieces of shanties and forebitters. A whore of his own? He remembered something about a woman: the price, the sound of a piano, but for some reason, nothing about her body. He was aware of being submerged in water and dragged out of it by rude hands; he was gritty with sand and felt the hard surface of wet wood against his cheek, a little dog nipping at his ankle, then nothing. The end of his shore leave.

"This mysterious, divine Pacific zones the world's whole bulk about; makes all coasts one bay to it; seems the tide-beating heart of earth," said Ishmael as the *Pequod*, heading for the Japanese hunting grounds, drew in for the kill. "That serene ocean rolled eastwards from me a thousand leagues of blue."

The *Beara Head* now made its true Pacific voyage. The Horn to Valparaiso had been a week's jump north from the cape, two hundred miles off the coast, to the desert shore. The passage to San Francisco, port to port on the same side of the Americas, would nevertheless take the ship many blue leagues into the heart of the great ocean, bending its course, in the way of the sailing ship, to the flow of winds and the set of currents.

It was a deep-water passage; the *Beara Head* saw no land from start to finish. The ship passed nine hundred miles southeast of the Galapagos; when it crossed the equator, the nearest coast was Mexico, sixteen hundred miles northeast. Before it bore off before the westerlies, at about thirty-five degrees north latitude, it was closer to the Hawaiian Islands than to the mainland. By luck, it sailed from Valparaiso towards the end of the hurricane season in the eastern North Pacific, and although there is the odd tropical storm in December, it usually forms close to the coast, far from the barque's devious route. The ship's distance from Vallipo to Frisco was about eighty-five hundred miles.

The *Beara Head* sailed with offsetting conditions: light in ballast but dragged down by a foul hull. It made an average passage of around two hundred miles a day, forty-two days in all. Mostly, the Pacific was as serene as Ishmael envisioned. The barque ran into a gale its third day out. Nothing unusual: a quick buster with fifty knots or so for a day, falling away to a full-sail southwesterly and a fair-weather sunset.

The second gale overtook the vessel six days before it made port, just after they had found the westerlies and bore off for Frisco. It was like running their easting down in the Southern Ocean with a roaring-forties gale at their backs: a cold wind and big seas building with the fetch of a quarter of the earth behind them. Except they were in ballast and the helmsmen had to take extra care; eight hundred tons in the hold, instead of three thousand, made the barque's responses quick and skittish, increasing the chances of a broach. The foul bottom helped them now, its drag damping down the light hull's yawing. Foul or not, in forty-eight hours, running hard dead downwind under three lower topsails, they covered 580 miles, their best two-day run. Unlike Southern Ocean storms, this gale had no successor right behind to give them another pasting without pause or rest. The wind settled down to a full-sail breeze out of the north, still cold, but fair.

In between the brackets of these storms, the barque made steady, tranquil progress; it was lucky in the southern horse latitudes, experiencing light air for a day (too light to move the fouled hull), but then finding enough wind to ghost through to the southeast trades. Days of glorious movement for the ship. Watch on watch of the real seaman's labour for the crew: chipping iron, painting, chipping, slushing down the standing rigging, soogie-moogieing the decks, chipping some more, overhauling clapped-out lines, yet more chipping, working the old iron up (the rust had accumulated in two months south of fifty).

They spent another day immobile in the doldrums, but then they picked up enough moving air to propel them at a few knots towards the northeast trades. These slung them northwest, right through the northern horse latitudes and into the westerlies with barely a pause, the trades falling light and uncertain and then backing to the west within half a day.

The seamen all felt the same about this sea-skip north: What was a biblical forty days and nights in the sunny, blue Pacific after they had endured three times that in the grim, green Atlantic and the abyss of the bloody Horn? The prospect of Frisco had aborted everyone's desertion plans. It would have been different if they were headed back round the Horn with nitrates or across the whole stretch of the Pacific to Sydney. But no one jumped in Vallipo if he could wait a few weeks and do it in Frisco, the Sailortown of Sailortowns, the celebrated Barbary Shore, so-called because it was the sailor's version of a Muslim paradise.

Arriving at the Frisco heads, the barque hove to a few miles off Point Lobos, waiting for the usual clammy afternoon fog bank over the bay and its entrance to dissipate. The next morning was clear, and the *Beara Head* ran with the flood tide before a mild west wind under topsails, jib and spanker through the Golden Gate, threaded its way among a hundred other ships and came up to its anchor midway between the Embarcadero and the fort on Alcatraz Island.

The mate, on the fo'c's'le head, oversaw the growth of the chain and made sure the anchor was well dug in. The seamen hauled down the fore-and-aft sails, clewed up and furled the topsails, then loosed and refurled every other sail in a neat harbour stow. They snugged down and coiled the braces, sheets, tacks, halyards, clewlines and buntlines, then hung them on their pins. As the men completed the various jobs of anchoring and making the gear Bristol, they gathered in their accustomed place at the break of the poop. When they were all assembled there, the mate came and stood at the rail above them. They looked up at their bucko officer.

"That'll do the crew," he said, and turned away.

EPILOGUE

Full fathom five thy father lies;
Of his bones are coral made;
Those are pearls that were his eyes:
Nothing of him that doth fade
But doth suffer a sea-change
Into something rich and strange.
WILLIAM SHAKESPEARE, The Tempest

They disappear.

The sea-voyage story ends. History begins again. Or the search for the sparse bits and pieces of ordinary lives—that history—begins again: a few letters preserved at random, the hazy recollections of descendants. How could they have foreseen this belated thirst for facts? Times and years dissolve and shift; sequences are jumbled. The records of the island, still in its early days of settlement, are uncertain and incomplete. The homestead, the grandson, the sea chest, the graves (although they are obscured by the wear of time and the moist island's green growth) are the sources of knowledge. They are eloquent in a way, yet mute; they are fragments and suppositions.

First, they plunged into the boozy, whorey streets of the Barbary Shore, and then who knows? Any trail would go cold; they were still only partway through "the dangerous experiment of forcing [their] way to the shores of America," as Fenimore Cooper put it. These weren't men who wrote regular letters home, kept journals or did things of any note or notice. Benjamin and Anderson could have done anything, gone anywhere in the New World: inland to the mountains and open country; north towards British territory; back to sea—the two men made more than one passage under sail. Rumours in the family: stories of smuggling booze and illegal Chinese labourers into the British West Coast colony. The old square-rigger shipmates have mutual friends along the shores of the Inland Passage and up into Alaska; their names turn up in letters Benjamin and Anderson exchanged in later years: Robinson, Charlie, Sailor Jack, "the boys." The possibilities for adventure are rife on the beautiful, wild coast.

One thing is certain: years later, when Benjamin landed on Salt Spring, he had enough money to pay for a bargeful of California redwood to build a house on the island. He didn't need much for the forty-nine acres he acquired there. He "pre-empted" that: took it over from the Crown. Provided that he continued to live on it, the land was his for the nominal amount of $1.25 an acre, payable in two years or so. But the house was different. It was a good size for its time, one and a half storeys, with a fine, large verandah and set on a rise looking down towards the Upper Ganges Road, not far from the island's main village. It took some wherewithal to buy the material for such a building, especially good redwood. Benjamin's land spread out behind and on both sides of the house, an area of gentle, undulating hills; fir, cedar, hemlock and arbutus grew everywhere.

There are a few other more or less sure things: Benjamin married Annie and settled down in his fine house, had two daughters and a son, grew apples and pears, planted potatoes and other crops, had rheumatism so badly in his ankles that sometimes he couldn't walk. Annie got rheumatism, too, and was almost crippled by it for a few weeks at least once. Some years were too dry, some too wet, he wrote Anderson; the crops were "light," the fruit "scarce." One year, it rained all through November, and the potatoes were so bad, they would have few of them for the winter (a terrible thing for an Irishman). He sent three boxes of apples to Anderson, who was up north, close to Alaska, working at a salmon cannery on the Skeena River; he was to keep one and give one each to Sailor Jack and Robinson. Benjamin's daughter Edith was very sorry he wouldn't be there for Christmas that year. "She wanted to know where I left Mr. Anderson," Benjamin wrote.

Anderson stayed out there in the cold—itinerant, wandering from job to job in canneries, aboard sealers, maybe fishing boats. He wrote to Benjamin about where he was working that winter or what his prospects were. In a letter written on Christmas Eve, year unknown, he says that he will ship out on the sealer *Borealis* before New Year's. He was supposed to have had a job all winter handling freight on scows between a steamer and shore, but it fell through. "If the *Borealis* does not turn over once too often, I will be back in May," he wrote. He sent a case of venison down to Salt Spring. He apologized about a Christmas present of some cloth; he was going to buy something else, but he got drunk and some sailor came along and sold him the cloth instead. He was down in Moor's Cove, he told Benjamin, "seeing the boys." They hunted ducks and geese. The deer were so easy to kill that he just clubbed them. They were no good for eating, though, only for the skins.

In February 1894, he wrote from Skeena River that it had snowed six days out of every seven. He hadn't found anything to do all winter, and there was no work coming up. He wished he'd stayed down south. "Friend Ben," he wrote, "I hope the next square-rigger that carries Royal Yards will carry me out of the country altogether."

Benjamin swallowed the anchor and came ashore for good, not far from the sea, to be sure—it's a fifteen-minute walk to Ganges harbour—but he became rooted, as solid as the land. His sea-change was complete: from the few square feet of the narrow little house in the Irish Quarter to the acres of his own earth on the comely western island.

In October 1901, Benjamin received a letter from John Williams, at the Carlisle Cannery in Skeena River. Williams had heard of Anderson's death but had "no further particulars." There were rumours that no money was found on his body. That was strange, Williams wrote, because he had been paid off by the cannery in the amount of seventy-five dollars and had collected some debts too. In any event, before his departure, he had placed in Williams's charge a net gun and cooking gear. What did Benjamin want him to do with them?

I checked into the motel on Salt Spring Island on a mild, sunny day in October. I printed my name, then signed the card and pushed it towards the owner. He looked at it, startled.

"But there's another Derek Lundy on the island," he said. "Are you related?"

I was on a book tour and had a couple of days to spare between events in Victoria and Whitehorse. I had decided to rent a car and drive to Salt Spring to look around. I would stay overnight and go back to the Victoria airport the next day. I knew vaguely about my family's island connections but had almost forgotten them. The news of other Lundys brought it all back. Were these my long-separated relations?

As it turned out, they weren't. There was a small tribe of Lundys on the island, including my namesake, but they were late arrivals from Ontario. They might have been very distant relations—it's not a common name—but they weren't Benjamin's family. His descendants had a different surname now; Benjamin's daughters had, naturally, adopted their husbands' names when they married. His son had died young of tuberculosis. His grandson lived on the other side of the Upper Ganges Road from the old Lundy homestead, which had been sold out of the family twenty or so years ago.

Two years later, my wife, daughter and I wanted to spend a year out of the city, a sabbatical for Christine and somewhere different for Sarah and me. We went to Salt Spring Island and lived there for ten months. In its small community, I became "the other Derek Lundy," as the original one, "the real Derrick Lundy"—he used a variation of the spelling—adjusted with endless patience and good humour to sharing his identity, and to receiving phone calls, meant for me, from the eastern time zone before dawn.

We lived on the sea through the mild, rainy seasons of the school year. We got to know Benjamin's family. Because one of the owners of the Lundy homestead was the kindergarten teacher in Sarah's school, we visited the old house. It has been renovated inside, although the original wood is gleaming and intact. The exterior is almost unchanged from Benjamin's day; the surrounding meadows are beautiful and full of fruit trees.

After ten months, we moved back to the city. But Salt Spring is a hard place to give up. We have decided to move back to the island. We will buy a house there. Benjamin will have more of his descendants nearby. We hope to visit the homestead from time to time. Like Benjamin, I began my life in a tiny worker's house on a narrow street in Ireland, and it's a good feeling to be inside the one he created in our new world.

ACKNOWLEDGMENTS

I gratefully acknowledge the help of the staff at the National Maritime Museum library in Greenwich, the Public Records Office in Kew, the Family Research Centre in London and the Public Records Office of Northern Ireland, as well as Billy Hurst of Harland and Wolff in Belfast, Dr. Brian Trainor of the Ulster Historical Foundation and Heather Wareham, of the Maritime History Archive at Memorial University of Newfoundland. My thanks to Jack David, Jim and Frances Deas, Christian de Marliave, Marilyn Ellis, Martin Hawthorne, Janis and Linda Kraulis, Alan Lundy, Mark Lundy, William Lundy, Pauline Madden, Mervyn McMeekin, Jeffrey Miller, Jonathan Raban, Rocky Sampson, Tom and Margaret Thompson, Barend Visser, the captain and crew of the barque Europa, Bertrand Dubois and Siv Follin Dubois of Baltazar, and especially Bob and Diane Hele on Salt Spring Island.

I'm most grateful to my editor, Diane Martin, my agent, Anne McDermid, and to Scott Richardson and Janice Weaver.

Finally, my thanks and love to Christine Mauro and Sarah Lundy, who, all by themselves, make it all worthwhile.

PERMISSIONS